Models and Methods for Biological Evolution

SCIENCES

Computer Science, Field Directors –
Valérie Berthé and Jean-Charles Pomerol

Bioinformatics, Subject Heads –
Anne Siegel and Hélène Touzet

Models and Methods for Biological Evolution

Mathematical Models and Algorithms to Study Evolution

Coordinated by
Gilles Didier
Stéphane Guindon

iSTE

WILEY

First published 2024 in Great Britain and the United States by ISTE Ltd and John Wiley & Sons, Inc.

ISTE Ltd
27-37 St George's Road
London SW19 4EU
UK

www.iste.co.uk

John Wiley & Sons, Inc.
111 River Street
Hoboken, NJ 07030
USA

www.wiley.com

Library of Congress Control Number: 2023932775

British Library Cataloguing-in-Publication Data
A CIP record for this book is available from the British Library
ISBN 978-1-78945-069-9

ERC code:
PE6 Computer Science and Informatics
PE6_13 Bioinformatics, biocomputing, and DNA and molecular computation
LS8 Ecology, Evolution and Environmental Biology
LS8_7 Macroevolution, paleobiology

Contents

Preface

Gilles DIDIER[1] and Stéphane GUINDON[2]
[1]*IMAG, CNRS, Université de Montpellier, France*
[2]*LIRMM, CNRS, Université de Montpellier, France*

Evolution, in the usual meaning it takes in life science as the phenomenon by which living species evolve over time, is an essential biological process. It can arguably be said that it is the most important process, since all other biological phenomena are in some way derived therefrom. Evolution is thus at the origin not only of the extraordinary diversity of living beings, but has also shaped every biological function that can be observed. Its general theory dates back to the 19th century with the work of Darwin. Although this theory is now widely accepted and shared, its study remains an extremely lively and fertile scientific ground, which has extensively developed in recent decades, and is thus, consequently, rather broad.

It is obvious that the aim of this book cannot be to present every single aspect of evolution, but is instead to focus on the large number of interactions between biology, mathematics and computer science that its study has fostered. The main reason why evolutionary studies make such an extensive use of mathematical models and algorithms is due to the fact that biological evolution is a process which has been going on for more than 4 billion years and which is not, with rare exceptions (see Chapter 11), directly observable on our time scale. As such, it can only be studied from the data that are available to us today, that is, present-day species and the fossil record. In order to test hypotheses about the mechanisms governing evolution, it is generally necessary to express them in terms of mathematical models. The latter are simplified representations of biological reality that can be used to (imperfectly) reconstruct the evolutionary history of contemporary species (as well as ancestral species in the case

Models and Methods for Biological Evolution,
coordinated by Gilles DIDIER and Stéphane GUINDON.

of fossils) that we observe. By fitting these models to the data, their relevance can be assessed and the hypotheses initially proposed validated or rejected. The design of models and the inference of their parameters require us to deal with mathematical, computational and statistical problems that have contributed to opening up new fields of research, both theoretical and applied, in these areas.

The central object in the study of evolution is the tree of life. The latter is mainly a natural representation of the diversification of species in the sense that it describes the relationships between species (or even between individuals). In modeling, we shall see that it is sometimes interpreted as a support for evolution, which we can try to reconstruct from the available data, and sometimes as a representation of the statistical dependency between the characters carried by the species. Trees are also theoretical objects that have been studied in computer science and mathematics, particularly from the point of view of their combinatorics. Chapter 1 briefly presents this aspect before describing different evolutionary models leading to trees and the probabilities associated therewith within these models.

Although trees represent the framework in which evolution takes place, evolution itself operates on the various characteristics (which should be taken in the broadest sense here) that can be found in living beings. Moreover, it is through these characters that it can be studied. The most used "character" within this framework is genetic material, namely the molecules/polymers from which the sequence of elementary bricks can be extracted in the form of (long) words on finite alphabets. In fact, the development of genetics, first identified as a support for evolution, and DNA sequencing techniques have revolutionized the study of biological evolution by changing the nature and causing an explosion in the amount of exploitable data in this field. Using these (and other) data to better understand evolution requires an increase in mathematical and computational resources to address the ever-changing amount and type of data.

Chapter 2 presents the main Markovian models of DNA sequence evolution. These models, generally considered as mechanistic, describe evolutionary processes at the molecular level, over sufficiently long periods of time so that intraspecies genetic variability is negligible compared to interspecies variability. The vast majority of these models consider that the different positions along the genetic sequences evolve independently of each other and follow the same continuous-time Markovian model. This same chapter also describes probabilistic models for taking into account the variability of evolutionary rates along sequences, an important phenomenon from a biological point of view, particularly with regard to the evolution of genome coding parts, which are constrained by the structure of the genetic code. Finally, models of the same type as those used for DNA sequences can also be used for modeling the evolution of discrete characters such as the presence or absence of a given morphological characteristic or the number of fingers, etc.

Evolution also concerns the physical characteristics of species, in particular the so-called quantitative characters such as height, weight and so on. Although these are less used for phylogenetic inference, understanding their evolution is essential to biology, for example, for testing hypotheses about morphometric and allometric relationships in ecology. Evolutionary models of continuous characters prove to also be a relevant tool for the detection of possible traces of natural selection on the evolution of morphological characters. Chapter 3 presents in detail the generic framework in which these models are implemented as well as a wide range of regularly used approaches to appropriately model the correlation between character values that derives from evolutionary relationships between compared species.

The models presented in Chapters 2 and 3 assume that the characters being considered are probabilistically independent (e.g. sequences sites and the different edges in a phylogeny are considered as independent). Although this independence avoids an explosion of the computation time and the size of the models, it might not be considered realistic in many cases. Chapter 4 presents different approaches which make it possible to highlight and study the evolution interdependence of several characters, discrete or continuous. Similarly to the models presented in Chapter 3, the co-evolution models take into consideration the phylogenetic tree as a nuisance parameter in order to evaluate the correlation part between morphological characters that is not explained by evolutionary relationships.

Genetic sequences do not only evolve by means of mutations, as presented in Chapter 2, but sometimes change in a more radical way. In particular, at the genome level, evolution sometimes proceeds by duplication or inversion of whole sections of chromosomes. These types of changes are very rich in terms of information on the evolutionary distances between genomes. They provide a more comprehensive view of evolution than the "simple" models of point substitutions between nucleotides. Nonetheless, genomic rearrangements are more difficult to model mathematically. Chapter 5 reviews the general approach to detecting these rearrangements and reconstructing evolution at the genome scale.

The first five chapters of our book provide an overview of evolutionary models. The next four chapters illustrate how some of these models are applied in the context of phylogenetic inference, that is, for determining evolutionary relationships between species or individuals and the time since they have diverged. They present several approaches commonly used to answer these questions.

The first approach to reconstructing the evolution of a group of species consists of considering a distance or dissimilarity matrix, that is, a measure of the "resemblance" between pairs of species. Here, the assumption is made that the further apart species are from an evolutionary perspective, the less similar they are from the point of view of the chosen measure. Under this hypothesis, the tree that best represents these distances and that we will try to determine is close to the one that traces their evolution. The dissimilarity, or distance, used can be calculated based on genetic sequences and morphological characters. Although a distance drastically summarizes all the characters inherent to the species, different methods presented in Chapter 6 enable the reconstruction of realistic evolutionary trees from this information. These approaches are even virtually the only ones that can be applied in practice to reconstruct the trees comprising a large number of species due to their computational speed.

While the methods presented in Chapter 6 prove to be very fast, they do not directly take the evolution of the species under study into consideration, since the evolutionary distances between species only provide an approximate summary of the raw data. Other approaches directly involve mechanisms or evolutionary models for phylogenic inference. One of the first to be considered is parsimony, which is based on Occam's razor principle and seeks phylogenies involving the fewest possible evolutionary events. Parsimony has gradually been supplanted by approaches based on probabilistic models such as those presented in Chapter 2. The first way to use such models for phylogenetic inference consists of looking for the tree maximizing the probability of the observed data under the chosen model. This is called maximum likelihood. The method is quite close to parsimony in spirit and these two approaches are described in Chapter 7.

Another way of using sequence evolution-based probabilistic models for phylogenetic inference can be conducted in the context of Bayesian sampling, where we no longer seek to determine only the tree that maximizes the probability of the observed data, but to associate with any tree its posterior probability, namely its probability conditional on the observed data. This approach takes into account the uncertainty inherent to the inference process, thereby overcoming a limitation of maximum likelihood approaches. Chapter 8 presents the general principles of Bayesian approaches by Markov chain Monte Carlo methods as well as their applications to phylogeny.

Bayesian approaches are not the only ones capable of quantifying uncertainty in the trees inferred by the various methods. Evaluating this uncertainty is essential, since in many situations, evolutionary history is difficult to reconstruct, to the point of causing controversy among people who study it. Chapter 9 presents several approaches for quantifying the uncertainty associated with each branch of a phylogenetic tree. The non-parametric bootstrap technique, well known to statisticians, has long been the preferred approach. Other approaches, which are faster, have recently emerged over the last two decades. These are described in detail, providing an overview of current solutions allowing for quantifying uncertainties in phylogenetic tree reconstructions.

While previous chapters of this book focused on the models and methods that underpin modern tools for studying evolution, the following chapters illustrate how these approaches can be used to expand further our knowledge of biology, which in turn can sometimes contribute to enhancing phylogenetic analyses.

For example, considering phylogenies containing fossil species benefits at several levels. Beyond their paleontological interest, fossil species are essential to dating phylogenetic trees, more precisely to the dating of their internal nodes representing speciations. Evolutionary models presented in Chapters 2 and 3 do not allow the determination of evolutionary durations in "absolute" time. These are only determined relatively to an evolutionary rate that is not identifiable without a temporal calibration point (e.g. the date of an evolutionary event). On the other hand, geological strata in which the fossils are found provide a means to date more or less precisely the period during which the associated species lived. Integrating fossil species into a phylogeny thus makes it possible to add time information and to estimate speciation dates in particular. Chapter 10 presents the challenges and particularities of this type of data and the difficulties of integrating them into phylogenetic analysis.

Another field of application for phylogenetic models is that of phylodynamics, as presented in Chapter 11. The point here is to trace the evolution of pathogens, viruses or bacteria, in order to understand the underlying population dynamics. The phylogenetic tree no longer describes a sequence of speciations but rather the transmission events of the pathogen between successive hosts. The pace at which these events occur, when analyzed within the context of an appropriate epidemiological model, is directly related to the dynamics of the epidemy. Chapter 11 gives an overview of modern statistical approaches (and unfortunately topical at the time of writing this preface) mixing phylogenetics and epidemiology.

Finally, it should be noted that in phylogenetic inference methods presented so far (with the exception of the previous chapter), the evolutionary unit is the species. Tracing the evolution of individuals constituting a species, or even a population, is another major challenge. Chapter 12 gives a broad overview of the advances made possible by the genealogy of individuals traced on the basis of genetic data combined with probabilistic models in population genetics. Such approaches allow us to infer variations in the size of ancestral human populations and to reconstruct, from the analysis of georeferenced genetic sequences alone, migratory events that occurred several thousand years ago.

The aim of this book is not to provide an exhaustive overview of techniques for studying evolution or their applications. For example, phylogenomic approaches, which aim at reconstructing evolution based on the analysis of multiple genes, are not presented. Similarly, techniques for detecting traces left by natural selection within genetic sequences, or those for molecular dating, are not discussed in detail. Our approach here has been to focus on some of the most frequently used approaches at the current time and to provide an in-depth description thereof. Many examples for applying these methods can of course be found and, again, we only give an overview of these. Nevertheless, we hope that after reading this book readers unfamiliar with the subject will be interested and curious about evolutionary research, while allowing students and researchers in the field to deepen their knowledge of these questions.

November 2023

1

Trees: Combinatorics and Models

Gilles DIDIER[1] and Stéphane GUINDON[2]
[1] *IMAG, CNRS, Université de Montpellier, France*
[2] *LIRMM, CNRS, Université de Montpellier, France*

1.1. Introduction

Trees describe evolutionary processes at different time scales. Therefore, for long time scales, each branch of the tree, referred to here as "phylogeny", generally represents the evolution of a set of organisms exchanging genetic material, that is, a *species*. The nodes called "internal" (the set of nodes distinct from the leaves) then correspond to the formation of new clades or "*cladogenesis*". The formation of a new clade is caused by (or results in) a decrease or even an interruption in the exchange of genetic material, and it defines a partition of the set of organisms constituting the species in question. Under certain conditions, such a partition occurring is synonymous with speciation: the ancestral species gives rise to one or more new species, thus generating a bi- or multi-furcation in the phylogenetic tree. Finally, a leaf in a phylogenetic tree appears when a species disappears. A leaf can also be the result of the fact that the corresponding species has been sampled in order to be incorporated into a corpus of data to be analyzed.

On a shorter time scale, a tree can describe the evolutionary relationships between organisms of the same species or population. A branch of such a tree (a tree usually referred to as a "genealogy") then describes a succession of individuals linked by

Models and Methods for Biological Evolution,
coordinated by Gilles DIDIER and Stéphane GUINDON.

direct kinship relations. An internal node in this genealogy corresponds to a reproduction event where a parent gives birth to one or more new lineages, a lineage being defined as the ancestry of an individual. Finally, a leaf is associated with the disappearance or sampling of a lineage.

Regardless of the time scale being considered, the branching phenomena mentioned below lend themselves to probabilistic modeling. These processes have been the subject of a considerable amount of mathematical work. As such, as early as 1845, the French probabilist and statistician Irénée-Jules Bienaymé proposed a stochastic model of branching for studying the disappearance of surnames. The proposed model would become the Galton–Watson model and would be the foundation for a new branch in mathematics, combining combinatorial and probabilistic aspects. Coalescence processes, which, by simplifying, are equivalent to branching processes when the arrow of time is reversed, are also the source of a multitude of mathematical studies (for review articles on this theme, see Aldous et al. (1999); Möhle (2000); Berestycki et al. (2009)). The latter have contributed to shed new light on important questions in population genetics through the establishment of equivalence relations between population biology models based on very different hypotheses.

This chapter focuses on the two main probabilistic tree generating models, namely the birth–death process and the Kingman coalescent. We will first of all give the elements of combinatorics necessary to count these trees. The probabilities of the corresponding trees, ignoring the ages of the nodes, will then be determined next. The probability densities of trees comprising ages at the nodes will finally be detailed. These derivations will allow us to expose the underlying assumptions of the two tree models and discuss the links with other popular models in population biology.

1.2. Preliminary definitions

While trees are used to model the support of biological evolution, they are also formal objects studied as such in computer science and mathematics. In these two disciplines, a tree is defined as a cycle-free (non-oriented) connected graph. In other words, a tree is made up of a set of elements called *nodes* or *vertices*, some of which are connected by undirected *edges* in such a way that there exists one and only one path connecting any two vertices. Two vertices directly connected by an edge in a tree, or more generally in a graph, are said to be *neighbors* and a *path* of the tree designates a succession of vertices $(a_1, a_2, \ldots, a_n)$ such that a_i and a_{i+1} are connected to one another for any $1 \leq i < n$. The vertices of a tree that are connected to only one other vertex are differentiated from the others and called *leaves* of the tree. The other vertices of the tree are called *internal nodes*.

Two types of trees will be presented later in the text: non-rooted trees whose definition is exactly the same as the one above, and rooted trees which are trees in which one of the vertices is distinguished and called *root* of the tree. Here, we are more specifically interested in rooted trees, since they have a more direct interpretation in terms of evolution, which will be reviewed later. Within a phylogenetic context, unrooted trees essentially appear to visually represent relationships between species. Within an evolutionary context, the datasets are often chosen in such a way that determining the root of the tree becomes more or less easy.

Going back to rooted trees, they are usually shown, somewhat counterintuitively, with their roots at the top. An important point is that the root of a tree can be used to orientate it, because by traveling along an edge (or more generally a path) of the tree one either moves away from or toward the root. As a matter of fact, as for any vertex a of the tree which is not its root, there always exists a unique path from the root to a (because the tree is connected) and this path crosses one and only one neighbor of a (otherwise there would be a cycle), which is necessarily the only neighbor of a closer to the root and which is called the *direct ancestor* or the *parent* of a. This makes it possible to define a partial ordering relation such as "is an ancestor of" (in a wider sense than "is a direct ancestor of") or symmetrically "is a descendant of" on the vertices of the tree, a vertex b being an ancestor of vertex a if the only path leading from the root to a passes through b.

A distinguished class of rooted trees are *binary trees* in which any node which is not a leaf has exactly two direct descendants. An important property of rooted trees is that, if the root node and all the edges starting therefrom are removed, we obtain as many rooted trees as the root had direct descendants, these having as root these direct descendants. This property is widely used both for computing characteristic quantities on trees and for proving properties by induction. The general principle is the following: in the (quite frequent) situation where the property to be proved is trivially true for the tree consisting of a single vertex and where it is true for the whole tree if it is true for all the subtrees starting from its root, the induction principle makes it possible to establish that the property is thus valid for any tree. Similarly, if we know how to compute a certain characteristic quantity for the tree consisting of a single vertex and combine the quantities of the subtrees of the root of a given tree to determine its own characteristic quantity, it is possible to devise thereof a recursive computation algorithm that can be applied to any tree and whose complexity is linear with the size of the tree considered. For example, to compute the number of leaves of a tree, one merely has to note that this number is one in the case of a tree comprising only one vertex, and then to observe that the number of leaves of any tree is equal to the sum of the numbers of leaves of all the subtrees starting from its root. To demonstrate that in a binary tree, the number of leaves is equal to the number of internal nodes plus one, it is clear that the property holds in the case of the single-vertex tree, and then that the property holds for any tree with $n+1$ vertices if the property holds for any tree with less than n vertices. The total number of internal nodes of the tree is actually equal to the sum of those for the two subtrees starting from the root plus one (the root) and the total number of leaves is equal to the sum of those of these same subtrees which, having necessarily less than n vertices, both verify the property.

1.3. Counting trees

A natural mathematical problem within the context that matters to us here consists of counting the number of different trees. This is a fairly old question since it goes back to at least the 19th century and some of the works by Arthur Cayley. It should be noted that the number of different trees depends both on the type of tree being considered (for instance, rooted or not) but also on what distinguishes them from each other, in particular whether their nodes are labeled or not. Two trees in which no nodes are labeled are identical if there exists a bijection between the vertices of each one which preserves the neighborhood relations (i.e. any neighbor of the image of a node is the image of a neighbor of the latter and vice versa). Intuitively, this means that the graphs present the same shape. If the nodes of the trees are labeled, even partially, two trees will be considered identical only if there exists a bijection between the vertices of each tree which preserves both the neighborhood relations and the labels. Obviously, two identical labeled trees will remain identical if their labels are deleted. In the following, it is implicitly assumed that two labeled nodes of the same tree have different labels.

1.3.1. *Fully labeled non-rooted trees*

The number of non-rooted trees in which all nodes are labeled (each with a unique label) with n nodes is n^{n-2}. This result is due to Cayley. Prüfer (1918) proposed a proof by establishing a bijection between labeled trees of size n and sequences comprising $n-2$ numbers taken from 1 to n.

This bijection associates with any tree of size n where the nodes are labeled (in this case, rather numbered) from 1 to n, the sequence s of length $n-2$ whose terms are taken between 1 and n constructed as follows. s is initialized to the empty sequence. As long as the tree contains more than two nodes, the number labeling the neighboring node of the leaf labeled with the smallest number is concatenated at the end of s. The tree is then updated by deleting the leaf labeled with the smallest number and the same process iterated.

Reciprocally, any sequence $s = s_1 s_2 \dots s_{n-2}$ of numbers taken from 1 to n is associated with the tree constructed in the following manner. We consider the set I

which is initialized as the set of numbers between 1 and n not appearing in s. The graph $\mathcal{G}$ is then built, initially empty, by iterating the following process. Let m be the smallest number of I. We add to the graph $\mathcal{G}$ a node labeled by m and a node labeled by the first term of s if they were not in there and an edge is added between these two nodes. We then update the set I by deleting m and adding the first term of s if and only if it was the last occurrence of this number in s. The sequence s is also updated by deleting its first term. These operations are repeated until obtaining an empty sequence. There are then exactly two numbers between 1 and n in I (at this point, $n - 2$ numbers have been removed between 1 and n from I and any number between 1 and n is, at any time, an element of I). Two nodes labeled by these numbers are added in $\mathcal{G}$ if they were not included therein and they are connected by an edge between them (see Figure 1.1).

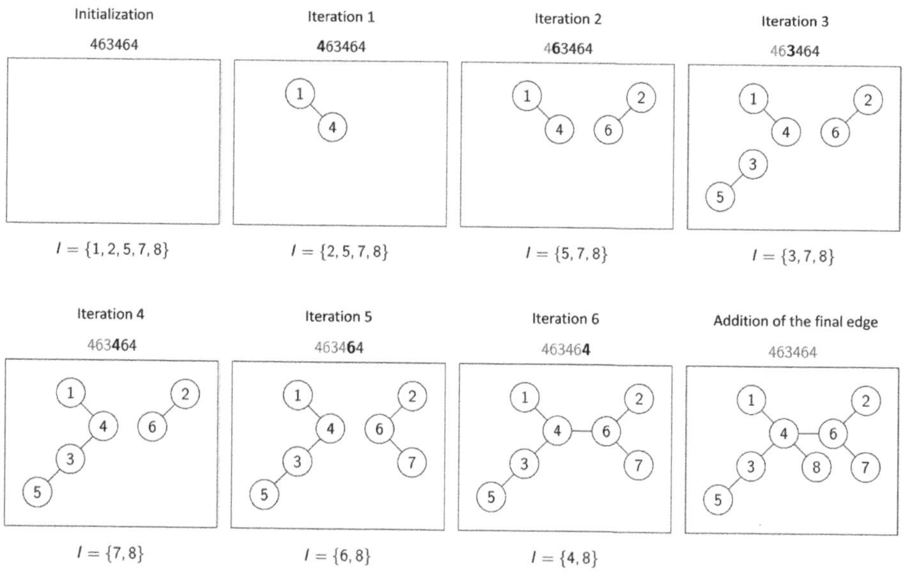

Figure 1.1. *Computation of the tree associated with the sequence 463464. For each iteration, the first term of the current sequence appears in bold and the states of the graph and of the set* I *are presented right after this iteration*

By construction, it follows that:

– the graph $\mathcal{G}$ contains exactly n nodes labeled from 1 to n after the last iteration;

– the graph $\mathcal{G}$ does not contain any cycle since, each time an edge is added, one of its two nodes will not be connected to any other node afterwards;

– after each iteration, any node of the graph $\mathcal{G}$ is connected either to a node labeled by an element of I, or to a node labeled by a term of the sequence s (updated). During the last iteration, any node of $\mathcal{G}$ is thus connected to any of the nodes labeled by the two numbers remaining in I and this last iteration connects the whole graph.

In summary, the graph $\mathcal{G}$ obtained from any sequence of $n-2$ numbers taken between 1 and n is indeed a connected and cycle-free graph; in other words, a tree of n vertices labeled from 1 to n.

Let us verify that the operation transforming a tree into a sequence, which will be called f, is indeed the reciprocal of the one transforming a sequence into a tree, which will be called g. Let s be any sequence of $n-2$ numbers taken between 1 and n. Let $\mathcal{T}$ be the tree obtained by applying transformation g to s and s' the sequence obtained by applying transformation f to $\mathcal{T}$. At the beginning of iteration 1 of transformation g applied to s, the set I contains the numbers labeling the leaves of $\mathcal{T}$, which is the current tree at the beginning of iteration 1 of transformation f applied to $\mathcal{T}$. By construction, the label of the neighbor of the leaf with the smallest label of $\mathcal{T}$ is s_1 and thus we correctly have $s'_1 = s_1$. Since the tree updated at the end of iteration 1 of f is the tree $\mathcal{T}$ deprived of its leaf with the smallest label which is also the image by g of the sequence $s_2 \dots s_{n-2}$ (up to renumbering the nodes and the sequence), one has in the same way $s'_2 = s_2$, etc. We can conclude that $f(g(s)) = s$ for any sequence s and a bijection has been established between the non-rooted labeled trees of size n and the sequences of length $n-2$ of numbers taken between 1 and n.

1.3.2. *Binary trees with labeled leaves*

We now consider the number of binary (therefore rooted) trees in which (all and only) the leaves are labeled, say from 1 to n, which is the most commonly found situation in phylogeny. We denote by $\mathcal{B}(n)$ the number of binary trees with n labeled leaves. We must bear in mind that n designates here the number of leaves and not the total number of nodes of the tree as previously. It has been previously seen that the number of internal nodes is then $n-1$ (the total number of nodes in the tree is thus $2n-1$). Let us consider a binary tree with n leaves labeled from 1 to n. The direct ancestor of the leaf labeled by n is either the root of the tree or another internal node. In the first case, if we delete leaf n and the root (its direct ancestor) as well as the two edges where they occur, a binary tree of $n-1$ leaves is obtained labeled from 1 to $n-1$. In the second case, the direct ancestor of leaf n is an internal node, which itself has a direct ancestor. Since the tree is binary, the direct ancestor of n also has a direct descendant other than n. If we delete leaf n and its direct ancestor as well as the three edges where either occurs and add an edge between the direct ancestor of the direct ancestor of n and the sibling node of n, again a binary tree of $n-1$ leaves labeled 1 to $n-1$ is obtained. It is deduced that any binary tree of $n-1$ leaves labeled from 1 to $n-1$ can be obtained by deleting leaf n from $2n-3$ binary trees of n leaves labeled from 1 to n. As a matter of fact, a binary tree of $n-1$ leaves has $2n-3$ nodes (leaves or internals, including the root). The direct ancestor of leaf n can be added as a direct ancestor of any of them so that the same binary tree is obtained by deleting it as previously. Moreover, if the binary trees obtained by deleting leaf n from two binary trees with n labeled leaves are different, the two binary trees with n labeled leaves are also different. It can be thus concluded that the number of trees with n labeled leaves is equal to the number of binary trees with $n-1$ labeled leaves multiplied by $2n-3$, that is, $\mathcal{B}(n) = (2n-3)\mathcal{B}(n-1)$. Since there is only one binary tree with two labeled leaves, it can be deduced that $\mathcal{B}(n) = (2n-3)(2n-5)\dots 1 = \prod_{i=2}^{n}(2i-3) = \frac{(2n-3)!}{2^{n-2}(n-2)!}$, a quantity that is denoted by $(2n-3)!!$

1.3.3. *Binary trees with labeled leaves and ordered internal nodes*

Within an evolutionary context, binary trees whose leaves are labeled and whose internal nodes are ordered with respect to one another in an order compatible with the tree are sometimes taken into consideration. An order on the nodes is *compatible* with their tree if it is such that the rank of any nodes is higher than that of any of their ancestors (and lower than that of any of their descendants). In particular, the root is always of rank 1 in a compatible order. Within a phylogenetic framework, a compatible order is interpreted as representing the temporal relationships of the evolutionary divergence events that led to this tree (see section 1.3.4).

Counting the number of ordered and labeled binary trees can be achieved in the same way as in the unordered case. Given an ordered binary tree with n leaves labeled from 1 to n. If leaf n and its direct ancestor are removed in the same way as before, an ordered binary tree is obtained with $n-1$ leaves labeled from 1 to $n-1$ in which the ranks of the nodes of higher order than the direct ancestor of n have been decreased by 1.

The number of possible positions of the direct ancestor of the leaf n depends on its rank among the internal nodes. We show by induction that if its rank is r, then the direct ancestor of leaf n has r possible positions. This trivially holds for $r = 1$, since the direct ancestor is then necessarily the root. Let us suppose that the property holds for some $1 < r < n-1$; in other words, there are exactly r branches (of the tree with $n-1$ leaves) that start from a node of rank strictly less than r and reach a node of rank greater than or equal to r or a leaf. Among these, a branch reaches the node of rank r and the $r-1$ others reach internal nodes of rank strictly higher than r or tips. Since the tree is binary, there are two branches leaving from the node of rank r. It is deduced that there are $r+1$ branches whose ancestor has a rank strictly lower than $r+1$, and the descendant has a rank higher than or equal to $r+1$ or is a leaf, and the property is thus true for any rank r. Finally, if its rank is r, there are as many possible positions of the direct ancestor of leaf n as there are branches starting from a node of rank strictly less than r and reaching a node of rank greater than or equal to r or a leaf, that is, r. As every ordered binary tree with $n-1$ labeled leaves, each possible rank and position for the direct ancestor of the leaf giving a different ordered tree and by denoting $\mathcal{O}(n)$ the number of ordered binary trees of size n, we have $\mathcal{O}(n) = \mathcal{O}(n-1)\sum_{r=1}^{n-1} r = \mathcal{O}(n-1)\frac{n(n-1)}{2}$. Since there is only one ordered tree with two labeled leaves, we conclude that $\mathcal{O}(n) = \frac{n!(n-1)!}{2^{n-1}}$.

1.3.4. *Number of orders of internal nodes of a given tree*

Since the order of the internal nodes constitute additional information, there are more ordered and labeled trees than only labeled trees. The following result shows how to compute the number of ordered trees corresponding to a given (non-ordered) tree, that is, the number of orders compatible with a given tree. The number of ordered trees determined in section 1.3.3 can be obtained by summing this number of compatible orders over all binary trees with labeled leaves.

THEOREM 1.1.– *The number $\mathcal{R}(\mathcal{T})$ of orders of internal nodes compatible with a binary, rooted tree $\mathcal{T}$ whose leaves are labeled is equal to:*

$$\mathcal{R}(\mathcal{T}) = \mathcal{R}(\mathcal{T}_a)\mathcal{R}(\mathcal{T}_b)\binom{\mathbf{L}_{\mathcal{T}_a} + \mathbf{L}_{\mathcal{T}_b} - 2}{\mathbf{L}_{\mathcal{T}_a} - 1}$$

where $\mathcal{T}_a$, $\mathcal{T}_b$, $\mathbf{L}_{\mathcal{T}_a}$ and $\mathbf{L}_{\mathcal{T}_b}$ are, respectively, the two sub-trees starting from the root of $\mathcal{T}$ and their numbers of leaves.

PROOF.– Let us first remark that the restriction of any order of internal nodes compatible with $\mathcal{T}$ to the internal nodes of $\mathcal{T}_a$ (respectively, $\mathcal{T}_b$) is an order compatible with $\mathcal{T}_a$ (respectively, $\mathcal{T}_b$). Conversely, any order obtained by merging an order compatible with $\mathcal{T}_a$ with an order compatible with $\mathcal{T}_b$ is an order compatible with $\mathcal{T}$. It follows that there are as many orders compatible with $\mathcal{S}$ as there are ways to merge an order of $\mathcal{T}_a$ with an order of $\mathcal{T}_b$. A merging is perfectly determined by the set of positions of the internal nodes of $\mathcal{T}_a$ in the order on $\mathcal{T}$. The positions of those of $\mathcal{T}_b$ are actually deduced therefrom and the relative order of the internal nodes of $\mathcal{T}_a$ (respectively, of $\mathcal{T}_b$) within its positions is given by the order on this subtree. In summary, we have:

- $\mathcal{O}(\mathcal{T}_a)$ orders compatible with $\mathcal{T}_a$;
- $\mathcal{O}(\mathcal{T}_b)$ orders compatible with $\mathcal{T}_b$;
- $\binom{\mathbf{L}_{\mathcal{T}_a}-1}{\mathbf{L}_{\mathcal{T}_a}+\mathbf{L}_{\mathcal{T}_b}-2}$ ways of merging an order of internal nodes of $\mathcal{T}_a$ with an order of internal nodes of $\mathcal{T}_b$.

Since all these possibilities can be independently combined to give an order compatible with $\mathcal{T}$, we obtain that $\mathcal{O}(\mathcal{T}) = \mathcal{O}(\mathcal{T}_a)\mathcal{O}(\mathcal{T}_b)\binom{\mathbf{L}_{\mathcal{T}_a}+\mathbf{L}_{\mathcal{T}_b}-2}{\mathbf{L}_{\mathcal{T}_a}-1}$. □

We observe that the number of compatible orders depends on the shape of the considered tree. In particular, it is linear with the number of leaves for the "caterpillar" tree but exponential in the case of a balanced tree.

1.3.5. *Directed binary trees*

A *directed* binary tree (not to be confused with the usual notion of graph orientation) is a binary tree in which the two direct descendants of any internal node can be distinguished. It is possible, for example, to distinguish between the eldest and the youngest, but the nature of this distinction does not matter here and we will speak of "left child" and "right child". Like the order of the internal nodes, the orientation of the direct descendants is an additional information. There are thus more directed binary trees than non-directed binary trees. To direct a binary tree whose leaves are labeled, it is necessary and sufficient to make as many independent binary choices ("what is its right child?") as there are internal nodes in the tree, that is, $n-1$ if the tree has n leaves. We thus have:

– $2^{n-1}\mathcal{B}(n) = 2^{n-1}(2n-3)!!$ unordered directed binary trees with n labeled leaves;

– $2^{n-1}\mathcal{O}(n) = n!(n-1)!$ ordered binary trees with n labeled leaves.

Similarly, there are $2^{n-1}\mathcal{R}(\mathcal{T})$ ordered directed binary trees corresponding to a binary tree $\mathcal{T}$ with n labeled leaves.

1.4. Probabilities of trees resulting from branching processes

Binary trees (directed, ordered or not) whose leaves are labeled appear naturally as the result of branching processes. In order to remain as general as possible, we consider here processes that involve elements that we shall call *lineages* and in which the only type of possible event at any time is that a lineage present at that time gives birth to a new lineage. The only other assumptions made are that the processes considered start with a single lineage and that they are *homogeneous in lineage*, that is, at any time t, no lineage alive at time t is more likely than another to create to a new lineage.

The evolution of such a process is naturally represented as a tree. More formally, any realization of this type of process is associated with the tree in which:

1) the internal nodes and the leaves of the tree are in bijection, respectively, with birth events and with the lineages of the process generating the tree;

2) for any leaf x, the direct ancestor node of the leaf associated with lineage x is the one corresponding to the last birth event involving x, which may be either its own birth or the last time it gave birth to a new lineage;

3) for any leaf x, the direct ancestor node of the internal node associated with the birth of lineage x is the one corresponding to the last event before the birth of x involving the parent lineage of x.

By construction, such a tree is binary and its leaves are labeled, but it is undirected and its internal nodes are unordered. If we add the following to the above points:

4) the internal nodes are ordered according to the temporal order of the events with which they are associated;

5) the left (respectively right) child of the internal node associated with the event "x gives birth to y" is the root of the subtree containing the leaf x (respectively, y).

A binary and oriented tree with ordered internal nodes and labeled leaves is obtained. If only point 4 (respectively, 5) is considered, the tree thus defined is only ordered (respectively, oriented).

In short, any realization of a branching process can be associated with a binary tree with labeled leaves that can be equipped with an orientation and/or an order on the internal nodes.

The birth order of the lineages (not to be confused with the order of the internal nodes) can be used to both direct and order the internal nodes of the tree (unordered and undirected) associated with a process realization. Starting with the orientation, the birth order of the lineages is used to recursively associate a lineage/leaf with any node of the tree in the following way. Each leaf is associated with itself and each node is associated with the oldest leaf/lineage among those associated with its two direct descendants (one of its descendants is thus always associated with the same leaf as itself). The orientation is then direct: the left child of an internal node is the direct descendant that is associated with the same leaf as itself and its right child is its other direct descendant. By construction, the internal node corresponding to the birth of lineage x is then the direct ancestor of the deepest node associated with x if it is not the root (otherwise x is the initial lineage). The birth order of the lineages then perfectly determines the order of the internal nodes of the tree defined by point 4 above.

Conversely, the orientation and order of the internal nodes of a tree enables us to determine without ambiguity the birth order of its lineages. Every lineage/leaf can actually be recursively associated with any node of the tree based on the orientation.

Again, we proceed recursively by associating with any internal node the leaf associated with its left child (a leaf being associated with itself). By construction, the internal node associated with the birth of the lineage x is then the direct ancestor of the deepest node associated with x if it exists (otherwise x is the initial lineage). The order of the nodes then perfectly determines the order of these events.

Since the way leaves, and therefore lineages, are labeled is arbitrary (in the sense that it depends neither on the tree nor on their birth order), a symmetry argument leads to the following remark.

REMARK 1.1.– If the lineages are arbitrarily labeled, all their birth orders are equiprobable.

Recall that a process is said to be *lineage homogeneous* if a birth taking place at a given time is attached to any lineage present at that time with uniform probability.

LEMMA 1.1.– *Given n lineages and their birth order resulting from a realization of a homogeneous birth process (without knowing their kinship relations), all trees have probability $\frac{1}{(n-1)!}$.*

PROOF.– Since the process is lineage homogeneous, the parent of the ith lineage is taken uniformly from the $(i-1)$ lineages alive at the time of its birth, that is, with probability $\frac{1}{i-1}$ for each one of them. It follows that, given the birth order of the lineages, the joint probability of a given set of parent–child relationships, therefore a tree, is $\frac{1}{(n-1)!}$. □

THEOREM 1.2.– *Let $\mathcal{T}^{o,r}$, $\mathcal{T}^{o}$, $\mathcal{T}^{r}$ and $\mathcal{T}$ be the binary trees with labeled leaves, respectively, both directed and ordered, directed but not ordered, ordered but not directed and neither directed nor ordered, resulting from a process of homogeneous birth in lineages. Conditional on the number n of final lineages, the probabilities of these trees are:*

$$\mathbf{P}(\mathcal{T}^{o,r} \mid \mathbf{L}_{\mathcal{T}^{o,r}} = n) = \frac{1}{(n-1)!n!}$$

$$\mathbf{P}(\mathcal{T}^{o} \mid \mathbf{L}_{\mathcal{T}^{o}} = n) = \frac{2^{n-1}}{(n-1)!n!}$$

$$\mathbf{P}(\mathcal{T}^{r} \mid \mathbf{L}_{\mathcal{T}^{r}} = n) = \frac{\mathcal{R}(\mathcal{T})}{(n-1)!n!}$$

$$\mathbf{P}(\mathcal{T} \mid \mathbf{L}_{\mathcal{T}} = n) = \frac{2^{n-1}\mathcal{R}(\mathcal{T})}{(n-1)!n!}$$

PROOF.– From remark 1.1, the probability of a birth order of n given lineages is $\frac{1}{n!}$. Combining this with lemma 1.1, the joint probability of a tree/birth order pair of n given lineages is $\frac{1}{(n-1)!n!}$. Since giving the birth order is equivalent to orientating the tree and ordering its nodes, we have indeed $\mathbf{P}(\mathcal{T}^{o,r} \mid \mathbf{L}_{\mathcal{T}^{o,r}} = n) = \frac{1}{(n-1)!n!}$.

Given a binary tree with labeled leaves, any pair "orientation of this tree" and "compatible order of its internal nodes" corresponds to a unique birth order of its lineages and vice versa. The probability of a directed but not ordered tree (respectively, ordered but not directed and neither directed nor ordered) is obtained by summing the joint probability of a tree/birth order pair over all possible directions (respectively, all possible orders of internal nodes and all possible directions and orders). Sections 1.3.5 and 1.3.4 then give us the probabilities $\mathbf{P}(\mathcal{T}^o \mid \mathbf{L}_{\mathcal{T}^o} = n)$, $\mathbf{P}(\mathcal{T}^r \mid \mathbf{L}_{\mathcal{T}^r} = n)$ and $\mathbf{P}(\mathcal{T} \mid \mathbf{L}_{\mathcal{T}} = n)$. □

The distribution of the shape of a binary tree conditional on its number of leaves, namely $\mathbf{P}(\mathcal{T} \mid \mathbf{L}_{\mathcal{T}} = n)$, is called the *Yule–Harding distribution* and was established by Harding (1971).

1.5. Birth–death processes

A birth–death process assumes that at some time t, every living individual reproduces at a rate λ to give a (single) new individual and disappears (dies) at a rate μ (Kendall 1948). In other words, denoting by N_t the random variable counting the number of individuals alive at time t, the birth–death model gives us the probabilities of evolution of the number of individuals during an infinitesimal time interval $\mathrm{d}t$. For any strictly positive integer n, we have:

$$\mathbf{P}(N_{t+\mathrm{d}t} = n+k \mid N_t = n) = o(\mathrm{d}t) \text{ for all } k \geq 2$$

$$\mathbf{P}(N_{t+\mathrm{d}t} = n+1 \mid N_t = n) = n\lambda\mathrm{d}t + o(\mathrm{d}t)$$

$$\mathbf{P}(N_{t+\mathrm{d}t} = n \mid N_t = n) = 1 - n(\lambda+\mu)\mathrm{d}t + o(\mathrm{d}t)$$

$$\mathbf{P}(N_{t+\mathrm{d}t} = n-1 \mid N_t = n) = n\mu\mathrm{d}t + o(\mathrm{d}t)$$

$$\mathbf{P}(N_{t+\mathrm{d}t} = n-k \mid N_t = n) = o(\mathrm{d}t) \text{ for all } k \geq 2$$

If, at any time t, one has $N_t = 0$, then individuals can no longer be born or die and one has $N_{t'} = 0$ for any time $t' \geq t$. Let I be the number of individuals present at time 0, which is the beginning of the process (in the most usual case $I = 1$). Let $p_n(t) = \mathbf{P}(N_t = n)$. We have in particular $p_I(0) = 1$ and $p_n(0) = 0$ for all $n \neq I$. For a given time $t \geq 0$, one has:

$$p_n(t+\mathrm{d}t) = p_{n+1}(t)(n+1)\mu\mathrm{d}t + p_{n-1}(t)(n-1)\lambda\mathrm{d}t$$
$$+(1-n(\lambda+\mu)\mathrm{d}t)p_n(t) + o(\mathrm{d}t)$$

for all $n \geq 1$and $p_0(t+\mathrm{d}t) = p_1(t)\mu\mathrm{d}t + p_0(t) + o(\mathrm{d}t)$.

It can be derived that:

$$\frac{\mathrm{d}p_n(t)}{\mathrm{d}t} = (n+1)\mu p_{n+1}(t) + (n-1)\lambda p_{n-1}(t) - n(\lambda+\mu)p_n(t)$$

for all $n \geq 1$, and:

$$\frac{\mathrm{d}p_0(t)}{\mathrm{d}t} = \mu p_1(t)$$

If we consider the generating function ϕ defined as:

$$\phi(z,t) = \sum_{n=-\infty}^{\infty} p_n(t)z^n$$

we get:

$$\frac{\phi(z,t)}{\partial t} = (z-1)(\lambda z - \mu)\frac{\phi(z,t)}{\partial z}$$

The above differential equation can be solved, for example, by the method of characteristics to obtain:

$$\phi(z,t) = \left(\frac{\mu(1-e^{-(\lambda-\mu)t}) - z(\mu - \lambda e^{-(\lambda-\mu)t})}{\lambda - \mu e^{-(\lambda-\mu)t} - z\lambda(1-e^{-(\lambda-\mu)t})}\right)^I$$

$$= \left(\frac{\mu(1-e^{-(\lambda-\mu)t}) - z(\mu - \lambda e^{-(\lambda-\mu)t})}{\lambda - \mu e^{-(\lambda-\mu)t} - \lambda(1-e^{-(\lambda-\mu)t})}\right)^I \left(\frac{1 - \frac{\lambda(1-e^{-(\lambda-\mu)t})}{\lambda-\mu e^{-(\lambda-\mu)t}}}{1 - z\frac{\lambda(1-e^{-(\lambda-\mu)t})}{\lambda-\mu e^{-(\lambda-\mu)t}}}\right)^I$$

$$= \left(\frac{\mu(1-e^{-(\lambda-\mu)t}) - z(\mu - \lambda e^{-(\lambda-\mu)t})}{(\lambda-\mu)e^{-(\lambda-\mu)t}}\right)^I \left(\frac{\frac{(\lambda-\mu)e^{-(\lambda-\mu)t}}{\lambda-\mu e^{-(\lambda-\mu)t}}}{1 - z\left(1 - \frac{(\lambda-\mu)e^{-(\lambda-\mu)t}}{\lambda-\mu e^{-(\lambda-\mu)t}}\right)}\right)^I$$

The negative binomial formula tells us that:

$$\left(\frac{p}{1-(1-p)z}\right)^n = \sum_{k=0}^{\infty}\binom{k+n-1}{n-1}p^n(1-p)^k z^k$$

Hence, it can be derived that:

$$\left(\frac{\frac{(\lambda-\mu)e^{-(\lambda-\mu)t}}{\lambda-\mu e^{-(\lambda-\mu)t}}}{1-z\left(1-\frac{(\lambda-\mu)e^{-(\lambda-\mu)t}}{\lambda-\mu e^{-(\lambda-\mu)t}}\right)}\right)^I =$$

$$\sum_{k=0}^{\infty}\binom{k+I-1}{I-1}\left(\frac{(\lambda-\mu)e^{-(\lambda-\mu)t}}{\lambda-\mu e^{-(\lambda-\mu)t}}\right)^I\left(\frac{\lambda(1-e^{-(\lambda-\mu)t})}{\lambda-\mu e^{-(\lambda-\mu)t}}\right)^k z^k$$

Since:

$$\left(\frac{\mu(1-e^{-(\lambda-\mu)t})-z(\mu-\lambda e^{-(\lambda-\mu)t})}{(\lambda-\mu)e^{-(\lambda-\mu)t}}\right)^I =$$

$$\frac{\sum_{j=0}^{I}\binom{I}{j}(\mu(1-e^{-(\lambda-\mu)t}))^{I-j}(-\mu+\lambda e^{-(\lambda-\mu)t})^j z^j}{\left((\lambda-\mu)e^{-(\lambda-\mu)t}\right)^I}$$

It finally yields that the coefficient of z^n in $\phi(z,t)$ is for all $n \geq 0$:

$$p_n(t) = \sum_{k=0}^{\min(I,n)}\binom{I}{k}\binom{I+n-k-1}{I-1}$$
$$\frac{\left(\mu(1-e^{-(\lambda-\mu)t})\right)^{I-k}\left(\lambda(1-e^{-(\lambda-\mu)t})\right)^{n-k}\left(-\mu+\lambda e^{-(\lambda-\mu)t}\right)^k}{(\lambda-\mu e^{-(\lambda-\mu)t})^{I+n-k}}$$

In particular for $I = 1$, the usual case considered in evolution, one has:

$$p_n(t) = \frac{e^{-(\lambda-\mu)t}(\lambda-\mu)^2\left(\lambda(1-e^{-(\lambda-\mu)t})\right)^{n-1}}{(\lambda-\mu e^{-(\lambda-\mu)t})^{n+1}} \text{ for all } n \geq 1$$

and:

$$p_0(t) = \frac{\mu(1-e^{-(\lambda-\mu)t})}{\lambda-\mu e^{-(\lambda-\mu)t}}$$

1.5.1. *Probability density of a birth–death tree*

We are interested here in the joint probability density of a set of random variables whose combination defines a rooted phylogenetic tree with dated nodes. The random variables in question are as follows: a directed tree without leaf labels (τ_o), the ages of the associated nodes ($t_1, \ldots, t_{n-1}$) and the number of leaves (n). The joint density of these three variables is conditionally defined to the time at which the observation of the n lineages is performed (t_0). This time is considered here as being unique.

We also consider as known the values of birth and death parameters, λ and μ, as well as the time at which the first lineage was born (t_n), or time of origin of the process. Finally, we assume that this time of origin is a random variable uniformly distributed in the interval $[t_0, t_b]$. We thus consider here that the realizations of the birth and death process are initiated after time t_b and before t_0. For practical reasons, the axis of time is reversed with respect to the progression of the process t_b which goes from t_b to t_0 (i.e. we have $t_b \geq t_0$ even though from the point of view of the process, t_b is earlier than t_0). The joint density of the tree is given by (see Figure 1.2):

$$p(\tau_o, t_1, \ldots, t_{n-1}, n|\lambda, \mu, t_n, t_0, t_b) = p_1(t_n - t_0) \prod_{i=1}^{n-1} \lambda p_1(t_i - t_0) \qquad [1.1]$$

The time of origin of the process t_n is generally not known. It is therefore necessary to integrate over all the values that this random variable can take. Assuming a uniform distribution on $[t_0, t_b]$, the density of interest becomes:

$$\begin{aligned}
&p(\tau_o, t_1, \ldots, t_{n-1}, n|\lambda, \mu, t_0, t_b)\\
&= \int_{t_0}^{t_b} p(\tau_o, t_1, \ldots, t_{n-1}, n|\lambda, \mu, t_n, t_0, t_b) p(t_n|\lambda, \mu, t_0, t_b) \mathrm{d}t_n\\
&= \lambda^{n-1} \prod_{i=1}^{n-1} p_1(t_i - t_0) \left(\frac{1}{t_b - t_0}\right) \int_{t_0}^{t_b} p_1(t_n - t_0) \mathrm{d}t_n\\
&= \lambda^{n-1} \prod_{i=1}^{n-1} p_1(t_i - t_0) \left(\frac{1}{t_b - t_0}\right)\\
&\frac{\lambda - \mu}{\mu} \left(\frac{1}{\lambda - \mu e^{-(\lambda-\mu)t_0}} - \frac{1}{\lambda - \mu e^{-(\lambda-\mu)t_b}}\right) \qquad [1.2]
\end{aligned}$$

When $t_0 = 0$ and $t_b \to \infty$, the integral in the previous expression is equal to $1/\lambda$. It then yields:

$$p(\tau_o, t_1, \ldots, t_{n-1}, n|\lambda, \mu, t_0 = 0, t_b \to \infty) \propto \lambda^{n-2} \prod_{i=1}^{n-1} p_1(t_i)$$

and the symbol "proportional to" ($\propto$) becomes necessary because $1/(t_b - t_0) \to 0$ when $t_b \to \infty$ and $t_0 = 0$.

Phylogenetic trees are generally undirected and leaf-labeled, often associated with the names of the genetic sequences being analyzed. There exists 2^{n-1} directed trees that differ exclusively in the color of their branches (see Figure 1.2). Each of these combinations is equiprobable. There are also $n!$ permutations of labels at the leaves of the tree. Each of these is, again, equiprobable. The joint probability density of an undirected tree, n, with labels at the leaves is given by the following expression:

$$p(\tau, t_1, \ldots, t_{n-1}, n|\lambda, \mu, t_0 = 0, t_b \to \infty) \propto \frac{2^{n-1}}{n!} \lambda^{n-2} \prod_{i=1}^{n-1} p_1(t_i)$$

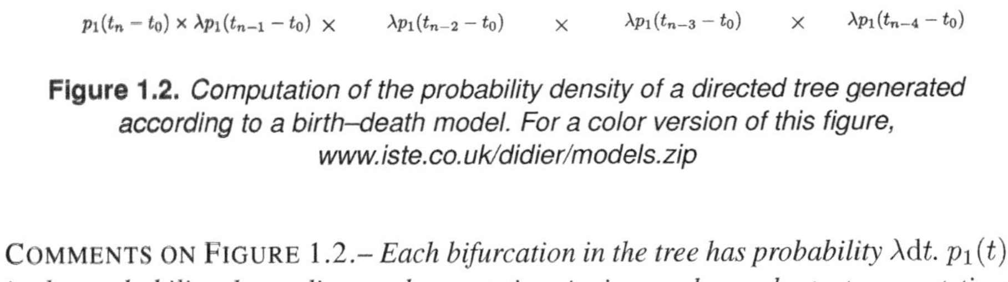

Figure 1.2. *Computation of the probability density of a directed tree generated according to a birth–death model. For a color version of this figure, www.iste.co.uk/didier/models.zip*

COMMENTS ON FIGURE 1.2.– *Each bifurcation in the tree has probability* λdt. $p_1(t)$ *is the probability that a lineage born at time* t *gives a descendant at present time* ($t_0 = 0$). *The term* $p_1(t_i)$ *given under each tree corresponds to the probability density associated with the bifurcation occurring at time* t_i. *The tree generated here is directed (each branch has a color) and is denoted* τ_o.

The number of leaves in the tree is usually determined by the sampling contingencies rather than by the stochastic process generating the data. It is then preferable to give the joint density of the tree conditionally on n. It thus follows that:

$$
\begin{aligned}
&p(\tau, t_1, \ldots, t_{n-1} | n, \lambda, \mu, t_0 = 0, t_b \to \infty) \\
&= \frac{p(\tau, t_1, \ldots, t_{n-1}, n | \lambda, \mu, t_0 = 0, t_b \to \infty)}{\Pr(n | \lambda, \mu, t_0 = 0, t_b \to \infty)} \\
&= \frac{p(\tau, t_1, \ldots, t_{n-1}, n | \lambda, \mu, t_0 = 0, t_b \to \infty)}{\int_{t_0}^{t_b} p_n(t_n) p(t_n | \lambda, \mu, t_0 = 0, t_b \to \infty) \mathrm{d}t_n} \\
&= \frac{\frac{2^{n-1}}{n!} \lambda^{n-2} \prod_{i=1}^{n-1} p_1(t_i)}{\int_{t_0}^{t_b} p_n(t_n) \mathrm{d}t_n} \\
&= \frac{\frac{2^{n-1}}{n!} \lambda^{n-2} \prod_{i=1}^{n-1} p_1(t_i)}{1/\lambda n} \\
&= \frac{2^{n-1}}{(n-1)!} \lambda^{n-1} \prod_{i=1}^{n-1} p_1(t_i) \\
&= \prod_{i=1}^{n-1} \left(\frac{2\lambda}{i} \right) p_1(t_i)
\end{aligned}
\qquad [1.3]
$$

It is also useful to give the joint density of the tree conditional on t_{n-1}, that is, the age of the first divergence. Molecular dating analyses often rely on a marginal distribution of the age of this node derived from fossil analysis. The product of the joint density of the tree knowing t_{n-1} according to the birth–death process by the marginal density derived from fossils thus allows molecular dating to be calibrated adequately. We have:

$$
\begin{aligned}
&p(\tau, t_1, \ldots, t_{n-2} | t_{n-1}, n, \lambda, \mu, t_0 = 0, t_b \to \infty) \\
&= \frac{p(\tau, t_1, \ldots, t_{n-1} | n, \lambda, \mu, t_0 = 0, t_b \to \infty)}{p(t_{n-1} | n, \lambda, \mu, t_0 = 0, t_b \to \infty)} \\
&= \frac{p(\tau, t_1, \ldots, t_{n-1} | n, \lambda, \mu, t_0 = 0, t_b \to \infty)}{\sum_\tau \int \cdots \int_{0 \le t_1 \le \ldots \le t_{n-1}} p(\tau, t_1, \ldots, t_{n-1} | n, \lambda, \mu, t_0 = 0, t_b \to \infty) \mathrm{d}t_1 \ldots \mathrm{d}t_{n-2}} \\
&= \frac{\prod_{i=1}^{n-1} p_1(t_i)}{\sum_\tau \frac{1}{\mu^{n-2}(n-2)!} p_0(t_{n-1})^{n-2} p_1(t_{n-1})} \\
&= \frac{(n-2)!}{\sum_\tau 1} \prod_{i=1}^{n-2} \mu \frac{p_1(t_i)}{p_0(t_{n-1})}
\end{aligned}
$$

$\sum_\tau 1$ corresponds to the number of undirected trees with leaf labels. This is equal to $\frac{(n-1)!n!}{2^{n-1}}$ (see section 1.3.3). The desired density is then given by the following expression:

$$p(\tau, t_1, \dots, t_{n-2} | t_{n-1}, n, \lambda, \mu, t_0 = 0, t_b \to \infty)$$

$$= \frac{2^{n-1}}{n!(n-1)} \prod_{i=1}^{n-2} \mu \frac{p_1(t_i)}{p_0(t_{n-1})} \quad [1.4]$$

The above expression is that given by Yang and Rannala (1997, p. 718, equation 6). It should be noted here that the tree density is derived conditionally from the number n of sampled lineages. When the value of n is determined by the tree generation process, it is then preferable to use equation [1.3] as support for the inference of birth and death parameters. For instance, in the early stages of an epidemic, the number of sampled lineages is sometimes close to or at least strongly correlated with the total number of circulating pathogens. The value of n then conveys information about the birth–death process and incorporating this information within the context of inference proves to be useful. When n is determined by sampling, and in particular by possible financial constraints limiting the number of sampled individuals, it makes sense to refer to equation [1.4] for inference. Therefore, the last equation is more appropriate when there is no reason to believe that n is close to or correlated with the total number of living lineages at the time of sampling, as might be the case when the abundance of the pathogen far exceeds the surveillance capabilities of the epidemic.

1.6. The coalescent

Two particles coalesce when they collide with each other to form a single one. In the case where the particles involved correspond to lineages, the division of one of them, following a gene or genome speciation or duplication event, corresponds to a "coalescence" when the time arrow is reversed; in other words, when two lineages coalesce to form one. A wide variety of coalescence processes can be found. Some of these processes imply the merging of more than two lineages, for example. Others allow for multiple mergers at the same time.

The coalescent proposed by Kingman is perhaps the simplest of all these processes and it is the one that we shall describe here. Despite its relative simplicity, this model has led to great advances in the field of population biology. This success is explained by its close connection with several classical models in population genetics such as the Wright–Fisher model and the Moran model as we will see later in this chapter.

Let $\mathcal{P}_n$ be the set of partitions of a set of n elements that correspond here to our n species or genes. For $n = 3$ for example, with $[n] = \{a, b, c\}$ denoting the names (or labels) of the three species, then this set of partitions is as follows:

$$\mathcal{P}_3 = \{\{a|b|c\}, \{a, b|c\}, \{a, c|b\}, \{b, c|a\}, \{a, b, c\}\}$$

The $\{a|b|c\}$ partition is composed of three "blocks", each block corresponding here to a lineage, whereas $\{a, b|c\}$ is composed of two blocks. Generally speaking, each block corresponds to a particle. The Kingman coalescent is a stochastic process denoted by $(\Pi_t^n, t \geq 0)$ with $\mathcal{P}_n$ for state space. It is defined in the following manner:

– the initial state Π_0^n is the one corresponding to the set of singletons ($\{a|b|c\}$ in the example above);

– $(\Pi_t^n, t > 0)$ is a Markovian process whose transition rates from x to y, denoted by $q(x, y)$ are null if y is not obtained by merging two blocks of x and equal to 1 otherwise. For example, $q(\{a, b|c\}, \{a, c|b\}) = 0$ and $q(\{a|b|c\}, \{a, c|b\}) = 1$.

In other words, the initial state of the stochastic process that matters to us here corresponds to a set of n lineages taken at time $t = 0$. Each of the $n(n-1)/2$ block pairs is associated with an alarm that is triggered after a random time. The process of merging block pairs being Markovian, the distribution of the latter is an exponential of parameter 1. The first alarm sounds at a time corresponding to the minimum of the $n(n-1)/2$ random times. This time is thus distributed according to an exponential of parameter $n(n-1)/2$. The number of blocks then decreases from n to $n-1$ and the merging of two blocks taken among $n-1$ takes place after a random time of exponential distribution of parameter $(n-1)(n-2)/2$. This process goes on until a single block is obtained grouping all of the n initial lineages.

1.6.1. *Links with "classical" models in population genetics*

The Kingman coalescent, or n-coalescent, is remarkable for its strong connections with various mathematical models used in population genetics, establishing an equivalence link between them. As such, the genealogies obtained with the Moran model are identical, from a probabilistic perspective, to those derived from the n-coalescent. According to the Moran model, each individual within a population of N individuals dies after a random time and is replaced by an individual taken at random uniformly from N. The distribution of the life span of each individual is an exponential of parameter 1. In Figure 1.3, this duration corresponds to the time elapsed between two successive dots along one of the $N = 5$ vertical lines.

1.6.2. *Moran's model*

To determine the distribution of coalescence times within a genealogy generated according to the Moran model, we focus on a sample of n individuals taken among N at time $t = 0$ and we follow the set of N lineages constituting the population by back tracking from the present time to the past. The first dot occurs after an exponentially distributed period of parameter N (since the time elapsed between two successive dots along one of the lineages is an exponential of parameter 1). This dot corresponds to a coalescence between the lineages of our sample with a probability equal to $\frac{n}{N} \times \frac{n-1}{N-1} \times (1 - \frac{1}{N}) = \frac{n(n-1)}{N^2}$. In fact, the probability that the lineage on which the first dot is located is from our sample is n/N. In addition, there are $n - 1$ lineages with which this lineage can merge to give a coalescence, among $N - 1$ lineages in total. Finally, the probability that this same dot does not correspond to replacing a lineage by itself is $(1 - \frac{1}{N})$. Thereby, the lineages in our sample coalesce at a rate equal to $n(n-1)/N^2 \times N = n(n-1)/N$ coalescences per unit time. This rate is different from the Kingman coalescent rate, which is $n(n-1)/2$. However, when the Moran model time unit is replaced by a new unit, the coalescent time unit, such that one coalescent time unit equals $N/2$ Moran time units, then the coalescence rate measured in expectation of the number of coalescences per coalescent time unit is $n(n-1)/2$. Consequently, the distribution of coalescence times according to the Moran model is that of the Kingman coalescent after a simple scaling of the time unit.

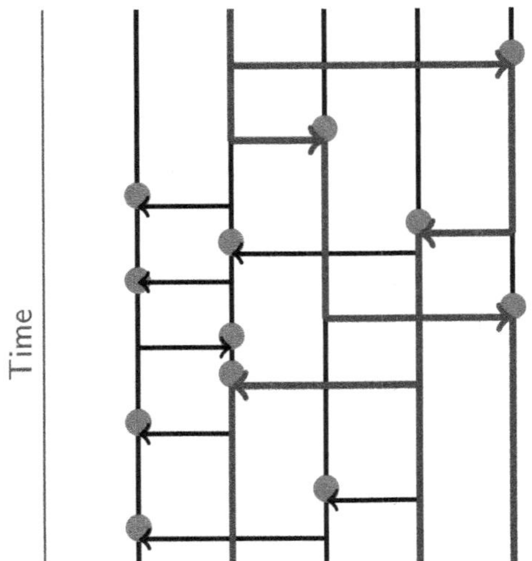

Figure 1.3. *Moran's model. Time flows from top to bottom. Each dot corresponds to the death of an individual. After such an event, the individual in question is immediately replaced by another. The red lines constitute a sample of size n = 3, taken from the population of size N = 5. For a color version of this figure, www.iste.co.uk/didier/models.zip*

It should be noted that when the time elapsed between two successive dots, taken from the set of the N lineages, is fixed and equal to a unit of time known as "Moran's time" (rather than a random quantity distributed according to an exponential of parameter N), then the probability that two lineages merge is $2/N^2$. Therefore, the probability that two lineages do not have a common ancestor until k units of Moran time is then equal to $(1 - 2/N^2)^k$. When the Moran time unit is scaled so that k units of that time are equivalent to $(N^2/2)t$ units of the "new" time (i.e. the time in Moran units is given by the time in coalescent units multiplied by $N^2/2$), then $\lim_{N\to\infty}(1 - 2/N^2)^{N^2t/2} = e^{-t}$. The Moran process thus still corresponds, this time asymptotically ($N \to \infty$), to a coalescent when one unit of the new time (the coalescent time) is equivalent to $2/N^2$ units of Moran time.

1.6.3. *The Wright–Fisher model*

The Wright–Fisher model is central to population genetics, because it makes it possible to quantify the phenomenon of genetic drift. As for the Moran model, we assume here that the population size is fixed and equal to N. Conversely, the generations here are non-overlapping: the N parents of generation t produce N descendants that replace their parents, thus becoming the N parents of generation $t + 1$. The relationship between individuals of two successive generations is defined by considering how each descendant "chooses" a parent. This is here a uniform random draw with replacement. Two descendants thus choose the same parent with a probability equal to $1/N$. The probability that two individuals have no common ancestor by back tracking g generations back in time is therefore equal to $(1 - 1/N)^g$. When the unit of time, the generation, is scaled such that g generations are equivalent to Nt units of "new" coalescent time, then $\lim_{N\to\infty}(1 - 1/N)^{Nt} = \exp(-t)$. The coalescence time is therefore distributed here, for a pair of individuals, according to an exponential of parameter 1. More generally, as for the Moran model, the distribution of coalescence times obtained according to the Wright–Fisher model is that of the coalescent when the time is scaled adequately.

1.6.4. *Generic model*

The Moran and Wright–Fisher models are strongly related to the Kingman coalescent in the sense that the distribution of times at which lineages merge within the genealogy of a sample is given by (in the case of the Moran model with an exponentially distributed lifetime of each individual) or tends toward (in the Wright–Fisher and Moran model with a fixed time between two successive reproduction/extinction events) the distribution defined by the Kingman stochastic coalescent process. It is therefore natural here to define more precisely the contours of a generic population model whose distribution of coalescence times is given by the n-coalescent.

Let X_i be the number of direct offspring of parent i. The generic model is such that $\sum_{i=1}^{N} X_i = N$, that is, the number of individuals in the population is fixed. The X_i, $i = 1, \ldots, N$ all have the same distribution. However, they are not independent since they are constrained by the size of the population. The generic model is said to be "exchangeable", because the order in which the N parents are considered in the sum $\sum_{i=1}^{N} X_i$ is not important. The fact that the X_i are identically distributed implies that the generic model is not aware of natural selection, and this carries on in each generation. Assuming that the generations are non-overlapping, the probability p that two offspring have the same parent is then given by the following expression:

$$
\begin{aligned}
p &= \mathbb{E}\Big(\sum_{i=1}^{N}\Big(\frac{X_i}{N}\Big)\Big(\frac{X_i - 1}{N-1}\Big)\Big) \\
&= \frac{1}{N(N-1)}\sum_{i=1}^{N}\mathbb{E}\Big(X_i(X_i-1)\Big) \\
&= \frac{1}{N(N-1)}\sum_{i=1}^{N}\mathbb{V}\Big(X_i\Big) \\
&= \frac{\mathbb{V}\Big(X_1\Big)}{N-1}
\end{aligned}
$$

Let us note here first that $\sum_{i=1}^{N} X_i = N$ implies that $\sum_{i=1}^{N} \mathbb{E}(X_i) = N$ and thus $\mathbb{E}(X_i) = 1$ (and $[\mathbb{E}(X_i)]^2 = 1$) because the X_i all have the same expected value. They also have the same variance and thereby $\sum_{i=1}^{N} \mathbb{V}\Big(X_i\Big) = N\mathbb{V}\Big(X_1\Big)$. When it is considered that the time elapses in a discrete way, the probability that two lineages do not have a common ancestor by going back k time units in the past is $(1 - \mathbb{V}(X_1)/(N-1))^k$ and the Kingman coalescent is thus obtained here asymptotically by changing the time scale such that k time units of the generic model are equivalent to $(N-1)t/\mathbb{V}(X_1)$ coalescent time units.

Consequently, for the generic model discussed in this section, convergence to the Kingman coalescent is achieved by accelerating time by a factor proportional to the population size and inversely proportional to the variance of the number of offspring produced by a parent. This factor, commonly referred to as "effective population size" and generally denoted by N_e, is thus a scalar that is used to scale up to the same scale (that of the Kingman coalescent) a variety of population models.

The effective population size is a central concept to population genetics because it makes it possible to quantify *genetic drift*. Therefore, when considering the forward-in-time Wright–Fisher model (that is if $g > h$, then g designates a point in time located in the future relative to h) for a population consisting of two types of individuals, the probability p_g of sampling two individuals of distinct type at generation $g > 0$ is equal to $p_g = p_0(1 - 1/N)^g$. This probability, called heterozygosity, decreases at the same rate, forward in time, as the coalescence rate measured this time by reversing the time arrow.

1.6.5. *Coalescent-generated tree probability density*

The process of lineage merging being Markovian, the derivation of the probability density of a tree generated according to the Kingman coalescent is almost immediate. It is nevertheless necessary to pay attention to the unit of time being used. We consider here that the node ages are expressed in calendar time units, as was already the case with the birth–death model. With the Wright–Fisher model, the number of generations multiplied by the effective population size N_e gives the correct unit of time for the Kingman coalescent. If ρ denotes the average duration of a generation, expressed in calendar time, then the coalescent time unit is given by the product of ρN_e by the calendar time. We thus have:

$$
\begin{aligned}
&p(\tau, t_1, \ldots, t_{n-1}|n, N_e, \rho) \\
&= \prod_{i=2}^{n} \frac{i(i-1)}{2N_e\rho} \exp\left[\frac{i(i-1)}{2N_e\rho}(t_{n-i+1} - t_{n-i})\right] \times \frac{2}{i(i-1)} \\
&= \left(\frac{1}{N_e\rho}\right)^{n-1} \exp\left[\sum_{i=2}^{n} \frac{i(i-1)}{2N_e\rho}(t_{n-i+1} - t_{n-i})\right]
\end{aligned}
$$

The previous equation involves the set of time periods between pairs of "successive" nodes, that is, pairs whose ages are (t_{i+1}, t_i) for any i ranging from 0 to $n-2$. Each term in this product corresponds to the multiplication of an exponential density of parameter $i(i-1)/2N_e$ by the probability that two lineages (with a label) in i do merge. Let us point out here that only the product ρN_e can be estimated from a tree. Therefore, inferring the effective size of a population is only possible on the condition that the average duration of a generation is known (and vice versa).

1.7. Conclusion

This chapter gives elements of combinatorics on trees and describes in detail the derivations of the probability densities of rooted trees generated according to the birth and death processes and according to the Kingman coalescent. Outlining the steps for these calculations makes it possible to highlight the main assumptions upon which these two models are based and therefrom to evaluate their biological relevance.

For example, both models assume that the sampled lineages are exchangeable: the probability density of the trees remains the same regardless of the permutation of the labels on the leaves of the tree. This assumption greatly simplifies the calculations. From a biological point of view, however, it implies that the different lineages are all "equal" and therefore does not allow natural selection to act.

The two tree generation models presented in this chapter also assume that their parameters, that is, the effective population size and birth and death rates, do not vary during evolution. By way of scaling the time unit, where the population size is expressed as a function of time (generally an exponential), it is possible to fit to the data a coalescent whose parameter varies during the evolution (while remaining constant between lineages taken at the same time). This type of model has also given rise to several variations, the best known being the following ones: "classic skyline", "generalized skyline", "Bayesian skyline", "Bayesian skyride" (see Pybus et al. (2000) for the first model of the "skyline" type and Ho and Shapiro (2011) for a review).

The variations in population size detected from this type of coalescent may nevertheless be mistaken for changes in the structuring of the population over time. Consequently, the transition from a panmictic population toward a population split into distinct demes leads to an increase in coalescence times, an increase that is all the greater as the migration rate between demes is low. Yet, the same phenomenon is also expected for a panmictic population whose effective size increases (for more details, see Mazet et al. (2016)). Similar limitations also affect birth–death models that allow the rates of both types of events to vary over time. Therefore, Louca and Pennell (2020) have shown that the likelihood of a given tree can be the same for an infinite number of distinct trajectories of birth and death rates (where a trajectory describes the variation of rates over time, the set of lineages at a given time all share the same rates). This result calls into question a number of macroevolutionary studies. Consequently, some of the studies focusing on possible correlations between diversification patterns and environmental factors should be reinterpreted with caution (for a more optimistic approach to this issue, see Pagel (2020) and Morlon et al. (2020)).

Furthermore, the two classes of models introduced in this chapter yield quite distinct predictions about the total number of lineages asymptotically. The Wright–Fisher and Moran models, and thus for the coalescent that is derived thereof, rely on a strict constraint of constant population size. For the birth–death model, the situation is more complex, because for infinitely long times the expectation of the total number of lineages tends either to an infinite quantity or to zero. From a biological point of view, it seems relevant to rely on a model according to which the population or the species under study necessarily becomes extinct in time. On the other hand, it is necessary to accept here the possibility that the population/species "explodes" sometimes to reach an infinite size, which is obviously more difficult to justify from a biological perspective.

A number of issues due to sampling also arise. When the number of lineages sampled is fixed by the experimental design, as is generally the case for the coalescent and, in some situations, for the birth–and–death model, then it makes sense to express the tree probability distribution conditional on the sample size. Nonetheless, in other situations, the number of sampled lineages must be considered as a random quantity whose value is derived from the stochastic process generating the tree. In terms of inferring demographic parameters, it is then the joint density of the tree and the number of lineages that best describes the data.

Although the influence of the type of sampling carried out is important from the point of view of inferring evolutionary parameters and testing biological hypotheses, these aspects have only been partially taken into account to date. Therefore, in the context of phylodynamic inference (see Chapter 11), only a few studies have observed and attempted to rectify biases in the inference of effective population sizes due to the time sampling of the sequences (Volz and Frost 2014; Karcher et al. 2016; Cappello and Palacios 2020; Parag et al. 2020). Stadler (2010) and Stadler et al. (2013) also describe a generalization of the birth–death model taking into account the sampling intensity and its variation in time.

1.8. References

Aldous, D.J. (1999). Deterministic and stochastic models for coalescence (aggregation and coagulation): A review of the mean-field theory for probabilists. *Bernoulli*, 5(1), 3–48.

Berestycki, N., Etheridge, A., Hutzenthaler, M. (2009). Survival, extinction and ergodicity in a spatially continuous population model. *Markov Processes and Related Fields*, 15, 265–288.

Cappello, L. and Palacios, J.A. (2020). Adaptive preferential sampling in phylodynamics. Report, Stanford University.

Harding, E.F. (1971). The probabilities of rooted tree-shapes generated by random bifurcation. *Advances in Applied Probability*, 3(1), 44–77.

Ho, S.Y. and Shapiro, B. (2011). Skyline-plot methods for estimating demographic history from nucleotide sequences. *Molecular Ecology Resources*, 11(3), 423–434.

Karcher, M.D., Palacios, J.A., Bedford, T., Suchard, M.A., Minin, V.N. (2016). Quantifying and mitigating the effect of preferential sampling on phylodynamic inference. *PLOS Computational Biology*, 12(3), e1004789.

Kendall, D. (1948). On the generalized "birth-and-death" process. *The Annals of Mathematical Statistics*, 19(1), 1–15.

Louca, S. and Pennell, M.W. (2020). Extant timetrees are consistent with a myriad of diversification histories. *Nature*, 580(7804), 502–505.

Mazet, O., Rodríguez, W., Grusea, S., Boitard, S., Chikhi, L. (2016). On the importance of being structured: Instantaneous coalescence rates and human evolution – Lessons for ancestral population size inference? *Heredity*, 116(4), 362–371.

Möhle, M. (2000). Ancestral processes in population genetics – The coalescent. *Journal of Theoretical Biology*, 204(4), 629–638.

Morlon, H., Hartig, F., Robin, S. (2020). Prior hypotheses or regularization allow inference of diversification histories from extant timetrees. *bioRxiv* [Online]. Available at: https://doi.org/10.1101/2020.07.03.185074.

Pagel, M. (2020). Evolutionary trees can't reveal speciation and extinction rates. *Nature*, 580(7804), 461–462.

Parag, K.V., du Plessis, L., Pybus, O.G. (2020). Jointly inferring the dynamics of population size and sampling intensity from molecular sequences. *Molecular Biology and Evolution*, 37(8), 2414–2429.

Prüfer, H. (1918). Neuer beweis eines satzes über permutationen. *Arch. Math. Phys.*, 27, 742–744.

Pybus, O.G., Rambaut, A., Harvey, P.H. (2000). An integrated framework for the inference of viral population history from reconstructed genealogies. *Genetics*, 155(3), 1429–1437.

Stadler, T. (2010). Sampling-through-time in birth–death trees. *Journal of Theoretical Biology*, 267(3), 396–404.

Stadler, T., Kühnert, D., Bonhoeffer, S., Drummond, A.J. (2013). Birth-death skyline plot reveals temporal changes of epidemic spread in HIV and hepatitis C virus (HCV). *Proceedings of the National Academy of Sciences*, 110(1), 228–233.

Volz, E.M. and Frost, S.D. (2014). Sampling through time and phylodynamic inference with coalescent and birth–death models. *Journal of the Royal Society Interface*, 11(101), 20140945.

Yang, Z. and Rannala, B. (1997). Bayesian phylogenetic inference using DNA sequences: A Markov chain Monte Carlo method. *Molecular Biology and Evolution*, 14, 717–724.

2

Models of Sequences and Discrete Traits Evolution

Étienne PARDOUX
I2M, CNRS, Aix-Marseille Université, France

2.1. Introduction

Models of sequence evolution are mainly used within the context of phylogenetic tree reconstruction. For instance, let us consider three species A, B and C that descend from a common ancestor. It is expected that there were two divergences in the past at different times, and we want to know what the shape of the phylogenetic tree of these three species is. If the genomes of A and B are closer to one another than each is to C, then C is predicted to have split from A and B before the latter two diverged.

However, to assert such a fact, it is always better to have a reliable model of genomic sequence evolution. In fact, it could be that the differences between the genome of C and those of species A and B involve nucleotides that evolve faster than those that differ between A and B, which might call into question the above statement.

It is therefore important to realistically model the accumulation of mutations in the genomes of two species, which are going to be called A and B, and that diverged at a certain date T in the past. The origin of time is placed at this divergence time, T becoming the present time. Let us denote by X_t the state of the genome of species A at any time $t \in [0, T]$, Y_t that of the genome of species B at the same time t. Since, at time $t = 0$, A and B are one and the same species, $X_0 = Y_0$. Between $t = 0$

Models and Methods for Biological Evolution,
coordinated by Gilles DIDIER and Stéphane GUINDON.

and $t = T$, the two species randomly accumulate mutations independently of each other. To shift from the genome of species A to that of species B, one must follow the evolution of X_t backwards in time from $t = T$ to $t = 0$, then follow the evolution of Y_t from $t = 0$ to $t = T$. We assume that the laws of the two processes $X_\cdot$ and $Y_\cdot$ are the same, and that in addition the evolution of X_t in backward time is the same as that of X_t in the forward time direction. Then shifting from the genome of A to that of B corresponds to the evolution of one of the two processes during a time $2T$.

So far we have talked about stochastic processes without specifying what kind of process we have in mind. In fact, it is quite natural to model X_t and Y_t as Markov processes. This class of processes is characterized by the following property. Knowing that $X_t = x$, the future evolution of X after t is independent of its past before t. Moreover, it will be seen that some of these processes are reversible in the sense that, if they are initialized by their invariant probability (which is totally natural since the evolutions are considered over very long durations), then for all T, the distributions of $\{X_t, 0 \le t \le T\}$ and $\{X_{T-t}, 0 \le t \le T\}$ are the same.

2.2. Discrete set-valued continuous-time Markov process

This is the only class of processes that we will consider here. Such a process is constant between its jumps, which occur at random times.

2.2.1. *Poisson processes*

The Poisson process is the prototype of Markov processes with values in a discrete set (finite or countable). It takes its values in the set $\mathbb{Z}_+ := \{0, 1, 2, \ldots\}$ of non-negative integers. It is a process $\{P_t,\ t \ge 0\}$ which starts from zero ($P_0 = 0$), and where all jumps are equal to $+1$. It thus remains to specify the probability distribution of the sequence of jump times; in other words of the sequence, $T_1 = \inf\{t > 0,\ P_t = 1\}$, $T_2 = \inf\{t > 0,\ P_t = 2\}$, ..., $T_n = \inf\{t > 0,\ P_t = n\}, \ldots$.

If P_t is a *standard* Poisson process (that is of intensity 1), then the infinite sequence of random variables $T_1, T_2 - T_1, \ldots, T_n - T_{n-1}, \ldots$ is i.i.d. (independent and identically distributed) of common distribution by the exponential distribution of parameter 1, that is to say, for all $t > 0$, $n \ge 1$ (one sets by convention $T_0 = 0$):

$$\mathbb{P}(T_n - T_{n-1} > t) = \exp(-t)$$

Showing that the random number of jumps of P_t in an interval $(s, r]$ (with $0 \le s < r$) follows the Poisson distribution of parameter $r - s$ is not too difficult to do.

Given a standard Poisson process P_t, for all $\lambda > 0$, the process $P_t^\lambda := P_{\lambda t}$ is called a Poisson process of intensity λ. Its waiting times between two consecutive jumps are i.i.d. common exponential distributions of parameter λ, that is, $\mathbb{P}(T_n - T_{n-1} > t) = \exp(-\lambda t)$. The distribution of the number of jumps of this process in the interval $(s, r]$ is the Poisson distribution with parameter $\lambda(r - s)$. Moreover, the Poisson process P_t has independent increments, that is, for all $k \geq 1$, $0 \leq t_0 < t_1 < \cdots < t_k$, the random variables $P_{t_1} - P_{t_0}, P_{t_2} - P_{t_1}, \ldots, P_{t_k} - P_{t_{k-1}}$ are independent. This property is closely related to the fact that the distribution of the waiting times between two consecutive jumps is exponential.

Since P_t has independent increments, the Markov property of the Poisson process can be easily established, which is stated in the following proposition.

PROPOSITION 2.1.– For all $0 < s < t$, $n \geq 1$, $0 < s_1 < s_2 < \cdots < s_n < s$, given the $n + 2$ integers $k_1 \leq k_2 \leq \cdots \leq k_n \leq k \leq j$,

$$\mathbb{P}(P_t = j \mid P_{s_1} = k_1, P_{s_2} = k_2, \ldots, P_{s_n} = k_n, P_s = k) = \mathbb{P}(P_t = j \mid P_s = k)$$

PROOF.– Let us consider the right-hand member:

$$\begin{aligned}\mathbb{P}(P_t = j \mid P_s = k) &= \frac{\mathbb{P}(P_s = k, P_t = j)}{\mathbb{P}(P_s = k)} \\ &= \frac{\mathbb{P}(P_s = k, P_t - P_s = j - k)}{\mathbb{P}(P_s = k)} \\ &= \mathbb{P}(P_t - P_s = j - k)\end{aligned}$$

where the last identity results from the independence of the random variables P_s and $P_t - P_s$. A similar computation shows that the left-hand side of the identity of the statement is also $\mathbb{P}(P_t - P_s = j - k)$, hence the result. □

2.2.2. *Finite set-valued continuous-time Markov process*

Again, we have a process which is constant between its jumps, which occur at random times. Once more, the waiting time between two consecutive jumps follows an exponential distribution, but its parameter depends on the state where the process is. Moreover, the jumps are now also random. Let us consider a Markov process X_t with values in the finite set E of cardinal d. The *infinitesimal generator* of this process is a matrix Q of dimension $d \times d$. All off-diagonal terms of matrix Q are non-negative, the diagonal terms being non-positive, and such that the sum of the terms of each row of the matrix Q is zero. There are two equivalent ways of describing the process X_t, from this matrix Q:

1) With every pair $(i, j) \in E \times E$ such that $i \neq j$, a Poisson process $P_t^{(i,j)}$ of intensity $Q_{i,j}$ is associated, such that these $d \times d - d$ Poisson processes are mutually independent. If at time $t \geq 0$ $X_t = i$, we wait for the first jump time after t of one of the processes $P_t^{(i,j)}$, $j \in E \backslash \{i\}$. If it is a jump of the process $P_t^{(i,j_0)}$, then at this instant the process X_t jumps from state i to state j_0.

2) If, at time $t \geq 0$, $X_t = i$, we wait for the first jump of a Poisson process of intensity $q_i := \sum_{j \neq i} Q_{i,j}$, and at this time the new state of process X_t is chosen according to the probability $\{q_i^{-1} Q_{i,j},\ j \in E\backslash\{i\}\}$.

REMARK 2.1.– It is easy to show that j_0 in the first description is almost surely well defined and the two descriptions are equivalent.

If we sample a jump Markov process in continuous time at its jump times, we obtain a Markov chain in discrete time, which is the sequence of values taken by the process. By combining this discrete-time chain with the sequence of sojourn times in the various states, the continuous-time process is reconstructed.

One has the following equation for the evolution of the marginal distributions of the process X_t (Pardoux 2007, see Corollary 7.2.2). In the following statement, $\mu(t)$ denotes the probability distribution of X_t, and this for all $t \geq 0$. It should be pointed out that for us a probability on E will always be a line vector (whose coordinates are non-negative of sum 1).

PROPOSITION 2.2.– The collection $\{\mu(t),\ t \geq 0\}$ of probabilities on E satisfies the progressive Kolmogorov equation (also called Fokker–Planck equation):

$$\frac{d\mu(t)}{dt} = \mu(t)Q, \quad t > 0, \qquad [2.1]$$

$$\mu(0) = \text{distribution of } X_0 \,. \qquad [2.2]$$

If, as implicitly assumed, the process is homogeneous, in the sense that its infinitesimal generator Q does not depend on time, then the Markov process X_t has at least one invariant probability which is a probability ν on E.

When we say that the probability ν is invariant for X_t, it means that for all $t > 0$, $\nu P(t) = \nu$ (where the left-hand side of this identity is the product of a $d \times d$ matrix by a row vector on the left side, and the matrix $P(t)$ is defined as follows: $P_{i,j}(t) = \mathbb{P}(X_t = j \mid X_0 = i)$). By differentiating this identity with respect to t at time $t = 0$, we deduce that $\nu Q = 0$ which is equivalent to the fact that ν is invariant.

An important question is whether this invariant probability is unique or not. This uniqueness is equivalent to the irreducibility of the matrix Q which is stated as follows.

Q is said to be irreducible if and only if for any $i, j \in E$, $i \neq j$ there exists $k \geq 1$ and a sequence $i_0 = i, i_1, \ldots, i_k = j$ of elements of E which are such that for all $1 \leq \ell \leq k$, $Q_{i_{\ell-1}, i_\ell} > 0$. In other words, the process X_t can move from the state i to the state j in k jumps. If the process X_t is irreducible (in the sense that its matrix Q is irreducible), then its unique invariant probability satisfies the following two properties when $t \to \infty$:

– for any initial probability μ, $\quad \mu P_t \to \nu$;

– for any function $f : E \mapsto \mathbb{R}$, $\quad \frac{1}{t}\int_0^t f(X_s)ds \to \langle \nu, f\rangle$,

where $\langle \nu, f\rangle = \sum_{i \in E} f(i)\nu_i$. The second property is called the ergodic theorem. It results in fact from the first property, which would not be true in full generality for an irreducible Markov chain in discrete time.

In the following, only irreducible processes will be considered, a property which is in particular satisfied when $Q_{i,j} > 0$ for all $i \neq j$.

Let us now consider our process X_t with as initial distribution its invariant probability ν. Then ν is the distribution of X_t for all $t \geq 0$. Let us consider the process $\{X_t,\ 0 \leq t \leq T\}$ and the same process in backward time $\{\widetilde{X}_t := X_{T-t},\ 0 \leq t \leq T\}$. Very easily, it can be shown that $\widetilde{X}_t$ is a Markov process. The marginal distributions (the distributions at every time t) are the same as those of X_t. What about its transitions? Could it be that, for any $0 \leq s < t \leq T$, $i, j \in E$, $\mathbb{P}(X_t = j \mid X_s = i) = \mathbb{P}(\widetilde{X}_t = j \mid \widetilde{X}_s = i)$? If this is the case, then the process X_t is said to be reversible. An important result (and with an easy proof) is that the process X_t is reversible (with respect to its invariant probability ν) if and only if the following detailed balance equation is satisfied:

$$\nu_i Q_{i,j} = \nu_j Q_{j,i}, \quad \text{for all } i \neq j\,.$$

It is easy (by summing up this relation with respect to i, or as well with respect to j) to show that if this equation is satisfied, then ν is necessarily invariant for X_t.

REMARK 2.2.– Let ν be a given probability on E. To verify that $\nu Q = 0$, d different sums of d numbers each must be calculated. To verify that the pair (ν, Q) satisfies the detailed balance equation, one merely has to verify $d(d-1)$ identities which do not require any computation. Thereby, this last relation is often much easier to verify than $\nu Q = 0$. If it is satisfied, then one knows that ν is the invariant probability and that the process X_t is reversible. If it is not satisfied, then either ν is not invariant, or it is invariant, but the process is not reversible.

2.3. Models of DNA sequence evolution

DNA can be seen as a (long) sequence of letters taken from a four-letter alphabet: A for adenine, G for guanine, C for cytosine and T for thymine, which are the four nucleotides, the building blocks of DNA. A sequence of nucleotides of length s can be considered as an element of a 4^s-element set, and this sequence evolves over time by random modifications, at random times, of an element of the sequence. The study of a process with values in a set of cardinal 4^s could be carried out, but even for moderate values of s, it would lead to complex patterns about which not much can be said.

Apart from a very small number of advanced research studies, all models of DNA sequence evolution assume that the evolutions of individual nucleotides in the sequence are independent, even if this assumption is not very realistic. One is therefore reduced to studying a model of evolution for each nucleotide at a given site.

The case where the evolution at the various sites follows the same distribution will first be considered, then models where the rate of evolution varies from one site to another. We will assume that the corresponding process is Markovian, which thus brings us back to the study of a Markov process with values in $E := \{A, G, C, T\}$, hence of cardinal 4. This results in a limited number of parameters to choose from, and we will now consider most of the models that have been proposed in the literature and that are used in phylogeny software programs.

In each of the following models, in order to simplify the formulas, the diagonal terms of the Q matrix will not be explicited. They are easily expressed in terms of the other terms of the same row by the formula $Q_{ii} = -\sum_{j \neq i} Q_{ij}$.

It should be emphasized that if Q is an infinitesimal generator, so is uQ for all $u > 0$. Changing Q to uQ is equivalent to changing the time t to ut. Many authors choose the parameter u in such a way that the average rate of mutations at equilibrium (i.e. when the marginal distributions are equal to the invariant probability) is equal to 1.

2.3.1. *The Jukes–Cantor model*

In this model, all mutations occur at the same rate $\alpha > 0$ (Jukes et al. 1969). In other words, we have:

$$Q = \begin{pmatrix} \cdot & \alpha & \alpha & \alpha \\ \alpha & \cdot & \alpha & \alpha \\ \alpha & \alpha & \cdot & \alpha \\ \alpha & \alpha & \alpha & \cdot \end{pmatrix}$$

Let $P_{i,j}(t) = \mathbb{P}(X_t = j \mid X_0 = i)$. Since $\frac{d}{dt}P(t) = P(t)Q$, we have:

$$\frac{d}{dt}P_{i,j}(t) = -3\alpha P_{i,j}(t) + \alpha \sum_{k \neq j} P_{i,k}(t)$$

$$= \alpha - 4\alpha P_{i,j}(t)$$

Therefore:

$$P_{i,j}(t) = \frac{1}{4} + \left(\delta_{i,j} - \frac{1}{4}\right) e^{-4\alpha t} \quad \text{where } \delta_{i,j} = \begin{cases} 1 & \text{if } i = j \\ 0 & \text{otherwise} \end{cases}$$

In particular, if $\mu(t)$ denotes the distribution of X_t, $\mu(t) \to \nu := (1/4, 1/4, 1/4, 1/4)$ when $t \to \infty$. Since ν is the uniform measure and $Q_{i,j} = Q_{j,i}$, the process is reversible. The average mutation rate is 1 if $\alpha = 1/3$.

2.3.2. *The Kimura model*

In fact, the four nucleotides naturally are grouped into two subclasses: A and G are purines, and C and T are pyrimidines. Transitions (change in nucleotide while remaining in the same class) occur at a higher rate than transversions (changing classes). Hence the following Kimura's (1980) model, with two parameters $0 < \beta < \alpha$:

$$Q = \begin{pmatrix} \cdot & \alpha & \beta & \beta \\ \alpha & \cdot & \beta & \beta \\ \beta & \beta & \cdot & \alpha \\ \beta & \beta & \alpha & \cdot \end{pmatrix}$$

Based on the detailed balance equation, we verify that the unique invariant probability is again the uniform probability $\nu = (1/4, 1/4, 1/4, 1/4)$ and that the model is reversible, since the matrix Q is symmetric. The average mutation rate is 1 if $\alpha + 2\beta = 1$, which leads in particular to $\beta < 1/3 < \alpha$.

2.3.3. *The Felsenstein model*

This model (Felsenstein 1981) is chosen to admit an arbitrary probability chosen in advance $\nu = (\nu_A, \nu_G, \nu_C, \nu_T)$ as invariant probability where $\nu_i > 0$ for

$i = A, G, C, T$. A parameter $u > 0$ is also given. The infinitesimal generator in this case is given by:

$$Q = u \begin{pmatrix} \cdot & \nu_G & \nu_C & \nu_T \\ \nu_A & \cdot & \nu_C & \nu_T \\ \nu_A & \nu_G & \cdot & \nu_T \\ \nu_A & \nu_G & \nu_C & \cdot \end{pmatrix}$$

For $i \neq j$, $Q_{i,j} = u\nu_j$. Therefore:

$$\nu_i Q_{i,j} = u\nu_i\nu_j = \nu_j Q_{j,i}$$

The process admits ν as its unique invariant probability and it is reversible.

The transition probabilities in time t can also be expressed as follows:

$$P_{i,i}(t) = e^{-ut} + (1 - e^{-ut})\nu_i$$
$$P_{i,j}(t) = (1 - e^{-ut})\nu_j, \quad \text{for } j \neq i$$

The average rate of mutations is $u(1 - \nu_A^2 - \nu_G^2 - \nu_C^2 - \nu_T^2)$, therefore it is 1 if $u = (1 - \nu_A^2 - \nu_G^2 - \nu_C^2 - \nu_T^2)^{-1}$.

2.3.4. *The HKY model*

This model, due to Hasegawa, Kishino and Yano, generalizes both the Kimura and Felsenstein models. It can be seen as the Felsenstein model in which the single parameter u is replaced by two parameters $0 < \beta < \alpha$ to differentiate the transition and transversion rates. The infinitesimal generator of the HKY model is written as:

$$Q = \begin{pmatrix} \cdot & \alpha\nu_G & \beta\nu_C & \beta\nu_T \\ \alpha\nu_A & \cdot & \beta\nu_C & \beta\nu_T \\ \beta\nu_A & \beta\nu_G & \cdot & \alpha\nu_T \\ \beta\nu_A & \beta\nu_G & \alpha\nu_C & \cdot \end{pmatrix}$$

Just as the invariant measure of the Kimura model is the same as that of the Jukes–Cantor model, the invariant measure of the HKY model is that of the Felsenstein model and the HKY process is reversible. Actually, $\nu_i Q_{i,j} = \nu_j Q_{j,i}$ as soon as $i \neq j$ since the two terms of the identity are $\alpha\nu_i\nu_j$ if i and j are both purines or both pyrimidines, and $\beta\nu_i\nu_j$ if i and j are a purine and a pyrimidine.

The average rate of mutations in this model is $2\alpha(\nu_A\nu_G + \nu_T\nu_C) + 2\beta(\nu_A\nu_C + \nu_A\nu_T + \nu_G\nu_C + \nu_G\nu_T)$.

2.3.5. *The general time reversible model*

One may ask what is the most general model in which the invariant probability is chosen and which is reversible. The answer is known; it is the so-called general time reversible (GTR) model which was introduced in the field of models of nucleotide evolution by Tavaré (1986). It is written as:

$$Q = \begin{pmatrix} \cdot & a\nu_G & b\nu_C & c\nu_T \\ a\nu_A & \cdot & d\nu_C & e\nu_T \\ b\nu_A & d\nu_G & \cdot & f\nu_T \\ c\nu_A & e\nu_G & f\nu_C & \cdot \end{pmatrix}$$

with a, b, c, d, e and f positive constants. The reversibility condition is easy to verify.

The average mutation rate is $2(a\nu_A\nu_G + b\nu_A\nu_C + c\nu_A\nu_T + d\nu_G\nu_C + e\nu_G\nu_T + f\nu_C\nu_T)$. Setting this rate to 1 means putting a linear constraint between the six parameters a, b, c, d, e and f. With this convention, the number of parameters to estimate for this model is $3 + 5 = 8$. In fact, estimating the invariant probability ν is not too difficult. It may be desirable to reduce the number of other parameters to be estimated. This explains why the GTR model is not widely used.

2.4. Models of rate evolution along the sequence

The many studies conducted on genomes using observed data indicate that some sites are evolving more rapidly than others. Failure to taking this variability into account can lead to serious estimation errors. For example, Wakeley (1994) has shown that ignoring such variations when they exist induces a bias in the estimation of the difference between the transition and transversion rates. It therefore makes sense to vary the parameter u from site to site. There are essentially two different ways to model such a rate change.

2.4.1. *Independent and identically distributed rates along the sequence*

The first possibility is to assume that the rates at the various sites are the realization of i.i.d. random variables with values in $\mathbb{R}_+$. Several types of distributions are used.

2.4.1.1. *Discrete distribution*

We choose an integer $K > 1$, K positive real numbers $u_1, \ldots, u_K$ and K positive weights $\varrho_1, \ldots, \varrho_K$ such that $\sum_{i=1}^{K} \varrho_i = 1$. The discrete distribution considered is then the probability $\sum_{i=1}^{K} \varrho_i \delta_{u_i}$. A random variable that follows this law equals u_i with probability ϱ_i, $1 \leq i \leq K$. If one imposes that the expectation of the distribution is fixed and equal to 1, there are $2(K-1)$ parameters to be estimated. Even with a very small K it gives a good number of parameters to determine.

2.4.1.2. *Continuous distribution*

A continuous distribution on $\mathbb{R}_+$ can be chosen. The most popular in this field is the Gamma distribution. The Gamma distribution is a two-parameter distribution $\Gamma(\alpha, \beta)$ whose density is the function defined on $\mathbb{R}_+$ by:

$$\frac{\beta^\alpha}{\Gamma(\alpha)} e^{-\beta x} x^{\alpha-1}$$

where $\Gamma(\alpha) := \int_0^\infty e^{-x} x^{\alpha-1} dx$. It is easily verified that the expectation of this distribution is α/β and the variance α/β^2. If we decide that the expectation must be 1, one chooses $\alpha = \beta$ and the variance is $1/\alpha$. For $\alpha < 1$, the density has a form of L (it tends to $+\infty$ in the vicinity of $x = 0$), while for $\alpha > 1$ it cancels out at $x = 0$ and has the shape of a bell curve that becomes less asymmetric when α becomes large.

2.4.1.3. *Discretized Gamma distribution*

The problem with general discrete distribution is the number of parameters to be estimated. This problem is solved in the case of the Gamma distribution, but at the expense of heavier computational efforts when using maximum likelihood, which limits the use of this type of model to small data sets. A solution proposed by Yang (1996) is the use of a discrete distribution that approximates the Gamma distribution, which makes it possible to obtain a family of one-parameter discrete distributions. Yang chooses a four-class-based approximation.

2.4.2. ***Hidden Markov model***

The approach presented in section 2.4.1 allows substitution rates to vary from site to site, and the values taken by these rates at the various sites are unknown. Moreover, this approach assumes that the rates are independent across sites.

There are good reasons to think that it is more realistic to allow correlation between rates at nearby sites. This is what the so-called hidden Markov chain-based approach of Felsenstein and Churchill (1996), and also using a different terminology of Yang (1995) allows.

In this approach, the rates at the various sites are assumed to be the realization of a Markov chain (called "hidden", because it is not directly observed) that evolves along the considered genomic sequence of length s. A finite set F consisting of the various possible values for the rates in our model is provided. This is the state space of our Markov chain $\{\tau_i,\ 1 \leq i \leq s\}$. If $k, \ell \in F$:

$$\mathbb{P}(\tau_{i+1} = k \mid \tau_i = \ell) = P_{\ell,k}$$

where P is the transition matrix of a Markov chain, that is a square $d \times d$ matrix (if $d = \text{card}(F)$), whose entries are non-negative, and the sum of the terms of each line is 1. For any $1 \leq i \leq s-1$, the line ℓ of the matrix P is the conditional distribution of τ_{i+1} given that $\tau_i = \ell$. We assume that the matrix P is irreducible (which is verified exactly as the irreducibility of the matrix Q in section 2.2.2). Then the chain of the τ_i admits a unique invariant probability π such that $\pi P = P$. If the distribution of τ_1 is π, then the distribution of τ_i is also π for any $2 \leq i \leq s$. By choosing π as the distribution of τ_1, the distribution of the sequence $\{\tau_i,\ 1 \leq i \leq s\}$ is entirely determined by the pair π and P. For all $a_1, \dots, a_s \in F$:

$$\mathbb{P}\left(\cap_{i=1}^{s}\{\tau_i = a_i\}\right) = \pi_{a_1} \prod_{i=1}^{s-1} P_{a_i, a_{i+1}}$$

This is the so-called a priori rate distribution, the a posteriori distribution being the conditional distribution of the τ_i knowing the observed data.

2.5. Models of discrete trait evolution

Continuous-time Markov processes are also used to model the evolution of discrete traits such as eye or coat color, the presence or absence of a given trait, the type of environment in which the species lives, the number of fingers, etc. The simplest of these is the Mk model that generalizes the Jukes–Cantor model to any number of states (Pagel 1994; Lewis 2001). The evolution of a discrete trait that can take k different values is determined by the following infinitesimal generator:

$$Q = q \begin{pmatrix} -(k-1) & 1 & \dots & 1 \\ 1 & -(k-1) & \dots & 1 \\ \vdots & \vdots & \ddots & \vdots \\ 1 & \dots & 1 & -(k-1) \end{pmatrix}$$

where it is seen that the model has only one parameter in addition to the number k of states, the evolution speed q. Under the Mk model, a state jumps to any other with the same probability. The process associated with the model Mk admits $\left(\frac{1}{k}, \dots, \frac{1}{k}\right)$ as its unique invariant probability and it is reversible.

The main advantage of this model is that the transition probabilities at time t have simple expressions. The same argument as in section 2.3.1 gives:

$$P_{i,j}(t) = \frac{1}{k} + \left(\delta_{i,j} - \frac{1}{k}\right) e^{-kqt}$$

The Mk model has been extended to allow non-uniform transition probabilities between state pairs, for example, the SYM and ARD models of Paradis et al. (2004).

2.6. References

Felsenstein, J. (1981). Evolutionary trees from DNA sequences: A maximum likelihood approach. *Journal of Molecular Evolution*, 17(6), 368–376.

Felsenstein, J. and Churchill, G.A. (1996). A hidden Markov model approach to variation among sites in rate of evolution. *Molecular Biology and Evolution*, 13(1), 93–104.

Hasegawa, M., Kishino, H., Yano, T.-A. (1985). Dating of the human-ape splitting by a molecular clock of mitochondrial DNA. *Journal of Molecular Evolution*, 22(2), 160–174.

Jukes, T.H. and Cantor, C.R. (1969). Evolution of protein molecules. In *Mammalian Protein Metabolism*, Munro, H.N. (ed.), Academic Press, New York.

Kimura, M. (1980). A simple method for estimating evolutionary rates of base substitutions through comparative studies of nucleotide sequences. *Journal of Molecular Evolution*, 16(2), 111–120.

Lewis, P.O. (2001). A likelihood approach to estimating phylogeny from discrete morphological character data. *Systematic Biology*, 50(6), 913–925.

Pagel, M. (1994). Detecting correlated evolution on phylogenies: A general method for the comparative analysis of discrete characters. *Proceedings of the Royal Society of London. Series B: Biological Sciences*, 255(1342), 37–45.

Paradis, E., Claude, J., Strimmer, K. (2004). APE: Analyses of phylogenetics and evolution in R language. *Bioinformatics*, 20(2), 289–290.

Pardoux, E. (2007). *Processus de Markov et applications. Algorithmes, réseaux, génome et finance*. Dunod, Malakoff.

Tavaré, S. (1986). Some probabilistic and statistical problems in the analysis of DNA sequences. *Lectures on Mathematics in the Life Sciences*, 17(2), 57–86.

Wakeley, J. (1994). Substitution-rate variation among sites and the estimation of transition bias. *Molecular Biology and Evolution*, 11(3), 436–442.

Yang, Z. (1995). A space-time process model for the evolution of DNA sequences. *Genetics*, 139(2), 993–1005.

Yang, Z. (1996). Among-site rate variation and its impact on phylogenetic analyses. *Trends in Ecology & Evolution*, 11(9), 367–372.

3

Evolutionary Models of Continuous Traits

Paul BASTIDE[1], Mahendra MARIADASSOU[2] and Stéphane ROBIN[3]

[1]*IMAG, CNRS, Université de Montpellier, France*

[2]*MaIAGE, INRAE, Université Paris-Saclay, Jouy-en-Josas, France*

[3]*LPSM, Sorbonne Université, Paris, France*

3.1. Motivations

The evolutionary models presented in Chapter 2 are mainly motivated by the study of biological macromolecules, namely, DNA, RNA or protein sequences. These models all have in common the use of a low cardinal state space or alphabet: four nucleotides for DNA and RNA and 20 amino acids for proteins. It is nevertheless usual to observe or measure *quantitative* traits in related organisms: height, weight, growth rate, life expectancy, length of a body part, etc. These traits can be discretized in order to be analyzed as discrete traits, for example, using the Mk model presented in Chapter 2; it is however more natural to consider them as intrinsically *continuous*.

As with DNA sequences, it is expected that closely related species (e.g. the chimpanzee and the bonobo) will possess more similar trait values than two more distantly related species (e.g. the chimpanzee and the marmoset). This link between sequence (or trait) similarity and relatedness is the basis of many phylogenetic tree reconstruction methods (see Chapters 7 and 8). The purpose of continuous trait evolution models is to describe this similarity in order to understand or correct it.

Models and Methods for Biological Evolution,
coordinated by Gilles DIDIER and Stéphane GUINDON.

3.1.1. *Comparative methods*

When measuring several traits, it is natural to study their relationships. Is height correlated with longevity in animals? Is sepal length correlated with petal length in flowers? A naive way to answer these questions consists of calculating the correlation coefficient between traits.

Unfortunately, and as illustrated in Figure 3.1, this naive view can lead to the detection of apparent correlations (a) where there is only a compositional effect (c) due to the species phylogeny (b). In this caricatural example, the structure induced by the phylogeny can be simply corrected by considering a *group* effect (in *gray* and *black* in Figure 3.1(c)) but the optimal correction is less obvious for more realistic phylogenies.

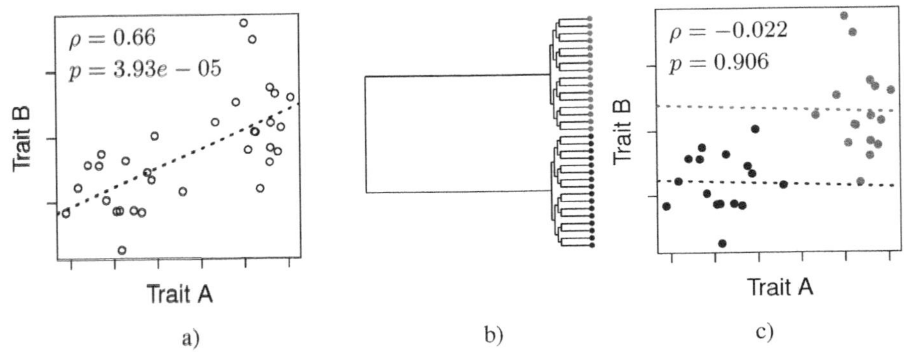

Figure 3.1. *Two traits (A and B) are measured on 32 species*

COMMENTS ON FIGURE 3.1.– *A naive analysis (a) reveals a strong ($\rho = 0.66$) and significant ($p = 3.93 \times 10^{-5}$) correlation but only induced by the evolutionary history of the species (b): the 32 species are divided into two groups (in gray or black) of 16 strongly related species. Once this structure is taken into account, there is no longer any correlation between traits A and B (c).*

In his seminal work on comparative methods in phylogeny, Felsenstein (1985) proposes a systematic correction based on the notion of independent phylogenetic contrasts. The particularly clever idea consists of no longer comparing *trait values* measured at the leaves, but *differences in trait values* observed between disjoint portions of the tree. This idea is illustrated in Figure 3.2. The contrast C_{ij} between a node i and a node j is given by $C_{ij} = (Y_j - Y_i)/\sqrt{d_{ij}}$, where Y_i (respectively, Y_j) is the trait value for node i (respectively, j) and d_{ij} is the distance in the tree between nodes i and j. This method requires that the trait value for the ancestral nodes be calculated beforehand. It can be shown that, under certain assumptions, it transforms the n correlated trait values (1 per species) into $n - 1$ independent and identically distributed contrasts (see section 3.3.2).

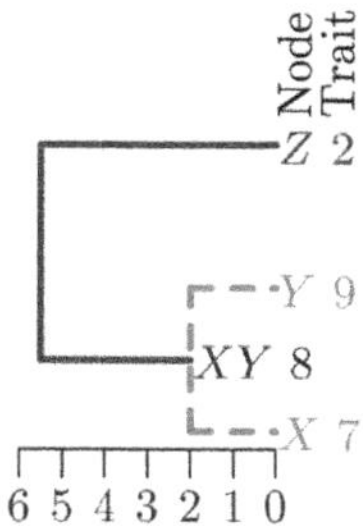

Nodes	Difference	d_{ij}	Contrast
$X - Y$	2	4	1
$XY - Z$	6	9	2

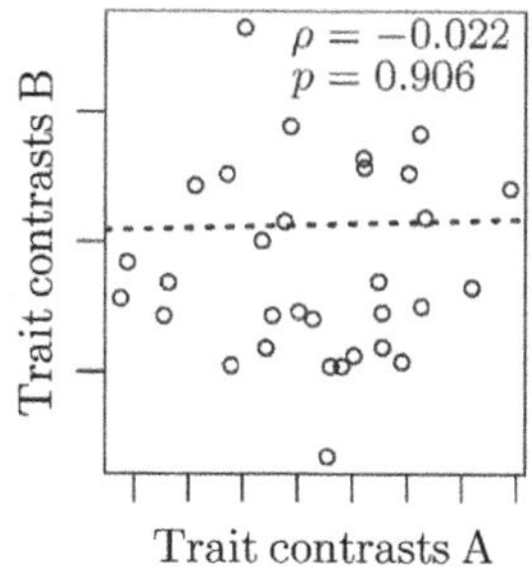

a) b)

a) b)

Figure 3.2. *Contrasts and phylogenetics*

COMMENTS ON FIGURE 3.2.– *(a) Illustration of the computational principle for phylogenetic contrasts. The difference in trait between nodes X and Y is* 2 *and these two nodes are at distance* 4 *apart in the tree. Their contrast is therefore* $2/\sqrt{4} = 1$. *It is clearly visible that the contrasts* $X - Y$ *(gray, dashed line) and* $XY - Z$ *(black, solid line) imply disjoint tree portions. (b) Contrast of traits A and B in the previous example. No correlation is visible between traits.*

This method is especially interesting for what it assumes without making it explicit: (i) the trait joint variation has greater interest than their joint values, (ii) trait variations in disjoint portions of the tree are independent and (iii) the characteristic scale of these variations increases with the evolutionary distance (here as the square root). These three aspects all evoke a *dynamic* view of the trait that can be formalized by an evolutionary process describing how the trait varies over time.

We will present in section 3.2 a simple example of an evolutionary model for one trait before generalizing it in section 3.3 to several traits. The multivariate model is particularly useful for computing the correlation between traits by correcting (i) the effect of phylogeny, as with contrasts, but also (ii) the effect of other variables in the more general context of phylogenetic comparative methods (PCM).

3.1.2. *Studies of evolutionary phenomena*

In the previous description, the tree is seen as a nuisance parameter that induces meaningless phylogenetic correlations that the formalism of tree-based evolutionary processes allows us to correct. However, the tree is a valuable source of information for identifying adaptive phenomena, such as the island gigantism of turtles (Jaffe et al. 2011), or the adaptation of the brain shape to changes in diet (Aristide et al. 2018).

In these situations, the subject of the study concerns less the comparison of traits than the intrinsic parameters of the model that enable capturing the speed of evolution of the trait, its ancestral value in certain branches of the tree, its optimal value, adaptive changes that may coincide with niche changes, mechanisms of competition between species, interactions between trait value and fitness of individuals, between trait value and geography, etc.

Our ability to understand the past dynamics of traits is obviously constrained by the complexity and realism of the models: it is obviously inconvenient to study the optimal value of a trait when this notion is not modeled by the evolutionary process! However, it is also constrained by our (in)ability to estimate the models: the most realistic models are both the most complex and the ones that require the most data for their fitting.

In section 3.4, we will see in details models that make it possible to study the optimal value of a trait, to detect adaptive events modifying this value (typically a change in niche) and to take into account the dispersion of the trait within a species. These models are all Gaussian and lend themselves well to estimation. Finally, in section 3.5, we will consider some non-Gaussian models that capture more sophisticated dynamics at the cost of an often much more complex estimation.

3.2. Brownian motion

A trait evolution model must, at the very least, describe fluctuations in trait value along a branch of the phylogenetic tree. These fluctuations are generally described by a continuous-time random process. To be of evolutionary interest, the model must also be able to describe how these fluctuations are shared among different branches of the same tree.

Although evolutionary models differ in the definition of the current process along a branch (Gaussian or not, univariate or not, etc.), most make the assumption of "branching" which states that the process gives rise to two independent copies starting from the current state of the process, at each node of the tree (see Figure 3.3). This representation allows us to model the correlation observed between the traits of related species through the intersection of their respective evolutionary trajectories.

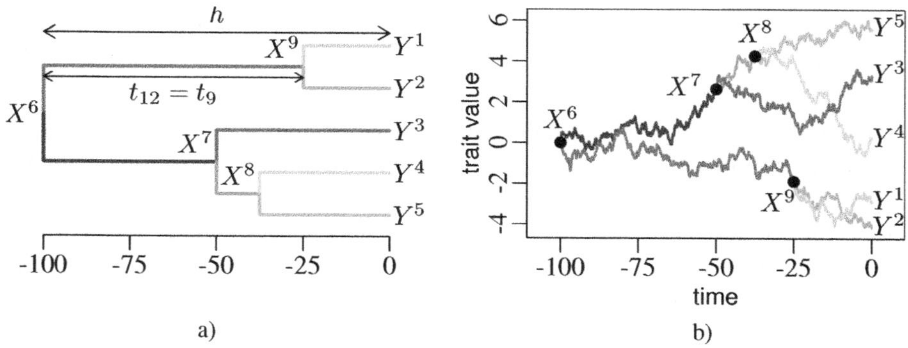

Figure 3.3. *Realization of a BM (with $\mu = 0$ and $\sigma^2 = 0.04$) on a time-calibrated phylogenetic tree. The colors of the branches of the tree (a) correspond to the colors of the stochastic processes (b). Only the leaves of the tree are observed (at time $t = 0$). For a color version of this figure, www.iste.co.uk/didier/models.zip*

3.2.1. *Description*

The simplest and most classical process to describe the evolution of a univariate trait is the Brownian motion (BM), governed by the equation:

$$W_0 = \mu, \ \mathrm{d}W_t = \sigma \mathrm{d}B_t, \ \forall t \in [0, h] \qquad [3.1]$$

where B_t is the Brownian motion of variance 1 defined as the only process with stationary and independent increments, almost surely continuous, and such that $B_t \sim \mathcal{N}(0, t)$ for any time t, $0 \leq t \leq h$ (for an introduction to this process within an ecological modeling framework, see Méléard (2016)). Assuming a fixed trait value at the root ($W_0 = \mu$) is tantamount to working conditionally on that value. The resulting model from this process combined with the branching hypothesis results in specifying the link between the values of the trait Y measured in two species i and j observed, respectively, at times t_i and t_j and which have been separated from each other at time t_{ij} ($0 \leq t_{ij} \leq \min(t_i, t_j)$) (see Figure 3.3(a)):

$$\mathbb{E}\left[Y^i\right] = \mu, \ \mathbb{V}\mathrm{ar}\left[Y^i\right] = \sigma^2 t_i, \ \mathbb{C}\mathrm{ov}\left[Y^i; Y^j\right] = \sigma^2 t_{ij} \qquad [3.2]$$

We refer to Felsenstein (2004, Chapter 23) for a formal and intuitive derivation of these equations. The term ultrametric tree is employed when all the species are contemporary and observed at the same time $t_i = h$ called "height of the tree". In this case, trait correlation observed in two species depends only on their shared evolutionary time by means of the formula $\mathbb{C}\mathrm{or}\left[Y^i; Y^j\right] = t_{ij}/h$.

3.2.2. *Phylogenetic regression and statistical transformations*

3.2.2.1. *Phylogenetic regression*

Similarly to the other uni- or multivariate Gaussian models that will be presented in sections 3.3 and 3.4, the BM model induces a Gaussian marginal distribution at the leaves of the tree with a variance-covariance structure specific to it. *Phylogenetic regression* (Grafen 1989, 1992) consists of describing the statistical relationships between the observed trait $\mathbf{Y}$ and a series of predictors (e.g. environmental variables) arranged in a matrix $\mathbf{X}$ of size $n \times k$ as a linear model whose residue distribution is precisely that given by the process being considered:

$$\mathbf{Y} = \mathbf{X}\boldsymbol{\theta} + \mathbf{E} \quad \text{with} \quad \mathbf{E} \sim \mathcal{N}\left(\mathbf{0},\, \sigma^2 \mathbf{C}_n\right) \qquad [3.3]$$

where $\mathbf{C}_n = [C_{ij}]_{1 \leq i,j \leq n}$ is the matrix representing the variance structure at the leaves, that is, for the BM model $C_{ij} = t_{ij}$ (see equation [3.2]).

The interest of this representation is twofold. First of all, the effects of the predictors on the trait value can be estimated without any confusion with those of the phylogenetic structure linking the different species. Second, it also provides a framework for measuring the strength of the effect of the phylogenetic structure on the diversity of the trait. As a matter of fact, the latter being entirely captured by the matrix $\mathbf{C}_n$, relaxing the structure is equivalent to "abating" the off-diagonal coefficients of this matrix to make it more similar to a diagonal matrix. This is tantamount to changing the structure of the underlying tree to give it the appearance of a star-shaped tree in which all species are independent. Since the matrix $\mathbf{C}_n$ depends actually only on the lengths of the branches of the tree under consideration, changing the variance structure amounts to imposing a transformation on the branches of the tree. This is the objective of the transformations proposed by Pagel (1999).

3.2.2.2. *Pagel's transformations*

A review of tree branch transformations due to Pagel (1999) is presented here, starting with the most commonly used one: the λ transformation. It turns out that the likelihoods resulting from these different transformations are explicit and depend only on one parameter, therefore they can be optimized numerically (Revell 2012; Pennell et al. 2014). Following the previous discussion, this parameter is often seen as measuring a "phylogenetic signal" defined as "a measure of the statistical dependence among species' trait values due to their phylogenetic relationships" (Revell et al. 2008). As underlined in the previous reference which interested readers are referred to, there are other ways of measuring this signal, one of the most popular being the K statistic (Blomberg et al. 2003), which is not discussed here.

3.2.2.2.1. Pagel's λ

The first idea is to mitigate the correlations between species by multiplying all the off-diagonal coefficients of the matrix $\mathbf{C}_n$ by a parameter λ between 0 and 1:

$$\begin{cases} C(\lambda)_{ij} = \lambda C_{ij} & \forall i \neq j \\ C(\lambda)_{ii} = C_{ii} & \forall 1 \leq i \leq n \end{cases} \qquad [3.4]$$

It is easy to check that this operation amounts to executing an BM on a tree whose branches have undergone the following transformation:

$$\ell_i(\lambda) = \begin{cases} \lambda \ell_i & \text{if } i \text{ is an internal node} \\ \ell_i + (1-\lambda)t_{\text{pa}(i)} = \lambda \ell_i + (1-\lambda)t_i & \text{if } i \text{ is a leaf} \end{cases} \qquad [3.5]$$

where ℓ_i is the length of the branch going from the single parent $\text{pa}(i)$ of i to i: $\ell_i = t_i - t_{\text{pa}(i)}$. This transformation is represented in Figure 3.4. It amounts to multiplying all the internal branches by λ, then to lengthening the outer branches to the fork so that the tree remains ultrametric and with the same total height h. The parameter λ can be seen as measuring the "phylogenetic signal": if $\lambda = 1$, the tree is not modified and the data distribution at the leaves remain unmodified (strong phylogenetic signal, the tree has a great influence on the data structure), whereas if $\lambda = 0$, the tree is reduced to a star tree and the data at the leaves are independent from one another (weak phylogenetic signal, the tree has no influence on the observed data).

As it will be seen in section 3.4.6, this λ parameter can be interpreted as a ratio of intra- and interspecific variances and is, for this reason, sometimes interpreted as a "phylogenetic heritability" parameter (Lynch 1991; Leventhal and Bonhoeffer 2016).

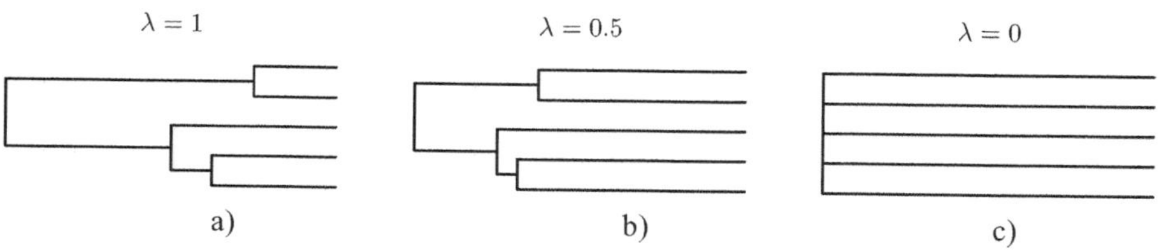

Figure 3.4. *Pagel's λ transformation. (a) $\lambda = 1$, original tree. (b) $\lambda < 1$, abated phylogenetic correlations. (c) $\lambda = 0$, independant observations*

This transformation is widely used in the literature with more than 2,300 citations[1] for the original article of Pagel (1999). The λ parameter is often seen as a nuisance parameter used for controlling in an ad hoc manner the importance of phylogeny in a phylogenetic regression (see (Revell 2010; Ceccarelli et al. 2018; Law et al. 2018) for examples of recent applications in evolutionary ecology, for the study of spider or musteloid groups). It can also be a parameter of interest in itself, for quantifying the relative significance of phylogeny with respect to independent environmental variations, for example, in virology (Vrancken et al. 2015).

3.2.2.2.2. Pagel's κ

This transformation consists of raising all lengths to a power $\kappa > 0$:

$$\ell_i(\kappa) = (\ell_i)^\kappa \quad \forall 1 \leq i \leq n \qquad [3.6]$$

The modified tree does not stay ultrametric (see Figure 3.5). If $\kappa > 1$, the heterogeneity between the branch lengths is increased and even more diversification is expected in long branches, and even less in short ones. Conversely, if $\kappa < 1$, the transformation tends to homogenize the length of all branches.

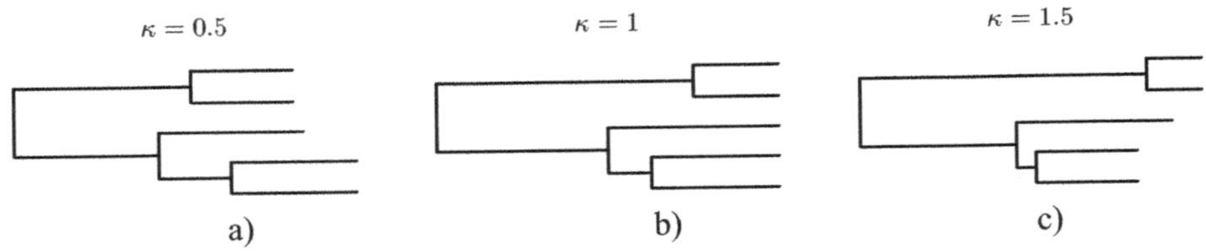

Figure 3.5. *Pagel's κ transformation. (a) $\kappa < 1$, homogeneous branch lengths. (b) $\kappa = 1$, original tree. (c) $\kappa > 1$, heterogeneous branch lengths*

3.2.2.2.3. Pagel's δ

In this last transformation, all node heights are raised to a power δ, while compensating by a factor $h^{1-\delta}$ so that the tree remains ultrametric and of total height h (see Figure 3.6):

$$t_i(\delta) = (t_i)^\delta \cdot h^{1-\delta} \quad \forall 1 \leq i \leq n \qquad [3.7]$$

This transformation has a similar interpretation to the ACDC model (see section 3.4.4.1). If $\delta < 1$, most of the evolution takes place in the ancestral branches (*decelerating*, or *early burst*), whereas if $\delta > 1$, the terminal branches are responsible for all the variance (*accelerating*).

1 From the Web of Science site (accessed 02/14/2020).

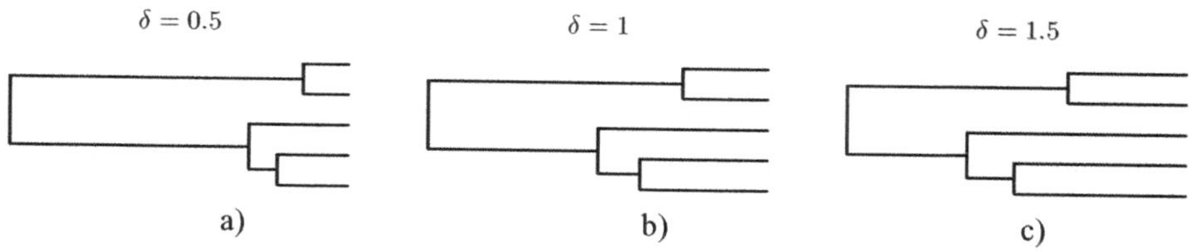

Figure 3.6. *Pagel's δ transformation. a) $\delta < 1$, evolution concentrated in the ancestral branches. b) $\delta = 1$, original tree. c) $\delta > 1$, evolution concentrated in the recent branches*

3.2.3. *Recursive algorithms for inference*

The phylogenetic regression formulation of equation [3.3] amounts to a simple linear model with known correlations (conditional on the phylogenetic tree, the shared evolutionary times and thereby the matrix $\mathbf{C}_n$ are known), such that the maximum likelihood estimators of the parameters $\boldsymbol{\theta}$ and σ^2 can be obtained as simple generalized least squares solutions (see, for example, Mardia et al. (1979)):

$$\begin{aligned}\hat{\boldsymbol{\theta}} &= (\mathbf{X}^T\mathbf{C}_n^{-1}\mathbf{X})^{-1}\mathbf{X}^T\mathbf{C}_n^{-1}\mathbf{Y}\\ \widehat{\sigma^2} &= (n-1)^{-1}(\mathbf{Y}-\mathbf{X}\hat{\boldsymbol{\theta}})^T\mathbf{C}_n^{-1}(\mathbf{Y}-\mathbf{X}\hat{\boldsymbol{\theta}})\end{aligned} \quad [3.8]$$

These formulas (and many others that will appear later) involve $\mathbf{C}_n^{-1}$. The naive approach which consists of inverting this matrix of size $n \times n$ has an algorithmic complexity of order $\mathcal{O}(n^{2.373})$ (Raz 2003) and thus becomes prohibitive for large trees. A more sophisticated approach makes use of the tree structure and employs a recursive *pruning* algorithm for computing these estimators without explicitly calculating $\mathbf{C}_n^{-1}$ and with a complexity of order $\mathcal{O}(n)$ in time and space.

This recursive algorithm, similar to the one used for the computation of the likelihood for discrete traits (see Chapter 7), is an algorithm of the "belief propagation" type (Kim and Perl 1983). Since its first description in this context by Felsenstein (1973) for the BM, a large number of adaptations (sometimes redundant) have been proposed (Hadfield and Nakagawa 2010; Fitzjohn 2012; Freckleton 2012; Pybus et al. 2012; Lartillot 2014; Cybis et al. 2015; Bastide et al. 2018a) for more complex Gaussian processes (see section 3.4). Mitov et al. (2019) propose a very general implementation of the algorithm that comprises the previous adaptations. The recursive algorithm can even in some cases (Landis et al. 2013; Hiscott et al. 2016; Duchen et al. 2017) be extended to non-Gaussian processes (see section 3.5). Finally, Ho and Ané (2014a) describe another linear algorithm, based on a slightly different formulation ("3-point structure"), which allows them to address generalized phylogenetic regressions (namely, logistic, Poissonian, etc).

3.3. Multivariate analysis

In the previous section, only the case where a single trait was observed at the tree leaves has been considered. The techniques described above can be immediately extended to cases where several features that evolve in an *independent* manner on the tree are observed. Nevertheless, as seen in the introductory example (see section 3.1.1), the main specific interest in the observed data sets lies in the correlation relationships between traits. This section will show how the multivariate model allows the correlated evolution of multiple traits on a phylogenetic tree to be modeled.

3.3.1. *Description*

3.3.1.1. *Definition*

The multivariate Brownian motion describes the evolution of a vector $\mathbf{W}_t$ of p traits over time t, with mean $\boldsymbol{\mu}$ and variance $\mathbf{R} = \mathbf{R}^{1/2}(\mathbf{R}^{1/2})^T$ by the following SDE:

$$\mathbf{W}_0 = \boldsymbol{\mu} \qquad \mathrm{d}\mathbf{W}_t = \mathbf{R}^{1/2}\mathrm{d}\mathbf{B}_t, \ \forall 0 \leq t \leq h \tag{3.9}$$

where $\mathbf{B}_t$ is the Brownian motion of variance $\mathbf{I}_p$ defined as the only process with stationary and independent increments, almost certainly continuous, and such that $\mathbf{B}_t \sim \mathcal{N}(\mathbf{0}_p, t\mathbf{I}_p)$ for any time t, $0 \leq t \leq h$. Again, assuming the trait vector at the root to be fixed is equivalent to working conditionally on the initial values of the traits.

3.3.1.2. *Distribution at the leaves*

The data now consists of a trait *matrix* $\mathbf{Y}$ of size $n \times p$: for each leaf i of the tree, $1 \leq i \leq n$, a trait vector $\mathbf{Y}^i$ of dimension p is observed, such that $\mathbf{Y}^T = (\mathbf{Y}^1, \dots, \mathbf{Y}^n)$. As for the univariate motion, the marginal distribution observed at the leaves can be written as follows (see also Felsenstein (2004, Chapter 23)):

$$\mathbb{E}\left[\mathbf{Y}^i\right] = \boldsymbol{\mu} \qquad \mathbb{C}\mathrm{ov}\left[\mathbf{Y}^i; \mathbf{Y}^j\right] = t_{ij}\mathbf{R} \tag{3.10}$$

The mean at the leaves is uniform and equal to the mean at the root $\boldsymbol{\mu}$, and the covariance between the trait q of leaf i and trait r of leaf j is simply the product of the variance R_{qr} between the two traits and the shared evolution time t_{ij} between the two leaves $\mathbb{C}\mathrm{ov}\left[Y_{iq}; Y_{jr}\right] = t_{ij}R_{qr}$.

This simple multiplicative structure implies that the contribution of the tree and the process can be naturally *factored*. This means that the distribution of $\mathbf{Y}$ can be directly written as a Gaussian matrix:

$$\mathrm{vec}(\mathbf{Y}) \sim \mathcal{N}\left(\mathrm{vec}(\mathbf{1}_n\boldsymbol{\mu}^T),\ \mathbf{R}\otimes\mathbf{C}_n\right) \qquad [3.11]$$

where vec is the vectorization operator which, to a matrix of size $n \times p$, associates a vector of dimension np by way of "stacking" the matrix column vectors one on top of the other, $\otimes$ is the Kronecker product, and $\mathbf{C}_n = [t_{ij}]_{1\leq i,j\leq n}$ is the matrix of shared evolution times (see section 3.2.2). The likelihood of the model is simply computed and makes it possible to write the maximum likelihood estimators of $\boldsymbol{\mu}$ and $\mathbf{R}$. The latter are detailed in section 3.3.3.

3.3.2. *Phylogenetic contrasts*

The contrasts introduced in section 3.1.1 find their full justification within this Gaussian framework. As a matter of fact, (i) the contrast of trait r between the nodes i and j defined by $C^r_{ij} = (Y_{ir} - Y_{jr})/\sqrt{d_{ij}}$ follows a Gaussian distribution $\mathcal{N}(0,\ R_{rr})$. Moreover, (ii) by the independent increments property of the BM, the contrasts C^r_{ij} and C^q_{kl} are independent as soon as the paths $i \leftrightarrow j$ connecting i to j in the tree, and $k \leftrightarrow l$ connecting k to l have no branches in common. Finally, equation [3.10] shows that (iii) the covariance between the same contrasts computed with different traits is $\mathbb{C}\mathrm{ov}\left[C^r_{ij}; C^q_{ij}\right] = R_{rq}$. Interested readers are referred to the book by Felsenstein (2004) (Chapters 23 and 24) for a more detailed derivation of these contrasts by their creator.

The contrasts thus enable (i) n correlated values to be replaced by $n - 1$ independent differences and (ii) the phylogenetic covariance matrix $\mathbf{C}_n$ as well as the ancestral value of the trait $\boldsymbol{\mu}$ to be simultaneously neutralized to simply study the covariance matrix of the traits $\mathbf{R}$. This simplification is achieved at the cost of a slight loss of power: shifting from n to $n - 1$ observations when computing the contrasts. The contrast method is presented for its historical interest only. It has now largely been replaced by PCM, and more specifically by Gaussian phylogenetic models, of which it is a particular case, and which lend themselves much better to generalizations such as the phylogenetic PCA.

3.3.3. *Phylogenetic PCA*

Equation [3.11], which describes the trait marginal distribution at the leaves, can be reduced to a simple Gaussian matrix of mean $\mathbf{1}_n\boldsymbol{\mu}^T$ of column variance $\mathbf{R}$ and row variance $\mathbf{C}_n$. Assuming that we work conditionally on the phylogeny, this matrix $\mathbf{C}_n$, being entirely characterized by the tree, is known. It is then possible to decorrelate the observations in rows by multiplying the observations on the right-hand side by the Cholesky matrix of $\mathbf{C}_n$. If $\mathbf{L}_n$ is such that $\mathbf{C}_n = \mathbf{L}_n\mathbf{L}_n^T$, then:

$$\text{vec}(\mathbf{L}_n^{-1}\mathbf{Y}) \sim \mathcal{N}\left(\text{vec}(\mathbf{L}_n^{-1}\mathbf{1}_n\boldsymbol{\mu}^T),\ \mathbf{R}\otimes\mathbf{I}_n\right) \qquad [3.12]$$

which brings us back to the classical multivariate analysis case where n independent vectors are observed. Explicit estimators for $\boldsymbol{\mu}$ and $\mathbf{R}$ are then available (see, for example, Mardia et al. 1979):

$$\begin{aligned}\hat{\boldsymbol{\mu}}^T &= (\mathbf{1}_n^T\mathbf{C}_n^{-1}\mathbf{1}_n)^{-1}\mathbf{1}_n^T\mathbf{C}_n^{-1}\mathbf{Y}\\ \hat{\mathbf{R}} &= (n-1)^{-1}(\mathbf{Y}-\mathbf{1}_n\hat{\boldsymbol{\mu}}^T)^T\mathbf{C}_n^{-1}(\mathbf{Y}-\mathbf{1}_n\hat{\boldsymbol{\mu}}^T)\end{aligned} \qquad [3.13]$$

The use of these notations makes it possible, as in classical multivariate analysis, to try to reduce the size of the observations through a said "phylogenetic" principal component analysis (pPCA) (Revell 2009). This transformation is tantamount to trying to eliminate the column correlations from the observations by using the eigen vectors of the variance estimator $\hat{\mathbf{R}} = \hat{\mathbf{V}}\hat{\mathbf{D}}^2\hat{\mathbf{V}}^T$ to calculate empirically independent scores $\mathbf{S} = (\mathbf{Y} - \mathbf{1}_n\hat{\boldsymbol{\mu}}^T)\hat{\mathbf{V}}$. This pPCA only differs from the classical PCA in that it takes into account the correlations induced by the tree as represented by $\mathbf{C}_n$. As shown in Figure 3.7, if the traits are indeed originating from a multivariate BM, it decorates the traits better than the classic PCA and allows us to correct the relationship effects between individuals.

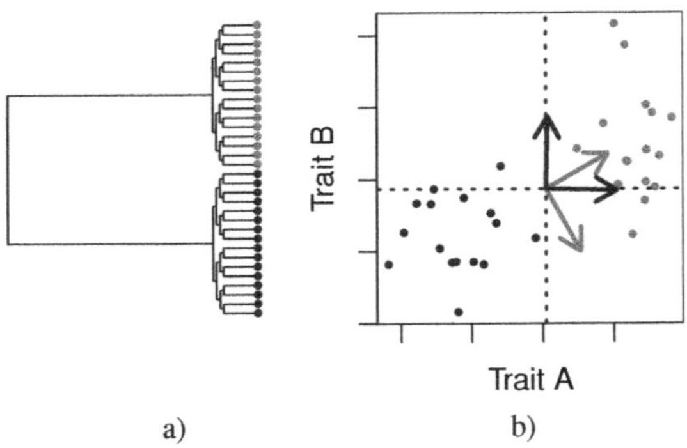

Figure 3.7. *Comparison between the PCA and the pPCA with the introduction example presented in section 3.1*

COMMENTS ON FIGURE 3.7.– *The tree (a) induces two strongly related groups, which biases the variance structure of the observed traits (b). The classical PCA (gray axes) is biased, whereas the phylogenetic PCA (black axes) approximates the true eigenvectors (dotted lines, no correlation).*

The pPCA has been used in the literature as a way to summarize and represent high-dimensional morphological data (for an application on spider morphologies, see Ceccarelli et al. (2018)). It is nonetheless important to bear in mind that the performance of this pPCA depends strongly on the underlying assumption of the Brownian evolution of traits. If this assumption is not verified, it is easy to show that this process induces important structural biases (Uyeda et al. 2015; Khabbazian et al. 2016; Bastide et al. 2018a) and it is not recommended to use it for pre-processing the data before a deeper analysis (Uyeda et al. 2015). The Brownian nature of evolution can be justified in some cases, but is not always relevant to the analysis of phylogenetic data. In the remainder of this chapter, other types of models will be presented to represent trait evolution in a more realistic way.

3.4. Gaussian models

3.4.1. *Some limits of the Brownian motion*

We have seen that all the phylogenetic comparative methods studied in the previous sections were based on the same evolutionary model of one (or more) traits following a Brownian motion branching on a phylogenetic tree. This very simple model allowed us to explicitly incorporate phylogenetic relationships in our statistical analyses. This framework considers that the BM represents the evolution of phenotype traits over time. This evolutionary assumption can be justified for certain types of traits under specific environmental conditions (see section 3.4.3). However, the BM can be seen as representing simple "white noise", which does not show any preference for any trait value over others and whose variance, which increases linearly with time, is not bounded. Such properties may contradict the intuitive idea that one has of a trait subject to environmental selection. More generally, what has made the strength and success of the Brownian motion, its simplicity, is also its limitation: such a process does not allow modeling or exploring the different mechanisms underlying phenotypic evolution.

From a theoretical point of view, this limitation is easy to remove. Maintaining the same underlying branching structure related to the phylogenetic tree, the BM has merely to be replaced by another well chosen stochastic process. This choice is generally based on two constraints: the process being used must be, first, consistent and able to usefully model some of the biological mechanisms in place, and, second rather simple to allow statistical analysis. The following explores several options that have been proposed in the literature.

3.4.2. *Ornstein–Uhlenbeck process*

3.4.2.1. *Univariate description*

The Ornstein–Uhlenbeck (OU) process has been proposed to model a *stabilizing evolution* around an optimum (Hansen and Martins 1996; Hansen 1997). Compared to the Brownian motion, this process presents, in addition to Gaussian noise, a mechanism of elastic recall toward a central value β interpreted as the optimal value of the trait in a given environment (see section 3.4.3 for more details). The stochastic equation defining this process is given below for a univariate trait W_t evolving over a time t between 0 and h (an example of a realization of this process is shown in Figure 3.8):

$$W_0 = \mu, \qquad \mathrm{d}W_t = -\alpha(W_t - \beta)\mathrm{d}t + \sigma \mathrm{d}B_t, \ \forall 0 \leq t \leq h \tag{3.14}$$

The speed with which the trait is attracted to its optimum is controlled by the *selection strength* α. To interpret this parameter, it is often replaced by the associated *phylogenetic half-life* $t_{1/2} = \ln(2)/\alpha$ (Hansen 1997). As it can seen in Figure 3.8, $t_{1/2}$ represents the time needed for the expectation of the trait to cover half the distance from its current value to its optimal value. This time $t_{1/2}$ can be compared to the total evolution time h given to the process, which often is the total height of the phylogenetic tree in the context of PCM. In order to make this relationship explicit, $t_{1/2}$ is often expressed as a percentage of total time h:

– If $t_{1/2}$ is *small* compared to h, it means that the process has enough time to reach its optimum, and therefore that the selection strength α can be considered *strong*. The deterministic part of elastic recall tends to outweigh the stochastic part.

– Otherwise, if $t_{1/2}$ is *large* compared to h, the process is "cut off" before it could reach its optimum, and the selection strength is *weak*. The stochastic Brownian part tends to dominate.

The distribution of the trait W_t at time t is still Gaussian:

$$W_t \sim \mathcal{N}\left(e^{-\alpha t}\mu + (1 - e^{-\alpha t})\beta, \ (1 - e^{-2\alpha t})s^2\right) \tag{3.15}$$

with $s^2 = \sigma^2/(2\alpha)$ the stationary variance that represents the limit variance of a process after a long time. The existence of a stationary distribution as well as the bounded nature of the variance are two properties that make the OU interesting to use compared to the BM.

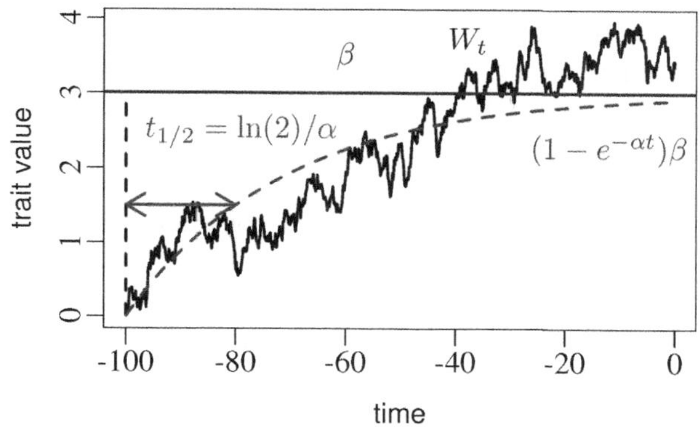

Figure 3.8. *Realization of a univariate OU W_t, with a fixed root value $\mu = 0$, a variance $\sigma^2 = 0.05$, an optimal value $\beta = 3$ (blue horizontal line) and a selection strength α such that $t_{1/2} = \ln(2)/\alpha = 20$ % of the allocated time ($h = 100$, double purple arrow). The exponential evolution of the expectation of W_t toward β is represented by the purple dotted curve. For a color version of this figure, www.iste.co.uk/didier/models.zip*

3.4.2.2. *Distribution at the leaves and identifiability*

Using again the notations from the previous sections, we obtain (see, for example, (Butler and King 2004) for a detailed derivation in a simple case):

$$\begin{aligned}\mathbb{E}\left[Y^i\right] &= e^{-\alpha t_i}\mu + (1 - e^{-\alpha t_i})\beta \\ \mathbb{C}\text{ov}\left[Y^i; Y^j\right] &= e^{-\alpha(t_i - t_{ij})} s^2 e^{-\alpha(t_j - t_{ij})} - e^{-\alpha t_i} s^2 e^{-\alpha t_j} \\ &= \frac{\sigma^2}{2\alpha}(1 - e^{-2\alpha t_{ij}})e^{-\alpha(t_i + t_j - 2t_{ij})}\end{aligned} \qquad [3.16]$$

The marginal distribution of the trait at the leaves is thus completely explicit, which allows us to make a direct inference (Butler and King 2004; Beaulieu et al. 2012; Pennell et al. 2014).

REMARK 3.1.– When α tends to 0, we have:

$$e^{-\alpha t_i}\mu + (1 - e^{-\alpha t_i})\beta \to \mu \quad \text{and} \quad \frac{\sigma^2}{2\alpha}(1 - e^{-2\alpha t_{ij}})e^{-\alpha(t_i + t_j - 2t_{ij})} \to \sigma^2 t_{ij}$$

so that the times of equation 3.16 converge toward those obtained by simple Brownian motion (see equation [3.2]). This confirms the previously given intuition: when $t_{1/2}$ tends toward infinity (i.e. when α tends to 0), a homogeneous OU on the tree behaves like a BM of parameters μ and σ^2.

For an ultrametric tree, we have $t_i = h$ for all the leaves of the tree, such that the trait has the same expectation $\lambda = e^{-\alpha h}\mu + (1 - e^{-\alpha h})\beta$ at all leaves and that only this parameter λ is identifiable. It is therefore impossible in the case of an ultrametric tree to independently infer the mean at the root of the tree and the optimum.

As shown in Figure 3.9, this has practical consequences from the biological interpretation perspective. In scenario A of Figure 3.9, the trait actually starts from its optimal value ($\mu = \beta = 1$) and stays there up to the leaves, such that the trait is seen as being at equilibrium in the population of species observed at the leaves. In contrast, in scenario B, the trait starts from a high value $\mu = 10$ and decreases toward its optimum $\beta = -2$ in an exponential way, however the total time allocated to its evolution (the tree height h) is too short to allow it to reach it, and all the traits are "cut" at an average value $\lambda = 1$. In this scenario, the observed species population is out of equilibrium, and it is expected that the trait continues to decrease in the future. However, since $\lambda = 1$ in both scenarios, they are indistinguishable from the data at the leaves alone. Only a temporal signal, obtained for instance by means of sampling fossils when they are available, could allow us to discriminate between these two very different scenarios.

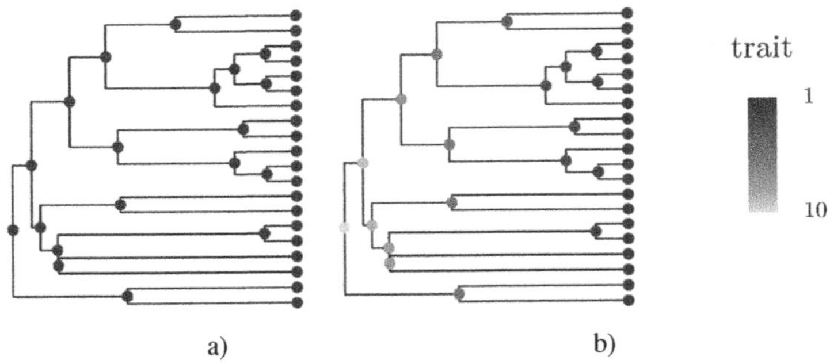

Figure 3.9. *Two equivalent OU processes such that $\lambda = 1$ and $t_{1/2} = 50\ \%$ of the total tree height h. Trait expectation at the nodes is represented by the color scale. (a) $\lambda = \mu = \beta = 1$. (b) $\lambda = 1$, $\mu = 10$ and $\beta = -2$. For a color version of this figure, www.iste.co.uk/didier/models.zip*

To address this identifiability issue, it is often assumed that $\lambda = \mu = \beta$, namely that the root of the tree is already at optimum (scenario A). This assumption can be justified if it is assumed, for example, that the trait has evolved a long time before reaching the root of the clade that matters to us. This difficulty, along with other problems relating to the inference of the OU are presented in Ho and Ané (2014b).

3.4.2.3. *Multivariate OU*

When the trait is multivariate, the dynamics of the trait return to its optimum is no longer described by a scalar α, but by a matrix $\mathbf{A}$ of size $p \times p$:

$$\mathbf{W}_0 = \boldsymbol{\mu}, \qquad d\mathbf{W}_t = -\mathbf{A}(\mathbf{W}_t - \boldsymbol{\beta})dt + \mathbf{R}^{1/2}d\mathbf{B}_t, \ \forall 0 \leq t \leq h \qquad [3.17]$$

Depending on the structure imposed to $\mathbf{A}$, this dynamics can be more or less complex (see Figure 3.10). For the trait to be attracted to its optimum, all the eigenvalues of $\mathbf{A}$ simply need to have strictly positive real parts (Bartoszek et al. 2012). In practice, the inference of this matrix requires a strong temporal signal, and it is often best to impose an a priori structure to it (Clavel et al. 2015).

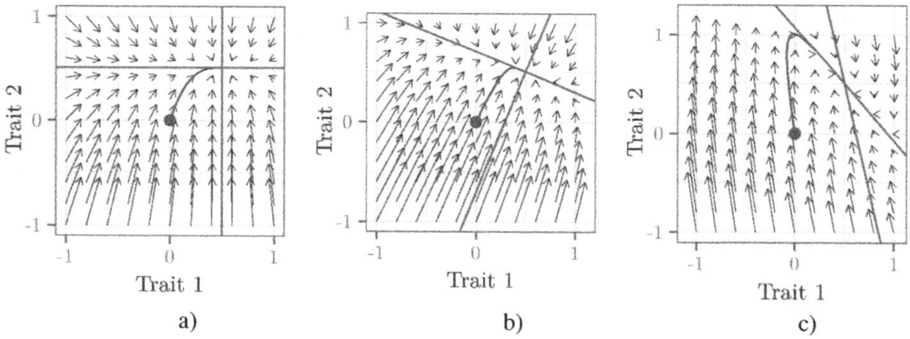

Figure 3.10. *Dynamics of return to the optimum for three bivariate OU with* **A** *a diagonal matrix (a); a symmetric matrix (b); or a matrix with real positive eigenvalues. For a color version of this figure, www.iste.co.uk/didier/models.zip*

COMMENTS ON FIGURE 3.10.– *The trajectory in blue represents the expectation evolution of the trait from* $\boldsymbol{\mu} = (0,0)^T$ *to the optimum* $\boldsymbol{\beta} = (0.5, 0.5)^T$. *The red axes follow the direction of the eigenvectors of the matrix and are centered in* $\boldsymbol{\beta}$. *The black arrows represent the vector field associated with the deterministic portion of equation [3.17].*

The marginal distribution of the trait at the leaves is written similarly to the univariate case (see, for example, Bartoszek et al. (2012), Appendix B for a detailed derivation):

$$\begin{aligned} \mathbb{E}\left[\mathbf{Y}^i\right] &= e^{-\mathbf{A}t_i}\boldsymbol{\mu} + (1 - e^{-\mathbf{A}t_i})\boldsymbol{\beta} \\ \mathbb{C}\text{ov}\left[\mathbf{Y}^i; \mathbf{Y}^j\right] &= e^{-\mathbf{A}(t_i - t_{ij})}\mathbf{S}e^{-\mathbf{A}^T(t_j - t_{ij})} - e^{-\mathbf{A}t_i}\mathbf{S}e^{-\mathbf{A}^T t_j} \end{aligned} \qquad [3.18]$$

where $\mathbf{S}$ is the multivariate stationary variance of the process given by the formula $\text{vec}(\mathbf{S}) = (\mathbf{A} \oplus \mathbf{A})^{-1} \text{vec}(\mathbf{R})$, with $\oplus$ being the Kronecker sum (Meucci 2009).

As in the univariate case, it is possible to directly use this distribution to infer the parameters of the multivariate OU (Bartoszek et al. 2012; Clavel et al. 2015). Compared to the Brownian case (see equations [3.10] and [3.11]), it is clear that we can no longer factor the contributions of the tree and the process itself to the variance matrix in an elementary way. This makes the inference of this process more complex and subject to numerical instabilities. In particular, no explicit estimator is available for $\mathbf{A}$ in the general case and its estimation depends critically on the parameterization chosen for this matrix (Clavel et al. 2015). The estimation is relatively easy when $\mathbf{A}$ and $\mathbf{R}^{1/2}$ commute: Khabbazian et al. (2016) considered the case where $\mathbf{A}$ and $\mathbf{R}^{1/2}$ are diagonal and Bastide et al. (2018a) the case where $\mathbf{A}$ is scalar and $\mathbf{R}^{1/2}$ arbitrary.

Since its introduction in phylogeny (Hansen and Martins 1996), the OU has experienced great success in the field, the number of publications using this model growing exponentially over a 10-year period between 2005 and 2014 (Cooper et al. 2016). This success is due in part to its intuitive and appealing interpretation of a trait subject to natural selection. However, this biological interpretation is not always justified, and it is important to keep in mind some of the subtleties of the modeling process, presented in section 3.4.3, when using this process.

3.4.3. *Biological interpretations and caveats*

The interpretation of the processes described in this chapter depends crucially on the *time scale* being considered. For time scales ranging from a few thousand to a few hundred thousand years, these processes enjoy well-established theoretical guaranties and may be related to specific biological mechanisms. Through the use of quantitative genetic tools, it is indeed possible to show (Lande 1976) that the evolution of a phenotype subjected to additive genetic drift in a constant adaptive landscape, once rescaled, can converge to a Brownian motion (for a flat adaptive landscape) or an Ornstein–Uhlenbeck process (for a landscape with a single adaptive peak). This type of phenomenon can be observed when a rather dense fossil record is available, carrying a sufficiently strong temporal signal. This is, for example, the case for the stickleback fish family (*Gasterosteidae*), in which it can be seen that the shape of the dorsal fin has evolved from one optimum to another in only a few thousand years, and despite a relatively weak strength of selection (Hunt et al. 2008; Hunt and Rabosky 2014). When selection pressures are stronger, a few years may even be enough to observe such an adaptation. In the family of island saurians (*Anolis*), some studies suggest, for example, that the introduction of an invasive species (*Anolis sagrei*) may have led to the adaptation of the endemic species (*Anolis carolinensis*) toward an arboreal-type niche in 15 years only between 1995 and 2010 (Stuart et al. 2014).

These very short time scales have nothing in common with the durations obtained during the calibration of phylogenetic trees of the ecological type, which are measured in the order of the tens or even hundreds of millions years. To cite a few examples, the common ancestors of the new world monkey clade (*Platyrrhini*), birds (*Aves*) and turtles (*Testudines*) date back to about 25, 110 and 210 million years before present time (Jaffe et al. 2011; Jetz et al. 2012; Aristide et al. 2016). For such durations, the assumptions used by Lande (1976) are no longer valid. In particular, it has been observed that the rate of phenotipical change inferred from the fossil record was generally much lower than the one that would be expected with such models, a phenomenon known as the "paradox of statis" (*paradox of stasis*; see, for example, Hansen and Houle (2004)). In such a context, the interpretation of the BM and OU processes must therefore be reviewed.

An idea often invoked is that, for this type of macro-evolutionary phenomena, the stochastic processes used reflect, rather than the consequences of genetic drift, those of the random evolution of successive ecological niches. The underlying assumption is that, for given environmental conditions, the trait under consideration reaches its optimum almost instantaneously (with respect to the macroevolutionary time scale), so that it is the trait optimum (or "secondary optimum"; Hansen 1997) that is followed over time rather than its value in itself. The BM can then be used for modeling, in a stochastic environment, the evolution of adaptive traits over phylogenies (Felsenstein 2004, Chapter 24). This interpretation would in particular explain the discrepancies observed between the evolutionary rates observed at the micro- and macro-evolutionary levels. Within the framework of an OU, the secondary optimum, subject to local environmental variations, itself evolves toward a "primary optimum" (Hansen 1997) supposed to represent an adaptive macroscopic balance. This balance is itself subject to undergo sudden changes, or *jumps* reflecting major ecological events, such as migration or climate disruption (Jaffe et al. 2011; see also section 3.4.5).

Although the OU is preferred to the BM to model the evolution of a stationary trait (Hansen and Orzack 2005; Hansen et al. 2008), when dealing with ecological data over a long period of time, it is important to keep in mind that the "naive" interpretation of a trait evolving to its optimum is at best very simplified, and at worst misleading (Cooper et al. 2016). These difficulties of interpretation may lead to adopting a more "pragmatic" approach for the OU which can be seen as a simple *statistical transformation* of phylogeny leading to relaxing the data tree structure imposed by a BM with an additional degree of freedom represented by α (see section 3.4.4.3).

3.4.4. *Further Gaussian processes*

3.4.4.1. *ACDC and EB models*

The ACDC model, for *accelerating/decelerating model,* was introduced by Blomberg et al. (2003) for modeling a trait evolving with a variance that grows or decreases exponentially in time. This is a simple variation on the BM, defined by the equation:

$$W_0 = \mu, \qquad \mathrm{d}W_t = \sigma_0 e^{rt/2} \mathrm{d}B_t, \ \forall 0 \leq t \leq h \tag{3.19}$$

with r a real that governs the evolutionary mode. If r is strictly negative, the variance decreases over time, and this model has been used to describe adaptive radiation known as *Early Burst* (EB) (Harmon et al. 2010). It is particularly used in paleontology for morphological traits for which fossil data are available, which is essential to get an idea of the temporal dynamics of trait evolution. Although attractive in appearance, this model is rarely selected in a statistically significant manner compared to the simple BM (Harmon et al. 2010; Slater and Pennell 2014).

As for the BM, trait expectation with this model remains constant and equal to μ, while the induced covariance between two leaves is given by (see Blomberg et al. (2003), Appendix 3) for a detailed derivation):

$$\mathbb{C}\mathrm{ov}\left[Y^i; Y^j\right] = \sigma_0^2 \left(\frac{e^{rt_{ij}} - 1}{r} \right) \tag{3.20}$$

This process can be seen as a transformation of the branches of the tree, and is equivalent, under certain conditions, to a univariate OU (see section 3.4.4.3).

3.4.4.2. *OUBM and OUOU models*

It is also possible to introduce an additional "layer" to the model by assuming that the optimal value $\beta(\mathbf{E}_t)$ of an OU depends on explanatory variables $\mathbf{E}_t$ (for example climatic data), which evolve themselves over time as a stochastic process, such as a BM or a OU. This is the principle of the OUBM and OUOU models, respectively, introduced by Hansen et al. (2008) and Bartoszek et al. (2012). In the simplest case of a univariate OUBM, the process is defined as:

$$\begin{cases} W_0 = \mu \\ \mathrm{d}W_t = -\alpha(W_t - \beta(E_t))\mathrm{d}t + \sigma_w \mathrm{d}B_t^w, \quad \forall 0 \leq t \leq h \\ \beta(E_t) = b_0 + b_1 E_t \\ \mathrm{d}E_t = \sigma_e \mathrm{d}B_t^e \end{cases} \tag{3.21}$$

The covariance structure obtained at the leaves can be explicitly derived and is given in the two previous references. This OUBM model has, for example, made it possible to identify a statistically significant link between the thermal preferences of certain species of saurians and the ambient temperature of their environment (Labra et al. 2009).

3.4.4.3. *OU and ACDC as statistical transformations*

When addressing an ultrametric tree, all leaves are at the same distance from the root ($t_i = h$, $\forall 1 \leq i \leq n$) and the covariance between two leaves induced by an OU given by equation [3.16] can be reduced to:

$$C_{ij}(\alpha) = e^{-2\alpha h} \left(\frac{e^{2\alpha t_{ij}} - 1}{2\alpha} \right) \qquad [3.22]$$

It is then easy to see that the OU is reduced to a simple BM running over the branches of a tree whose internal node heights have been transformed as relation (see Figure 3.11):

$$t_i(\alpha) = e^{-2\alpha h} \left(\frac{e^{2\alpha t_i} - 1}{2\alpha} \right) \quad \forall 1 \leq i \leq n \qquad [3.23]$$

This relation has been observed many times in the literature, and can be the basis for efficient inference procedures of an OU on an ultrametric tree (Blomberg et al. 2003; Ho and Ané 2014a; Pennell et al. 2014; Bastide et al. 2018a). It can also be noted that, when α tends to 0, $t_i(\alpha)$ tends to t_i and the tree is not modified. This is in line with the previously given interpretation (see remark 3.1): when α tends to 0, the homogeneous univariate OU converges to a BM.

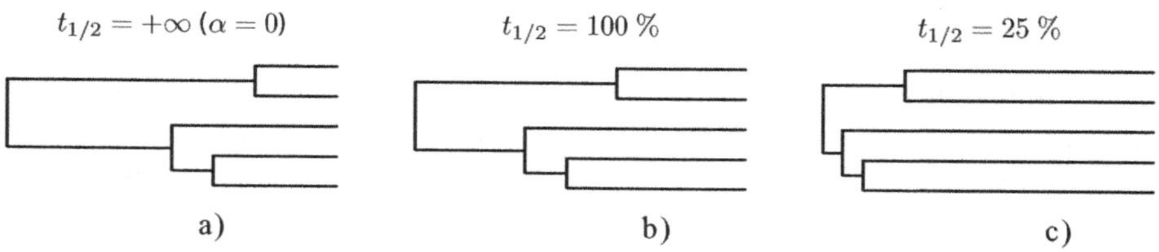

Figure 3.11. *Tree transformations equivalent to an OU. For $\alpha = 0$, the tree remains unmodified and the OU is reduced to a BM (a). As selection increases (namely when the half-life decreases as a percentage of the total tree height), the shape of the tree changes to increasingly look like a star tree (b and c)*

Similarly, the covariance structure induced by the ACDC (equation [3.20]) can be obtained by a BM of variance σ_0^2 executing over a tree whose branch lengths have been transformed by the relation

$$t_i(\alpha) = \frac{e^{rt_i} - 1}{r} \quad \forall 1 \leq i \leq n \tag{3.24}$$

It can be seen that this transformation is very similar to that of the OU. It is indeed possible to show that, under certain conditions, these two processes are equivalent.

PROPOSITION 3.1 (Equivalence of the OU and the ACDC; Uyeda et al. 2015).– With an ultrametric tree of height h, an OU process of variance at the root $\gamma^2 = 0$ of variance σ^2 and strength of selection α induces the same marginal trait distribution at the tree leaves as an ACDC process of parameters:

$$\begin{cases} r = 2\alpha \\ \sigma_0^2 = \sigma^2 e^{-2\alpha h} \end{cases} \tag{3.25}$$

It should be noted that when taking the strict definition of OU, one has $\alpha > 0$, and only the AC process ($r > 0$) can be obtained. If allowing a "generalized OU" where α can be negative (i.e. a process such that the trait is *repelled* by the value β), the DC or EB processes ($r < 0$) can also be used. This observation, that make it possible to see the ACDC as a special case of the OU, can be useful for the statistical analysis of a dataset (Aristide et al. 2018).

3.4.5. *Heterogeneous evolution*

In all the models previously described, it is assumed that a single process regulates the evolution of continuous traits throughout the phylogenetic tree. This assumption may seem reasonable for small trees, but its justification becomes more difficult for larger phylogenies covering entire clades of organisms such as mammals or birds (Meredith et al. 2011; Jetz et al. 2012). It then becomes necessary to relax it by allowing, for example, certain process parameters to change over time. We then define different *regimes* on certain tree clades, each having its own evolutionary mode.

Figure 3.12 shows an example of such a study focusing on the group of new world monkeys. The traits studied here correspond to two statistical scores chosen to best represent the variations in brain shape of these monkeys (Aristide et al. 2016). Such functional traits are expected to be influenced by the ecological niche in which the species under study evolve (Butler and King 2004). In the mentioned study, these evolutionary niches are defined by way of a combination of ecological parameters including the diet (comprising leaves, fruits, seeds or insects), the mode of locomotion (arboreal quadrupedalism, brachiation, etc.) and the typical size of communities (Aristide et al. 2016). The different colors represented in the tree in Figure 3.12 largely overlap these different evolutionary niches (Bastide et al. 2018a): all species of the same color share similar lifestyles.

These evolutionary regimes can be a priori defined according to appropriate biological assumptions or, which is more complex from a statistical perspective, be directly inferred from the data.

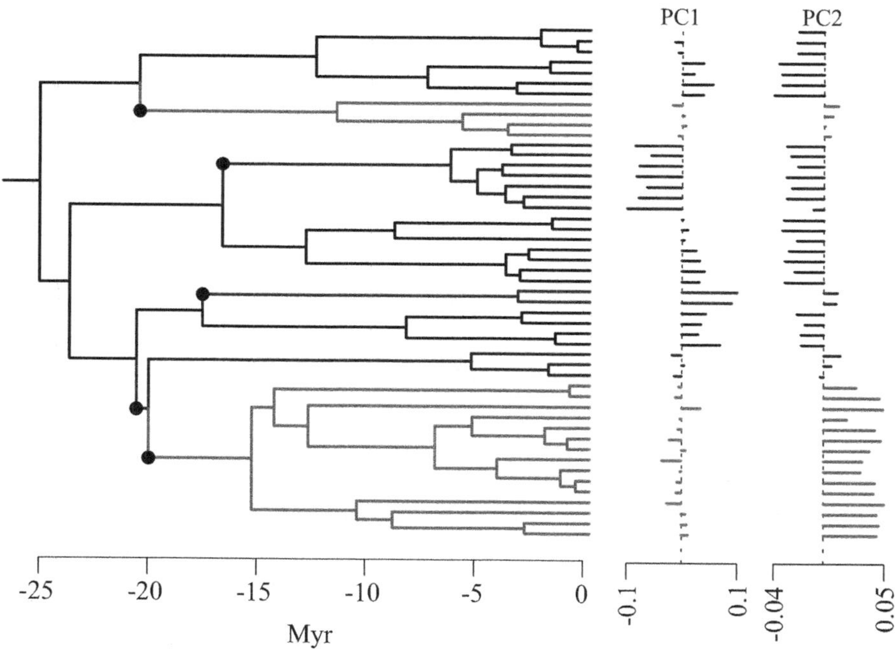

Figure 3.12. *Phylogenetic tree and traits representing brain shapes (principal components 1 and 2) for the new world monkey group (Aristide et al. 2016). The colored clades represent different evolutionary regimes, as inferred by PhylogeneticEM (source: Bastide et al. 2018a). For a color version of this figure, www.iste.co.uk/didier/models.zip*

3.4.5.1. *Regimes set a priori*

If the regimes are set in advance and we consider a process with parameters that are constant over each regime, it is possible to write and maximize the likelihood of the complete model. It is then possible to compare the performance of several well-chosen biological hypotheses for the positions of the regimes on the tree (see section 3.4.7 about selecting the model).

To the authors' knowledge, the most comprehensive tool is the R package mvMORPH software package (Clavel et al. 2015). This package provides all the processes and transformations cited, within a multivariate framework and with the possibility of a priori defining regimes on the tree. This tool is the culmination of a long succession of similar models, the main steps of which are listed below: vary only the optimum β of a univariate OU (Butler and King 2004), the variance σ^2 of a univariate BM (O'Meara et al. 2006) and the optimum and the variance of a univariate OU (Beaulieu et al. 2012).

The dataset presented in the previous section and illustrated in Figure 3.12(a), for example, was studied in such a framework by Aristide et al. (2016). A comparative study thus allows them to favor a model of OU with jumps, namely a scenario in which each group of species, such as defined in Figure 3.12, has its own optimal brain shape adapted to its ecological niche.

This model proves to be more consistent with the data collected than a simple stochastic model such as EB (see section 3.4.4.1), where the trait evolves with great variance at the beginning of the tree to then stabilize and thus give rise to clades of species with similar traits but not necessarily particularly well adapted to their specific environment.

In a slightly different perspective, Lemey et al. (2010) allow each branch b of the tree to have its own variance $\sigma_b^2 = \phi_b \sigma^2$ in a univariate BM. For a Bayesian inference problem, the heterogeneity parameters ϕ_b are then drawn following a well-chosen given emission distribution (log-normal or Cauchy for example).

3.4.5.2. *Automatic detection*

Automatic detection of the position of regimes in the tree is a more complex problem due to the large number of different models to be considered (all identifiable partitions of the tree into k different regimes, for $1 \leq k \leq n + m$ where m is the number of internal nodes in the tree; see Bastide et al. 2017).

Mahler et al. (2013) have proposed the first method for detecting the position of jumps in the tree in the parameter β. This approach is based on a heuristic and also allows the study of evolutionary convergence. It is limited to univariate or independent traits. Eastman et al. (2011, 2013) and Uyeda and Harmon (2014) proposed a Bayesian method based on a reversible-jump Markov chain Monte Carlo algorithm (RJMCMC), for detecting jumps, respectively, in the σ^2 of a BM, the mean μ of a BM and the optimum β of a univariate OU. By exploiting a linear model formulation and using the transformation described in section 3.4.4.3, Khabbazian et al. (2016) proposed a method based on the LASSO (Tibshirani 1996) for independent traits. Using the same tools, Bastide et al. (2018a) proposed a maximum likelihood method for the detection of jumps in this same parameter β for a correlated multivariate OU, but such that the strength of selection matrix $\mathbf{A}$ is proportional to the identity.

The latter method was used to produce the classification presented in Figure 3.12 (Bastide et al. 2018a). It is clear, a posteriori, that this classification is in agreement with the distribution into ecological niches as previously described (Aristide et al. 2016). Nevertheless, this relation is not taken into account by the model which makes use of the value of the continuous traits only as data, such that no conclusions can be drawn about the real impact of these niches on trait evolution. A more accurate integrative model capable of addressing these types of issues is yet to be developed.

3.4.5.3. *Identifiability*

In most comparative studies, such as the one on new world monkeys previously presented, generally only the traits measured at the tips of the tree, for current species, are generally accessible. For an ultrametric tree, we thus try to obtain information on an dynamic evolutionary process (such as the stochastic process with jumps in section 3.4.5.2) using measurements performed at a single time point. Such a situation is likely to give rise to identifiability problems. Figure 3.13 shows an example of such a situation. If we are modeling trait evolution as an OU for which each clade of a different color in the tree has a distinct optimum β, it is then possible to show that the five colorings presented in Figure 3.13 induce exactly the same marginal trait distribution observed at the leaves. In other words, these five jump allocations will have the same likelihood and will therefore not be distinguishable from observations alone.

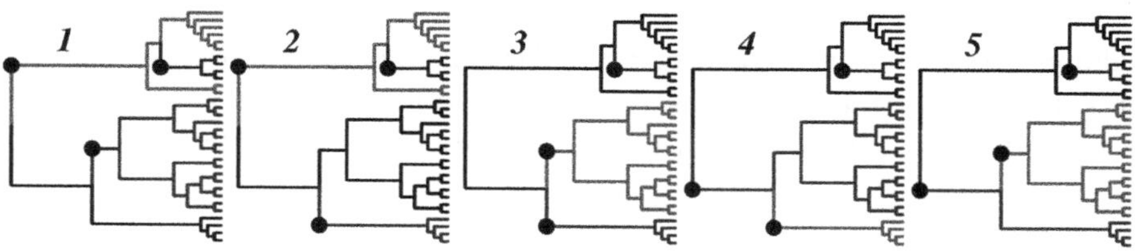

Figure 3.13. *Five equivalent allocations of three jumps (black dots) on a tree of 32 species. The partition induced at the leaves and actually observed is always the same, although the evolutionary scenarios represented by the position of the jumps can be very different. For a color version of this figure, www.iste.co.uk/didier/models.zip*

These scenarios may be quite different from a biological point of view. For example, in scenario 1, the four species at the bottom of the tree (in black) are in ancestral state: during the whole evolutionary period, their ecological niche has never changed. In contrast, in scenario 4, the optimum value of these same species (in green) underwent two successive changes that may represent ecological events such as migration or resulting from climate change. However, it is impossible to differentiate between these two scenarios on the basis of the traits observed nowadays. Only fossil data, if available, would, by introducing asynchronous measurements, discern between these scenarios. This problem, and its implications for statistical inference of the position of jumps in the tree as presented in the previous section, was studied in Bastide et al. (2017).

3.4.6. *Observation models*

In all previous sections, it was assumed that the only source of variability between observations was due to the stochastic process on the tree. The interpretation that was given to the BM and the OU (see section 3.4.3) assumes explicitly that, in the macro-evolutionary case, the secondary trait optimum of each species is directly observed which can be approximated by the mean of the trait within a species. In practice, this assumption is often far from being verified and may lead to significant biases in the results of a PCM (Silvestro et al. 2015; Cooper et al. 2016). This bias is particularly pronounced for heterogeneous data sets, originating from large collections and making use of multiple sources. For example, Rose et al. (2016) make a morphological study of the family of mosses (Bryophyta) based on specimens collected on site, but also on an extensive study of the scientific literature from which more than a thousand manuscript plates of botanical illustrations are extracted. Such a disparity in the source of the data must be taken into account in the statistical analyses.

The first simple way to take this variability into account consists of introducing, in addition to the phylogenetic variance $\sigma_p^2\mathbf{C}_n$ due to the stochastic process on the tree, a residual variance $\sigma_e^2\mathbf{I}_n$ that represents the independent effect of the environment on each measure. The total variance matrix at the leaves $\mathbf{V}_n$ is then written as:

$$\mathbf{V}_n = \sigma_p^2\mathbf{C}_n + \sigma_e^2\mathbf{I}_n \qquad [3.26]$$

By analogy with quantitative genetic models, the first variance is seen as a result of so-called "heritable" effects, while the second one represents the "non-heritable" variability. This model appears in the literature as the *phylogenetic mixed model* or PMM (Lynch 1991; Housworth et al. 2004).

In the case of a Brownian motion with an ultrametric tree of height h, the ratio h_p^2 of phylogenetic variance to total variance is sometimes referred to as "phylogenetic inheritance" (Lynch 1991):

$$h_p^2 = \frac{\sigma_p^2 h}{\sigma_p^2 h + \sigma_e^2} \qquad [3.27]$$

Due to its biological interpretation that seems attractive at first, this quantity has received a lot of attention, especially recently in virology (Alizon et al. 2010; Leventhal and Bonhoeffer 2016; Blanquart et al. 2017; Mitov and Stadler 2018). Within this simple framework, it should be noted that this model is equivalent to a Pagel λ transformation (see section 3.2.2.2.1). Under this model, the total variance at the leaves $\mathbf{V}_n^\lambda$ is in effect written as (Leventhal and Bonhoeffer 2016):

$$\mathbf{V}_n^\lambda = \sigma^2 \left[\lambda \mathbf{C}_n + (1-\lambda) h \mathbf{I}_n\right] \qquad [3.28]$$

This is tantamount to the model of equation [3.26] with $\sigma^2 = \sigma_p^2 + \sigma_e^2/h$ and $\lambda = h_p^2$. This equivalence enables us to see that the parameter λ, often interpreted as measuring a "phylogenetic signal", can also be interpreted as a ratio of variances representing "phylogenetic heritability". It is clearly visible that this formal equivalence between these two notions only holds for a simple Brownian motion and is only valid for the "statistical" definition of phylogenetic signal as given by Revell et al. (2008) and recalled in section 3.2.2.2.

In a multivariate framework, a similar model has been proposed by Ives et al. (2007) with an environmental variance assumed to be known. When multiple observations are available for each species, Felsenstein (2008) suggests that these individual observations be seen as additional leaves in the tree, connected to the node of their species by branches of zero length. This allows him to propose an extension of phylogenetic contrasts (see section 3.3.2) to this more complex situation. Within a similar framework, Goolsby et al. (2017) propose a fast inference procedure of intraspecific variance based on an algorithm developed by Ho and Ané (2014a).

More generally, Hadfield and Nakagawa (2010) revisit and extend the analogy between the PCM and quantitative genetic models based on the study of pedigrees. More specifically, in the case of a BM, they show the mathematical equivalence between the computation of the phylogenetic correlation matrix $\mathbf{C}_n$ and the kinship matrix $\mathbf{A}$ (Crow and Kimura 1970, Chapter 4) computed with consanguinity coefficients equal to the branch lengths (see also Bastide et al. (2018b) for an explanation of this relation in the more general context of a phylogenetic network). The general model described in Hadfield and Nakagawa (2010) is then written as:

$$\mathbf{Y} = \boldsymbol{\mu} + \mathbf{P} + \mathbf{E} + \mathbf{M} \qquad [3.29]$$

where:

– $\mathbf{Y}$ is the observation matrix of size $n \times p$ where n is the number of observations and p the trait dimension;

– $\boldsymbol{\mu}$ is the matrix of fixed effects ($n \times p$);

– $\mathbf{P}$ is the matrix of random effects whose covariance is dictated by the stochastic model of evolution on the tree as previously seen;

– $\mathbf{E}$ is the matrix of residual variations of variance σ_e^2 independent for all observations and traits;

– $\mathbf{M}$ is the heterogeneity matrix representing the errors specific to each group of measures related to their origin. Observations can be grouped by species, but not necessarily. It makes it possible to take into account the effect of "meta-analysis" by homogenizing data originating from various sources.

The inference of such a model can be done by borrowing techniques from quantitative genetics (for a review, see Henderson (1976) and Thompson (2000)) and can be adapted to non-Gaussian models by resorting to MCMC-type techniques (Hadfield and Nakagawa 2010).

3.4.7. *Model selection*

Many models have been presented in this section. Each captures different features of the evolutionary processes involved, and it can be difficult in practice to choose the *best* model for a given problem. This problem is not specific to the study of continuous traits. It has already been encountered for discrete traits: for the choice of evolutionary model for nucleotides (JC, HKY, GTR, etc.) of the transition matrix for amino acids (BLOSUM, LG, etc.) including a $+\Gamma$ component (see Chapter 2).

As in the discrete case, the most common approach consists of using model selection criteria based on an AIC-type penalized likelihood, *Akaike's information criterion* (Akaike 1974), or BIC (*Bayesian information criterion*). These criteria select the optimal trade-off between model complexity and data fitting and are used to choose the best model among a list chosen in advance (e.g. OU, BM, EB or ACDC). More sophisticated versions of these criteria are also employed to select the optimal number of regimes for the automatic detection of regime changes (see section 3.4.5). Interested readers may refer to Giraud (2014) for an introduction to these model selection techniques.

3.5. Extensions and generalizations

Section 3.4 focuses on Gaussian models with a moderate number of traits, without interaction between traits or between traits and phylogeny, and for which it is generally possible to write the likelihood and fit the model efficiently. We present here some extensions to large classes of non-Gaussian processes, with interactions or in high dimension.

3.5.1. *Non-Gaussian models*

The Lévy process can be seen as a BM (possibly with tendency) perturbed by a Poisson process. The Poisson process induces position and random intensity jumps that are expected to capture Simpsonian-based evolution. This model was used within a Bayesian framework to analyze morphological data from great apes (Landis et al. 2013) and identify the most likely position of adaptive jumps. The model likelihood is factored over the branches of the tree in a similar way to Gaussian models and can therefore be computed via the recursive algorithm described in section 3.2.3. However, and unlike Gaussian processes, the likelihood of the Lévy process along a branch is not analytical and requires a numerical integration step.

Boucher et al. (2018) propose an extension to more general models in which trait evolution is no longer governed by simple processes (such as the BM, the OU or the Lévy process) but by Fokker–Planck type partial differential equations. The latter are widely used in population genetics and can be used for modeling complex phenomena (evolutionary trends, rapid adaptations, existence of multiple optimal values, etc.). Just like Lévy processes, these models suffer from technical limitations (only available for univariate traits, long estimation) but constitute an active field of research.

3.5.2. *Tree–trait interactions*

All the models presented so far assume that the trait evolves on a *fixed* and a priori known tree. The family of *xxSSE* (for *State Speciation Extinction*) models relaxes this assumption: the trait value now influences the speciation and extinction rates and thus, *in fine*, the topology of the tree. Maddison et al. (2007) introduce BiSSE (for *Binary SSE*) the first representative of the family. In this model, a binary latent trait $(0/1)$ evolves on a known and complete tree according to a continuous time Markov chain with transition rates q_{01} and q_{10}. A lineage at state i ($i \in \{0, 1\}$) has a speciation rate λ_i and an extinction rate μ_i. Conditional on the binary process, the trait is independent of speciation and extinction events. Maddison et al. (2007) describe a system of differential equations that can be numerically integrated for computing the data likelihood using the same recursive algorithms as previously.

Many representatives have enhanced the family since then: Fitzjohn et al. (2009) showed how to fit the model even if the tree is not completely observed, before extending it to categorical latent traits (Fitzjohn 2010) or continuous traits (Fitzjohn 2012), and Goldberg et al. (2011) proposed a biogeographic version. Finally, Rabosky and Huang (2016) developed a robust test, which can be applied to models beyond xxSSE to detect dependencies between trait value and speciation rate.

From another perspective, it is possible to jointly model the evolution of molecular sequences and continuous traits within an integrative Bayesian framework. The phylogenetic tree is then inferred using both sources of information together, and a two-step study is not necessary as it has been done so far (first inferring the tree from the sequences, and then using this tree as data to study quantitative traits). These approaches are sometimes referred to as *total evidence* and are based on the fundamental assumption that, conditional on the phylogenetic tree, the two evolutionary sequence and continuous trait processes are independent (Lemey et al. 2010; Pybus et al. 2012; Cybis et al. 2015).

3.5.3. *Interactions between species*

In Gaussian, Lévy or *xxSSE* models, the influence of other species on a species of interest is implicit. It can be read in changes of trait value but is not explicitly coded. It is in particular impossible to model explicitly traits whose optimal value depends on the traits of other species.

Manceau et al. (2016) have proposed a very general conceptual framework for the estimation of species interactions that considerably extends Gaussian models. This framework considers that the traits of the *set of species* existing at a time t in the past evolve as a multivariate Gaussian process. The idea is attractive and makes it possible to model co-evolution between species. Nonetheless, this model suffers, similarly to those mentioned in this section, from a high computational complexity. The time since the root must be divided into segments (delimited by speciation and extinction events) and the likelihood must be calculated differently on each segment. Moreover, if n species exist in a given segment, replacing n independent features of size p by a single feature of size np increases the complexity from $O(np^{2.373})$ to $O((np)^{2.373})$.

The study of this type of process with interactions is however an active field of research, and several methods have been proposed to model specific, mutualistic or competitive interaction types (Nuismer and Harmon 2015; Drury et al. 2016; Bartoszek et al. 2017; Drury et al. 2018; Aristide and Morlon 2019).

3.5.4. *Trait of high dimension*

Recursive likelihood algorithms are linear in n but supra-quadratic in p and thus require adaptations for high dimensional traits. We present two of them: the use of latent factors and pseudo-likelihood approaches.

3.5.4.1. *Factor model and mixed data*

When a large number of traits are measured, it is likely that some are highly correlated and redundant. One method to reduce the size of the process on the tree is to use a "phylogenetic factor analysis" (Tolkoff et al. 2018). In this model, it is assumed that the p original traits are linearly deduced from a restricted number $q \ll p$ of factors that evolve as a BM on the tree using a weight matrix.

From an algorithmic point of view, this amounts to performing the recursion on the q latent factors and results in reducing the complexity to $\mathcal{O}(nq^{2.373} + qp)$. The number of latent traits q is chosen by a marginal likelihood method in Tolkoff et al. (2018). The latent factor-based formulation also allows for the discrete and continuous data to be processed in a unified manner (Cybis et al. 2015) using a threshold model also known as *latent liability model* (Felsenstein 2005, 2012). The trait being observed is then either a linear combination of factors (continuous case) or a discretization of this combination (discrete case). The discrete trait only changes value if the latent combination crosses a threshold. Such a process makes it possible to model within a latent Markovian framework a discrete trait "with memory", which is a non-Markovian trait (Goldberg and Foo 2019).

3.5.4.2. *Pseudo-likelihoods*

Other approaches have recently been proposed to directly address large-sized data without involving latent traits. Goolsby (2016) uses a pairwise composite pseudo-likelihood with traits so that only two-dimensional data have to be addressed in the calculations. A statistical test procedure for choosing a model under these conditions is proposed. For their part, Clavel et al. (2019) propose a regularized likelihood procedure based on penalties of the L_1 or L_2 type. Among other things, this allows them to infer parsimonious covariance matrices $\mathbf{R}$, which impose a low number of interactions between traits.

3.6. Useful references

Table 3.1 references all the R software packages cited in this chapter. Nevertheless, this list is not intended to be exhaustive, and interested readers are referred to the *Task*

View of the CRAN dedicated to the phylogenetic comparative methods for a more complete overview of the tools[2].

Sofware Package	Models and Methods	Publication
ape	Tree manipulation (used by all the other packages listed), contrasts	(Paradis et al. 2004)
phytools	pPCA and other classic PCM	(Revell 2012)
geiger	Univariate PCM (including BM, OU, ACDC and Pagel transformations)	(Pennell et al. 2014)
mvMORPH	Multivariate PCM (with knowns jumps)	(Clavel et al. 2015)
phylolm	Generalized phylogenetic regression	(Ho and Ané 2014a)
Rphylopars	BM and OU with intra-specific variance	(Goolsby et al. 2017)
ouch	OU with (known) jumps in β	(Butler and King 2004)
OUwie	OU with (known) jumps in β or σ^2	(Beaulieu et al. 2012)
mvSLOUCH	Multivariate OU, OUBM and OUOU	(Bartoszek et al. 2012)
bayou	OU with (unknown) jumps in β	(Uyeda et al. 2015)
l1ou	independent multivariate OU with (unknown) jumps in β	(Khabbazian et al. 2016)
PhylogeneticEM	multivariate BM, OU with (unknown) jumps in β	(Bastide et al. 2017)
PCMbase	Gaussian models, with (known) jumps in the various parameters	(Mitov et al. 2019)
Diversitree	BiSSE, MultSSE, GeoSSE	(Fitzjohn 2012)
BAMMtools	ACDC with (unknown) jumps in σ_0 and SSE variant	(Rabosky 2014)

Table 3.1. *R software packages cited in this chapter*

The following two books also provide good additional resources on the subject:

– Felsenstein (2004) for a general introduction of the evolutionary models of continuous traits (see Chapters 23–25);

– Ravell and Harmon (2022) for a more up-to-date presentation of the latest methods.

3.7. Acknowledgements

The authors would like to thank Héloïse Bastide and Amir Yassin, the first readers of this text, who contributed greatly to its linguistic and scientific improvements. PB received funding from the *Fonds Wetenschappelijk Onderzoek* (FWO, Belgium) during his post-doctorate at the KU Leuven University (*Department of Microbiology and Immunology*, Rega Institute).

2 Available at: cran.r-project.org/web/views/Phylogenetics.html.

3.8. References

Akaike, H. (1974). A new look at the statistical model identification. *IEEE Transactions on Automatic Control*, 19(6), 716–723.

Alizon, S., von Wyl, V., Stadler, T., Kouyos, D.R., Yerly, S., Hirschel, B., Böni, J., Shah, C., Klimkait, T., Furrer, H. et al. (2010). Phylogenetic approach reveals that virus genotype largely determines HIV set-point viral load. *PLOS Pathogens*, 6(9).

Aristide, L. and Morlon, H. (2019). Understanding the effect of competition during evolutionary radiations: An integrated model of phenotypic and species diversification. *Ecology Letters*, 22(12), 2006–2017.

Aristide, L., dos Reis, S.F., Machado, A.C., Lima, I., Lopes, R.T., Perez, S.I. (2016). Brain shape convergence in the adaptive radiation of New World monkeys. *Proceedings of the National Academy of Sciences*, 113(8), 2158–2163.

Aristide, L., Bastide, P., dos Reis, S.F., Pires dos Santos, T.M., Lopes, R.T., Perez, S.I. (2018). Multiple factors behind early diversification of skull morphology in the continental radiation of New World monkeys. *Evolution*, 72(12), 2697–2711.

Bartoszek, K., Pienaar, J., Mostad, P., Andersson, S., Hansen, T.F. (2012). A phylogenetic comparative method for studying multivariate adaptation. *Journal of Theoretical Biology*, 314, 204–215.

Bartoszek, K., Glémin, S., Kaj, I., Lascoux, M. (2017). Using the Ornstein–Uhlenbeck process to model the evolution of interacting populations. *Journal of Theoretical Biology*, 429, 35–45.

Bastide, P., Mariadassou, M., Robin, S. (2017). Detection of adaptive shifts on phylogenies by using shifted stochastic processes on a tree. *Journal of the Royal Statistical Society: Series B (Statistical Methodology)*, 79(4), 1067–1093.

Bastide, P., Ané, C., Robin, S., Mariadassou, M. (2018a). Inference of adaptive shifts for multivariate correlated traits. *Systematic Biology*, 67(4), 662–680.

Bastide, P., Solís-Lemus, C., Kriebel, R., Sparks, K.W., Ané, C. (2018b). Phylogenetic comparative methods on phylogenetic networks with reticulations. *Systematic Biology*, 67(5), 800–820.

Beaulieu, J.M., Jhwueng, D.-C., Boettiger, C., O'Meara, B.C. (2012). Modeling stabilizing selection: Expanding the Ornstein–Uhlenbeck model of adaptive evolution. *Evolution*, 66(8), 2369–2383.

Blanquart, F., Wymant, C., Cornelissen, M., Gall, A., Bakker, M., Bezemer, D., Hall, M., Hillebregt, M., Ong, S.H., Albert, J. et al. (2017). Viral genetic variation accounts for a third of variability in HIV-1 set-point viral load in Europe. *PLOS Biology*, 15(6), 1–26.

Blomberg, S.P., Garland, T., Ives, A.R. (2003). Testing for phylogenetic signal in comparative data: Behavioral traits are more labile. *Evolution*, 57(4), 717–745.

Boucher, F.C., Démery, V., Conti, E., Harmon, L.J., Uyeda, J. (2018). A general model for estimating macroevolutionary landscapes. *Systematic Biology*, 67(2), 304–319.

Butler, M.A. and King, A.A. (2004). Phylogenetic comparative analysis: A modeling approach for adaptive evolution. *The American Naturalist*, 164(6), 683–695.

Ceccarelli, F., Koch, N.M., Soto, E.M., Barone, M.L., Arnedo, M.A., Ramírez, M.J. (2018). The grass was greener: Repeated evolution of specialized morphologies and habitat shifts in ghost spiders following grassland expansion in South America. *Systematic Biology*, 68(1), 63–77.

Clavel, J., Escarguel, G., Merceron, G. (2015). mvmorph: An r package for fitting multivariate evolutionary models to morphometric data. *Methods in Ecology and Evolution*, 6(11), 1311–1319.

Clavel, J., Aristide, L., Morlon, H. (2019). A penalized likelihood framework for high-dimensional phylogenetic comparative methods and an application to new-world monkeys brain evolution. *Systematic Biology*, 68(1), 93–116.

Cooper, N., Thomas, G.H., Venditti, C., Meade, A., Freckleton, R.P. (2016). A cautionary note on the use of Ornstein–Uhlenbeck models in macroevolutionary studies. *Biological Journal of the Linnean Society*, 118(1), 64–77.

Crow, J.F., Kimura, M. (1970). *An Introduction to Population Genetics Theory*. Harper & Row, New York.

Cybis, G.B., Sinsheimer, J.S., Bedford, T., Mather, A.E., Lemey, P., Suchard, M.A. (2015). Assessing phenotypic correlation through the multivariate phylogenetic latent liability model. *The Annals of Applied Statistics*, 9(2), 969–991.

Drury, J., Clavel, J., Manceau, M., Morlon, H. (2016). Estimating the effect of competition on trait evolution using maximum likelihood inference. *Systematic Biology*, 65(4), 700–710.

Drury, J.P., Grether, G.F., Garland, T., Morlon, H. (2018). An assessment of phylogenetic tools for analyzing the interplay between interspecific interactions and phenotypic evolution. *Systematic Biology*, 67(3), 413–427.

Duchen, P., Leuenberger, C., Szilágyi, S.M., Harmon, L.J., Eastman, J.M., Schweizer, M., Wegmann, D. (2017). Inference of evolutionary jumps in large phylogenies using Lévy processes. *Systematic Biology*, 66(6), 1–14.

Eastman, J.M., Alfaro, M.E., Joyce, P., Hipp, A.L., Harmon, L.J. (2011). A novel comparative method for identifying shifts in the rate of character evolution on trees. *Evolution*, 65(12), 3578–3589.

Eastman, J.M., Wegmann, D., Leuenberger, C., Harmon, L.J. (2013). Simpsonian "evolution by jumps' in an adaptive radiation of anolis lizards. *arXiv*, 1305.4216.

Felsenstein, J. (1973). Maximum likelihood and minimum-steps methods for estimating evolutionary trees from data on discrete characters. *Systematic Biology*, 22(3), 240–249.

Felsenstein, J. (1985). Phylogenies and the comparative method. *The American Naturalist*, 125(1), 1–15.

Felsenstein, J. (2004). *Inferring Phylogenies*. Sinauer Associates, Sunderland.

Felsenstein, J. (2005). Using the quantitative genetic threshold model for inferences between and within species. *Philosophical Transactions of the Royal Society B: Biological Sciences*, 360(1459), 1427–1434.

Felsenstein, J. (2008). Comparative methods with sampling error and within-species variation: Contrasts revisited and revised. *The American Naturalist*, 171(6), 713–725.

Felsenstein, J. (2012). A comparative method for both discrete and continuous characters using the threshold model. *The American Naturalist*, 179(2), 145–156.

Fitzjohn, R.G. (2010). Quantitative traits and diversification. *Systematic Biology*, 59(6), 619–633.

Fitzjohn, R.G. (2012). Diversitree: Comparative phylogenetic analyses of diversification in R. *Methods in Ecology and Evolution*, 3(6), 1084–1092.

Fitzjohn, R.G., Maddison, W.P., Otto, S.P. (2009). Estimating trait-dependent speciation and extinction rates from incompletely resolved phylogenies. *Systematic Biology*, 58(6), 595–611.

Freckleton, R.P. (2012). Fast likelihood calculations for comparative analyses. *Methods in Ecology and Evolution*, 3(5), 940–947.

Giraud, C. (2014). *Introduction to High-Dimensional Statistics*. CRC Press, Hoboken.

Goldberg, E.E. and Foo, J. (2019). Memory in trait macroevolution. *The American Naturalist*, The University of Chicago Press, 195(2), 300–314.

Goldberg, E.E., Lancaster, L.T., Ree, R.H. (2011). Phylogenetic inference of reciprocal effects between geographic range evolution and diversification. *Systematic Biology*, 60(4), 451–465.

Goolsby, E.W. (2016). Likelihood-based parameter estimation for high-dimensional phylogenetic comparative models: Overcoming the limitations of "distance-based" methods. *Systematic Biology*, 65(5), 852–870.

Goolsby, E.W., Bruggeman, J., Ané, C. (2017). Rphylopars: Fast multivariate phylogenetic comparative methods for missing data and within-species variation. *Methods in Ecology and Evolution*, 8(1), 22–27.

Grafen, A. (1989). The phylogenetic regression. *Philosophical Transactions of the Royal Society B: Biological Sciences*, 326(1233), 119–157.

Grafen, A. (1992). The uniqueness of the phylogenetic regression. *Journal of Theoretical Biology*, 156(4), 405–423.

Green, P.J. (1995). Reversible jump Markov chain Monte Carlo computation and Bayesian model determination. *Biometrika*, 82(4), 711–732.

Hadfield, J.D. and Nakagawa, S. (2010). General quantitative genetic methods for comparative biology: Phylogenies, taxonomies and multi-trait models for continuous and categorical characters. *Journal of Evolutionary Biology*, 23(3), 494–508.

Hansen, T.F. (1997). Stabilizing selection and the comparative analysis of adaptation. *Evolution*, 51(5), 1341.

Hansen, T.F. and Houle, D. (2004). Evolvability, stabilizing selection, and the problem of stasis. In *Phenotypic Integration: Studying the Ecology and Evolution of Complex Phenotypes*, Pigliucci, M. and Preston, K. (eds). Oxford University Press, New York.

Hansen, T.F. and Martins, E.P. (1996). Translating between microevolutionary process and macroevolutionary patterns: The correlation structure of interspecific data. *Evolution*, 50(4), 1404.

Hansen, T.F. and Orzack, S.H. (2005). Assessing current adaptation and phylogenetic inertia as explanations of trait evolution: The need for controlled comparisons. *Evolution*, 59(10), 2063–2072.

Hansen, T.F., Pienaar, J., Orzack, S.H. (2008). A comparative method for studying adaptation to a randomly evolving environment. *Evolution*, 62(8), 1965–1977.

Harmon, L.J. (2019). *Phylogenetic comparative methods*. [Online]. Available at: https://lukejharmon.github.io/pcm/.

Harmon, L.J., Losos, J.B., Jonathan, D.T., Gillespie, R.G., Gittleman, J.L., Bryan, J.W., Kozak, K.H., McPeek, M.A., Moreno-Roark, F., Near, T.J. et al. (2010). Early bursts of body size and shape evolution are rare in comparative data. *Evolution*, 64(8), 2385–2396.

Henderson, C.R. (1976). A simple method for computing the inverse of a numerator relationship matrix used in prediction of breeding values. *Biometrics*, 32(1), 69–83.

Hiscott, G., Fox, C., Parry, M., Bryant, D. (2016). Efficient recycled algorithms for quantitative trait models on phylogenies. *Genome Biology and Evolution*, 8(5), 1338–1350.

Ho, L.S.T. and Ané, C. (2014a). A linear-time algorithm for Gaussian and non-Gaussian trait evolution models. *Systematic Biology*, 63(3), 397–408.

Ho, L.S.T. and Ané, C. (2014b). Intrinsic inference difficulties for trait evolution with Ornstein–Uhlenbeck models. *Methods in Ecology and Evolution*, 5(11), 1133–1146.

Housworth, E.A., Martins, E.P., Lynch, M. (2004). The phylogenetic mixed model. *The American Naturalist*, 163(1), 84–96.

Hunt, G. and Rabosky, D.L. (2014). Phenotypic evolution in fossil species: Pattern and process. *Annual Review of Earth and Planetary Sciences*, 42(1), 421–441.

Hunt, G., Bell, M.A., Travis, M.P. (2008). Evolution toward a new adaptive optimum: Phenotypic evolution in a fossil stickleback lineage. *Evolution*, 62(3), 700–710.

Ives, A.R., Midford, P.E., Garland, T., Oakley, T. (2007). Within-species variation and measurement error in phylogenetic comparative methods. *Systematic Biology*, 56(2), 252–270.

Jaffe, A.L., Slater, G.J., Alfaro, M.E. (2011). The evolution of island gigantism and body size variation in tortoises and turtles. *Biology Letters*, 7(4), 558–561.

Jetz, W., Thomas, G., Joy, J., Hartmann, K., Mooers, A. (2012). The global diversity of birds in space and time. *Nature*, 491(7424), 444–448.

Khabbazian, M., Kriebel, R., Rohe, K., Ané, C. (2016). Fast and accurate detection of evolutionary shifts in Ornstein–Uhlenbeck models. *Methods in Ecology and Evolution*, 7(7), 811–824.

Kim, H. and Perl, J. (1983). A computational model for combined causal and diagnostic reasoning in inference systems. In *Proceedings of the Eighth International Joint Conference on Artificial Intelligence*. Morgan-Kaufmann, San Mateo.

Labra, A., Pienaar, J., Hansen, T.F. (2009). Evolution of thermal physiology in Liolaemus lizards: Adaptation, phylogenetic inertia, and niche tracking. *The American Naturalist*, 174(2), 204–220.

Lande, R. (1976). Natural selection and random genetic drift in phenotypic evolution. *Evolution*, 30(2), 314.

Landis, M.J., Schraiber, J.G., Liang, M. (2013). Phylogenetic analysis using Lévy processes: Finding jumps in the evolution of continuous traits. *Systematic Biology*, 62(2), 193–204.

Lartillot, N. (2014). A phylogenetic Kalman filter for ancestral trait reconstruction using molecular data. *Bioinformatics*, 30(4), 488–496.

Law, C.J., Slater, G.J., Mehta, R.S. (2018). Lineage diversity and size disparity in musteloidea: Testing patterns of adaptive radiation using molecular and fossil-based methods. *Systematic Biology*, 67(1), 127–144.

Lemey, P., Rambaut, A., Welch, J.J., Suchard, M.A. (2010). Phylogeography takes a relaxed random walk in continuous space and time. *Molecular Biology and Evolution*, 27(8), 1877–1885.

Leventhal, G.E. and Bonhoeffer, S. (2016). Potential pitfalls in estimating viral load heritability. *Trends in Microbiology*, 24(9), 687–698.

Lynch, M. (1991). Methods for the analysis of comparative data in evolutionary biology. *Evolution*, 45(5), 1065–1080.

Maddison, W.P., Midford, P.E., Otto, S.P. (2007). Estimating a binary character's effect on speciation and extinction. *Systematic Biology*, 56(5), 701–710.

Mahler, D.L., Ingram, T., Revell, L.J., Losos, J.B. (2013). Exceptional convergence on the macroevolutionary landscape in island lizard radiations. *Science*, 341(6143), 292–295.

Manceau, M., Lambert, A., Morlon, H. (2016). A unifying comparative phylogenetic framework including traits coevolving across interacting lineages. *Systematic Biology*, 66(4), syw115.

Mardia, K.V., Kent, J.T., Bibby, J.M. (1979). *Multivariate Analysis, Probability and Mathematical Statistics*. Academic Press, New York.

Méléard, S. (2016). *Modèles aléatoires en écologie et évolution. Mathématiques et applications*. Springer, Berlin/Heidelberg.

Meredith, R.W., Janečka, J.E., Gatesy, J., Ryder, O.A., Fisher, C.A., Teeling, E.C., Goodbla, A., Eizirik, E., Simão, T.L.L., Stadler, T. et al. (2011). Impacts of the cretaceous terrestrial revolution and KPg extinction on mammal diversification. *Science*, 334(6055), 521–524.

Meucci, A. (2009). Review of statistical arbitrage, cointegration, and multivariate Ornstein–Uhlenbeck. *SSRN Electronic Journal*, 20.

Mitov, V., Bartoszek, K., Stadler, T. (2019). Automatic generation of evolutionary hypotheses using mixed Gaussian phylogenetic models. *Proceedings of the National Academy of Sciences*, 201813823.

Mitov, V. and Stadler, T. (2018). A practical guide to estimating the heritability of pathogen traits. *Molecular Biology and Evolution*, 35(3), 756–772.

Nuismer, S.L. and Harmon, L.J. (2015). Predicting rates of interspecific interaction from phylogenetic trees. *Ecology Letters*, 18(1), 17–27.

O'Meara, B.C., Ané, C., Sanderson, M.J., Wainwright, P.C. (2006). Testing for different rates of continuous trait evolution using likelihood. *Evolution*, 60(5), 922–933.

Pagel, M. (1999). Inferring the historical patterns of biological evolution. *Nature*, 401(6756), 877–884.

Paradis, E., Claude, J., Strimmer, K. (2004). APE: Analyses of phylogenetics and evolution in R language. *Bioinformatics*, 20(2), 289–290.

Pennell, M.W., Eastman, J.M., Slater, G.J., Brown, J.W., Uyeda, J.C., FitzJohn, R.G., Alfaro, M.E., Harmon, L.J. (2014). geiger v2.0: An expanded suite of methods for fitting macroevolutionary models to phylogenetic trees. *Bioinformatics*, 30(15), 2216–2218.

Pybus, O.G., Suchard, M.A., Lemey, P., Bernardin, F.J., Rambaut, A., Crawford, F.W., Gray, R.R., Arinaminpathy, N., Stramer, S.L., Busch, M.P. et al. (2012). Unifying the spatial epidemiology and molecular evolution of emerging epidemics. *Proceedings of the National Academy of Sciences*, 109(37), 15066–15071.

Rabosky, D.L. (2014). Automatic detection of key innovations, rate shifts, and diversity-dependence on phylogenetic trees. *PLOS ONE*, 9(2).

Rabosky, D.L. and Huang, H. (2016). A robust semi-parametric test for detecting trait-dependent diversification. *Systematic Biology*, 65(2), 181–193.

Raz, R. (2003). On the complexity of matrix product. *SIAM Journal on Computing*, 32(5), 1356–1369.

Revell, L.J. (2009). Size-correction and principal components for interspecific comparative studies. *Evolution*, 63(12), 3258–3268.

Revell, L.J. (2010). Phylogenetic signal and linear regression on species data. *Methods in Ecology and Evolution*, 1(4), 319–329.

Revell, L.J. (2012). Phytools: An R package for phylogenetic comparative biology (and other things). *Methods in Ecology and Evolution*, 3(2), 217–223.

Revell, L.J. and Harmon, L.J. (2022). *Phylogenetic Comparative Methods*. Princeton University Press.

Revell, L.J., Harmon, L.J., Collar, D.C. (2008). Phylogenetic signal, evolutionary process, and rate. *Systematic Biology*, 57(4), 591–601.

Rose, J.P., Kriebel, R., Sytsma, K.J. (2016). Shape analysis of moss (Bryophyta) sporophytes: Insights into land plant evolution. *American Journal of Botany*, 103(4), 652–662.

Silvestro, D., Kostikova, A., Litsios, G., Pearman, P.B., Salamin, N. (2015). Measurement errors should always be incorporated in phylogenetic comparative analysis. *Methods in Ecology and Evolution*, 6(3), 340–346.

Slater, G.J. and Pennell, M.W. (2014). Robust regression and posterior predictive simulation increase power to detect early bursts of trait evolution. *Systematic Biology*, 63(3), 293–308.

Stuart, Y.E., Campbell, T.S., Hohenlohe, P.A., Reynolds, R.G., Revell, L.J., Losos, J.B. (2014). Rapid evolution of a native species following invasion by a congener. *Science*, 346(6208), 463–466.

Thompson, E.A. (2000). *Statistical Inference from Genetic Data on Pedigrees: V 6 (Nsf-Cbms Confererence Series in Probability & Statistics Volume 6)*. Institute of Mathematical Statistics, Beachwood.

Tibshirani, R. (1996). Regression selection and shrinkage via the lasso. *Journal of the Royal Statistical Society. Series B (Methodological)*, 58(1), 267–288.

Tolkoff, M.R., Alfaro, M.E., Baele, G., Lemey, P., Suchard, M.A. (2018). Phylogenetic factor analysis. *Systematic Biology*, 67(3), 384–399.

Uyeda, J.C. and Harmon, L.J. (2014). A novel Bayesian method for inferring and interpreting the dynamics of adaptive landscapes from phylogenetic comparative data. *Systematic Biology*, 63(6), 902–918.

Uyeda, J.C., Caetano, D.S., Pennell, M.W. (2015). Comparative analysis of principal components can be misleading. *Systematic Biology*, 64(4), 677–689.

Vrancken, B., Lemey, P., Rambaut, A., Bedford, T., Longdon, B., Günthard, H.F., Suchard, M.A. (2015). Simultaneously estimating evolutionary history and repeated traits phylogenetic signal: Applications to viral and host phenotypic evolution. *Methods in Ecology and Evolution*, 6(1), 67–82.

4

Correlated Evolution: Models and Methods

Guillaume ACHAZ[1,2] and Julien Y. DUTHEIL[3]
[1] *SMILE Group, CIRB, CNRS, INSERM Collège de France, Paris, France*
[2] *Éco-anthropologie, CNRS MNHN Université Paris Cité, France*
[3] *RG Molecular Systems Evolution, MPI for Evolutionary Biology, Plön, Germany*

4.1. Introduction

As soon as the word *co-evolution* is dropped, the attention of evolutionary biologists rises. Some pictures illuminating patterns of co-evolution between species such the joint evolution of the beak of the beautiful pollinating hummingbirds and its associated flowers (Kay et al. 2005) or the terrifying arm races between ferocious pathogens and their hosts (Woolhouse et al. 2002); soon after, many will think about Alice running on the spot with the Red Queen through the looking glass (Van Valen 1973). Our fascination about complex evolving entities that are in constant dynamical interactions often lies at the cross-road between ecology and evolution, both being arguably parts of the field coined population biology (Lewontin 2004).

There is, however, another common usage of the term *co-evolution*, popular among molecular evolutionary biologists, where the co-evolving entities are not interacting species of an ecological community, but traits of the same individual, species or genome. This chapter reviews a selection of models and methods related to this second meaning. We use equally the terms *co-evolution* and *correlated evolution* to name the joint evolution of two or more traits; *trait* refers to *any* character of individuals, ranging from morphological traits (e.g. height) to genetic variant (e.g. an allele at a locus). We are naturally inclined to accept patterns of co-evolution between the height and the body mass of individuals or, at the molecular level, between paired nucleotides of the same RNA molecule. We show in this review how, once abstracted, the models and methods that have been used to characterize the correlated evolution between morphological traits or between nucleotides share several common features. They are designed in the same general framework and only differ by details of their implementation.

To demonstrate correlated evolution between two traits, one has to reject the statistical independence between the two processes that govern the evolution of the two traits. To do so, one can look at the state of the two traits in many individuals, species or genomes and characterize the co-occurrence or co-abundance of the two traits among the cohort. Alternatively, one can compare the evolutionary paths of the two trait states and scan for signs of co-variations. The hope is that the patterns of correlated evolution is caused by biologically relevant interactions between the two traits. However, as for any pattern of correlation, the non-independence can be indirectly caused by correlations to a third hidden variable. For instance, an environmental variable may drive synchronized changes in several traits that are not interacting directly. As always, *correlation does not imply causation.*

When the evolution of the two traits are actually directly interacting, it is usually the case that their joint mutations have a different impact on fitness than what is expected by the mutation of each one. For example, they can have a synergistic effect when mutating both traits together result in a larger fitness improvement than the sum of each mutation. At the molecular level, interactions between loci, a phenomenon known as epistasis, were the subject of long-standing interests for geneticists (Phillips 2008). In its simplest form, let us consider two binary traits. An individual can then have four possible trait states: 00, 01, 00, 11. Each of the combination can be associated with a fitness value. The map of all possible combinations of trait states to a fitness value is known as a fitness landscape (Wright 1932). For the simple case of two traits with two states, the fitness landscape can be easily sketched as in Figure 4.1.

Depending on which fitness landscape both traits are evolving in, one expects different patterns of correlated evolution. Probably the strongest patterns of co-evolution would arise when only the double mutant is not associated with a fitness loss (Figure 4.1(d)). It corresponds to cases of compensation where the second mutation will compensate, or even over-compensate, the first one. In such a case, we would only observe individuals with trait combinations 00 or 11, and more rarely the intermediate states. Similarly, we expect to only witness synchronized mutations in the evolutionary paths of the two traits, where one mutation in a trait is immediately followed by a mutation in the other. Fitness landscapes have fascinated evolutionary biologists (among others) for almost a century and are still nowadays an object of vivid interest (Achaz et al. 2014; de Visser and Krug 2014; Yi and Dean 2019).

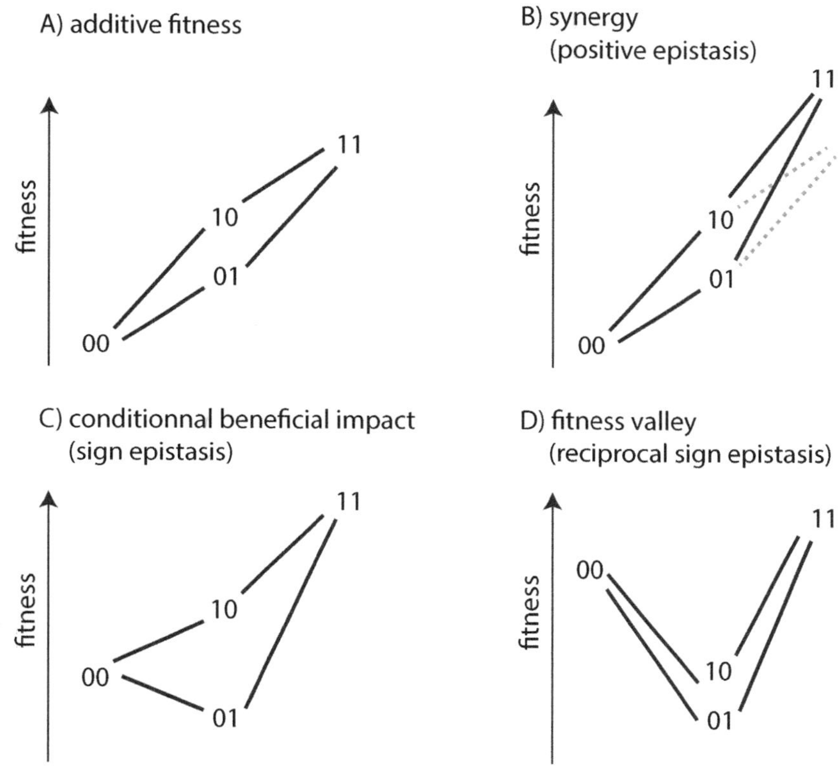

Figure 4.1. *Four examples of fitness landscapes with with binary traits, resulting in four possible combinations. The height of a combination represents its fitness. (A) When the fitness difference of the double mutant compared to the reference is simply the sum of each mutation, there is no dependency between the two mutations. The landscape is named additive and the two traits evolve independently. (B) When the fitness improvement of the double mutant is more than the additive case, there is a synergistic effect of both mutations. (C) The mutation of the second trait is deleterious in the reference ($00 \rightarrow 01$) but beneficial once the first trait has mutated ($10 \rightarrow 11$). (D) Both traits are deleterious when mutated on their own but compensate each other in the double mutant*

In this chapter, we will first explore models of correlated evolution between continuous and discrete traits (section 4.2). These models are historically used to characterize how traits are co-varying in the natural world as species evolve through time. We will then continue our journey with models of correlated evolution within genomes (section 4.3) that have many times been successfully applied to understand co-variations within the same molecule (RNA or proteins), but also at a larger scale to assess co-evolution between different genes or even between regulatory elements and genes. Finally, we will show how standard classical genetics can be rethought in terms of correlated evolution (section 4.4) between phenotypic trait(s) and molecular trait(s) in the genome.

In this chapter, we review the broad corpus of models and methods that have been developed to study coevolution. We more particularly explain the motivations and basic principles of a few chosen models and their related methodologies. We hope that this review will provide to the readers an original overview of the field of *correlated evolution* by integrating evolution, genetics, structural biology, statistical physics, modeling, and statistics.

4.2. Correlated evolution between traits

4.2.1. *Species are not independent*

Co-evolution is expected to result in correlations of character states when measured in several species. Comparative analysis, therefore, constitutes a natural method of investigation to exhibit coevolving traits. The simple comparative analysis of observed states, however, could be misleading as biological species are not statistically independent. Quite on the contrary, as all species share ancestry, the quest of correlated evolution requires dedicated methods. Ignoring the phylogeny results in biases when simple counting methods are applied to species traits. We illustrate this phenomenon, sometimes termed "phylogenetic inertia" (Harvey and Pagel 1991), with a correlation test.

We must first acknowledge the history partially shared by species, that is, classically represented by a phylogenetic tree. We will, therefore, account for the phylogenetic relationships between the species by modeling the evolution of traits along the tree. The most simple model of continuous trait evolution was introduced by Felsenstein (1973). Under this model, a trait Y mutation is only described by its intrinsic variance σ_Y^2. The evolution of the trait is modeled as a Brownian motion: the distribution $D_Y(t)$ of the evolved trait after a certain time t is a normal distribution with mean y_0 (trait at time 0) and a variance equal to $\sigma_Y^2 \times t$. To simulate the evolution of a trait along a tree, we can apply this procedure recursively on each branch of the tree, propagating the trait values down the tree.

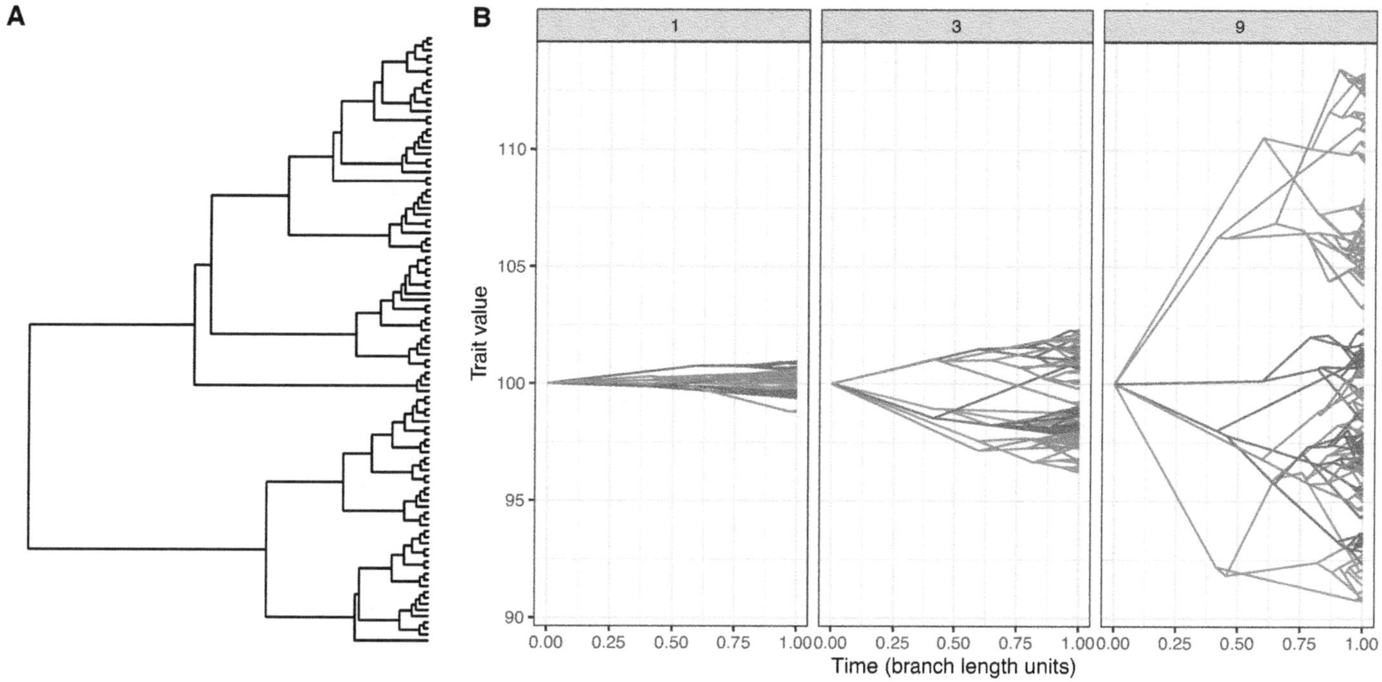

Figure 4.2. *Evolution of a continuous trait under a Brownian motion along a phylogeny. (A) A 100 species random phylogeny used for simulation. (B) Simulations under the phylogeny depicted in (A) using a Brownian motion with variance equal to 1, 3 or 9. Three independent replicates are displayed with three distinct colors in each case. Note the emergent co-variation in the traits, that is only due to the shared phylogeny. For a color version of this figure, www.iste.co.uk/didier/models.zip*

We note y_i the (ancestral) traits at node i in the tree; y_0 is the ancestral trait state at the root (node 0) of the tree. We further note the branch lengths $\{t_i\}_{1<i<n}$ between a node i and its parent p in a rooted tree with n branches. Starting from the root of the tree, a value for each descendant node i is then drawn randomly from the distribution

$$D_Y(t_i) = \mathcal{N}(y_p, \sigma_Y^2 \times t_i), \qquad [4.1]$$

where y_p is the value of the trait at the parent node. Figure 4.2 shows an example of three traits simulated with Brownian motions with distinct intrinsic variances.

Traits can be then measured at the leaves of the tree, representing a sample of species with partially shared histories. Once the traits are measured in the different species, one can compute simple correlations between the trait values. Here, we have chosen the Kendall's τ test because it is a non-parametric rank correlation test between two numerical variables, without any assumption regarding the distribution of the data. Like most correlation tests, it considers each pair of trait values as an independent sampling. Under the null hypothesis of independence, if one draw randomly two sets of values from independent distributions and compute their correlation, one has 5% chances to obtain a P value below 0.05. More generally, if one computes the distribution of P values over many such samples, we obtain a uniform distribution over $[0, 1]$.

When simulating two independent traits along the same phylogeny and assessing their correlation, we observe that the distribution of P values is skewed toward small values (Figure 4.4). This means that the two traits appear intrinsically correlated, despite being independently simulated along the phylogeny.

This implies that the correlation between two traits in a sample of species is the sum of an intrinsic correlation and a phylogenetic correlation. As a result, there is a need for methods that specifically test for the significance of the intrinsic correlation, eliminating for the phylogenetic one. One possibility is to only sample species that are very distant and/or have emerged from a rapid radiation and exhibit a star-like phylogeny. Another is to specifically incorporate the phylogeny in the method.

4.2.2. *The phylogenetically independent contrasts*

Using properties of the Brownian motion, Felsenstein (1985) introduced one of the first comparative method for continuous traits, named the phylogenetic independent contrasts (PIC). The idea behind the contrasts method is a transformation of the input variables, so that the values of each transformed variable become statistically independent. Considering a given phylogenetic tree, it occurs that the evolution of a trait y along two branches leading to a pair of node siblings i and j (e.g. 1 and 2) sharing the same parent node k (e.g. 8) is independent. As a result, the differences $y_1 - y_2, y_4 - y_5, y_6 - y_7, y_8 - y_3, y_{10} - y_{11}$ and $y_9 - y_{12}$ are all statistically independent (Figure 4.3). Following this logic, it is possible to transform a vector of n trait values $\{y_i\}_{1 \leq i \leq n}$ into $n - 1$ phylogenetically independent contrasts $\gamma_{ij} = y_i - y_j$.

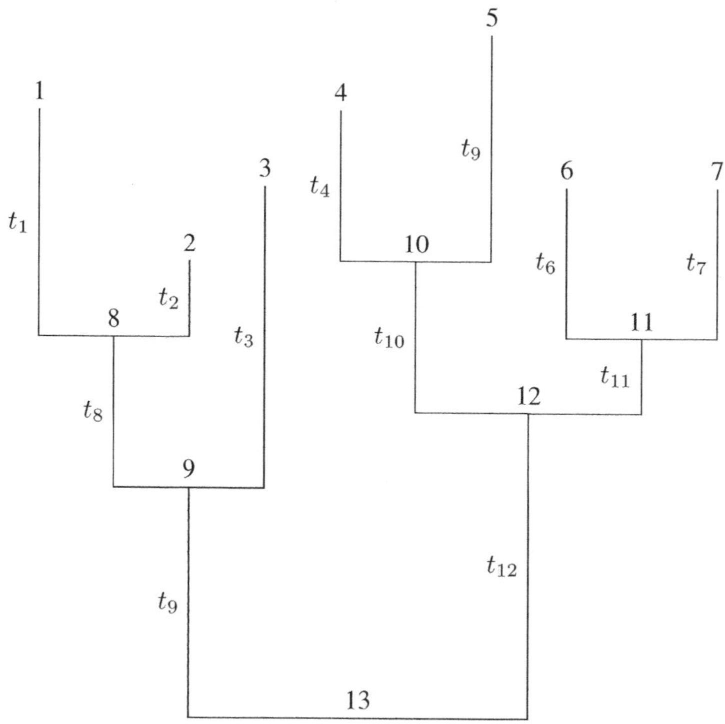

Figure 4.3. *Example rooted phylogeny with node labels and branch lengths*

Assuming a Brownian motion, the expectation and variance of the contrasts are:

$$E(\gamma_{ij}) = 0 \quad [4.2]$$

$$V(\gamma_{ij}) = \sigma_Y^2 \cdot (t_i + t_j). \quad [4.3]$$

To compute contrasts at internal nodes, we need to reconstruct ancestral trait states. The ancestral trait value y_k for any ancestral node k with two daughter branches i and

j is obtained by taking the weighted mean of the trait values at the direct descendant nodes:

$$y_k = \frac{\frac{1}{t_i} y_i - \frac{1}{t_j} y_j}{\frac{1}{t_i} + \frac{1}{t_j}}. \qquad [4.4]$$

We can therefore compute all contrasts and ancestral traits recursively along the tree, collapsing nodes after computing their associated contrasts. For instance, using the tree depicted in Figure 4.3, one computes first $\gamma_8 = y1 - y_2$ and $y_8 = \left(t_1^{-1} y_i - t_2^{-2} y_j\right) / \left(t_1^{-1} + t_2^{-1}\right)$ and then removes leaves 1 and 2. Branch length t_8 is then updated in order to accommodate the fact that y_8 was estimated and not observed, and has an additional estimation variance. This is equivalent to an additional time for y_8 to have evolved from y_9, which depends on branch lengths t_1 and t_2 (Felsenstein 1985):

$$t_8' = t_8 + \frac{t_1 \cdot t_2}{t_1 + t_2}. \qquad [4.5]$$

y_8 and t_8' are then used to compute γ_9 and y_9, using formula [4.4].

A simple method to test for the correlation between two continuous traits while accounting for the species phylogeny consists of first transforming each variable into independent contrasts, before testing for a correlation using standard classical tests. Figure 4.4 shows that this method successfully removes the phylogenetic correlation, at least when the true phylogeny is known and when the underlying trait evolves under a Brownian motion.

4.2.3. *Extending the linear model to account for phylogeny*

The PIC offer a simple way to compare different species traits. It proves, however, to be limited when testing hypotheses about trait evolution and environment variables, which are typically unknown for the ancestral species. As a result, these analyses are conducted within the linear modeling framework, where a given response variable y with n values $\{y_i\}_{1 \leq i \leq n}$ is modeled as a linear combination of m explanatory variables $\{x_j\}_{1 \leq j \leq m}$ with corresponding values $\{x_{ji}\}_{1 \leq j \leq m, 1 \leq i \leq n}$:

$$y_i = \hat{y}_i + \epsilon_i \qquad [4.6]$$

$$\hat{y}_i = a_0 + \sum_{j=1}^{m} a_j x_{j,i}, \qquad [4.7]$$

where a_0 is a constant (aka intercept) and the $\{a_j\}_{1 \leq j \leq m}$ are the model coefficients. The $\{x_{ji}\}$ terms may denote observed variables or some of their interactions. For two measured traits, there can be up to three possible variables (the two traits and their pairwise interaction), for three measured traits, there can be up to seven variables (the three observed traits, three possible pairwise interactions and one triple interaction), etc. In a typical evolutionary comparative analysis, y would be a biological trait of interest (e.g. body mass, age of sexual maturity, sexual dimorphism, and so on) and the x_j variables are either other biological variables such as life history traits, and/or environmental factors such as temperature, amount of precipitation, and so on.

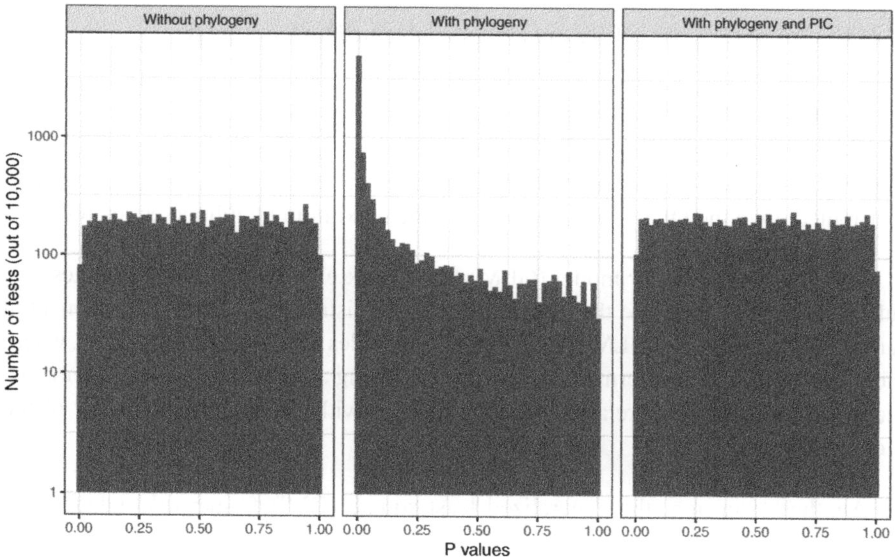

Figure 4.4. *P value distribution of Kendall's correlation test on two continuous traits simulated independently. Left: without phylogeny or a star-like phylogeny (no shared history). Center: with phylogeny. Right: with phylogeny and correction using phylogenetic independent contrasts (PIC). Note that 10,000 simulations were performed in each case, with a 100 pairs of points*

In the linear modeling framework, the model parameters $\{a_j\}$ are estimated using a maximum likelihood approach. The Gauss–Markov theorem states that, if the error terms $\{\epsilon_i\}_{1 \leq i \leq n}$ are uncorrelated and with identical variance (homoscedasticity), the best unbiased linear estimators are the ordinary least square (OLS) estimators, that is, the values of $\{a_j\}$ for which $\sum_{i=1}^{n} (y_i - \hat{y}_i)^2 = \sum_{i=1}^{n} \epsilon_i^2$ is minimal. Noting

$Y = (y_i)_{1 \leq i \leq n}$ as the response vector and $X = (x_{i,j})_{1 \leq i \leq n, 1 \leq j \leq m}$ the data matrix, such as

$$X = \begin{pmatrix} 1 & x_{1,1} & \cdots & x_{1,j} & \cdots & x_{1,m} \\ \vdots & \vdots & \cdots & \vdots & \cdots & \vdots \\ 1 & x_{i,1} & \cdots & x_{1,j} & \cdots & x_{i,m} \\ \vdots & \vdots & \cdots & \vdots & \cdots & \vdots \\ 1 & x_{n,1} & \cdots & x_{n,j} & \cdots & x_{n,m} \end{pmatrix}, \qquad [4.8]$$

and $A = (a_j)_{0 \leq j \leq n}$, the OLS estimator $\hat{A}$ is obtained by the formula

$$\hat{A}_{OLS} = (X^T X)^{-1} X^T Y, \qquad [4.9]$$

where X^T is the transposed matrix of X. If the error terms are, in addition, normally distributed, then the OLS estimators are also the maximum likelihood estimators.

The generalized linear model (GLM) framework extends the OLS estimates to a full class of models where the error terms follow distributions from the exponential family (Figure 4.5). The GLM framework also includes heteroscedastic models, where the variance of the error term is not uniform. A strong assumption, however, remains: the error terms are uncorrelated, which is at odds with the tree-based history of the biological data.

Another extension, the generalized least square (GLS) framework, integrates a correlation structure into the OLS estimators. Basically, it includes a variance–covariance matrix of individuals in the trait space, noted V. V is a square matrix of size n, the sample size. The GLS estimates of A are then simply given by the formula

$$\hat{A}_{GLS} = (X^T V^{-1} X)^{-1} X^T V^{-1} Y. \qquad [4.10]$$

An interesting property of the GLS procedure is that the variance-covariance matrix, V, does not need to be fully specified, and free parameters can be estimated jointly with the regression coefficients within the maximum likelihood framework.

Assuming a Brownian motion to model the evolution of continuous traits, the matrix of variances and covariances between species can be obtained directly from the branch lengths of the corresponding phylogenetic tree (Figure 4.6): the variance for each species is given by the total branch lengths from the root of the tree, while the covariance between two species is equal to the total length between the root and the most recent common ancestor of the two species. Grafen (1989) was the first to propose the use of a GLS procedure as a way to account for phylogeny in the comparative analysis. The method was later extended by Martins (1994) and Martins and Hansen (1997) and re-branded as "phylogenetic generalized least squares" (PGLS). When a purely bifurcating tree is used with a Brownian motion model, the PGLS boils down to the simpler PIC method (Felsenstein 2003). Paradis and Claude (2002) later further extended the framework using generalized estimating equations (GEE, Figure 4.5). A powerful property of the GLS and GEE frameworks is the possibility to test different models of evolution along a phylogeny, beyond the simple Brownian motion. Blomberg et al. (2003), for instance, extended the original Brownian motion to include heterogeneity in the rate of evolution.

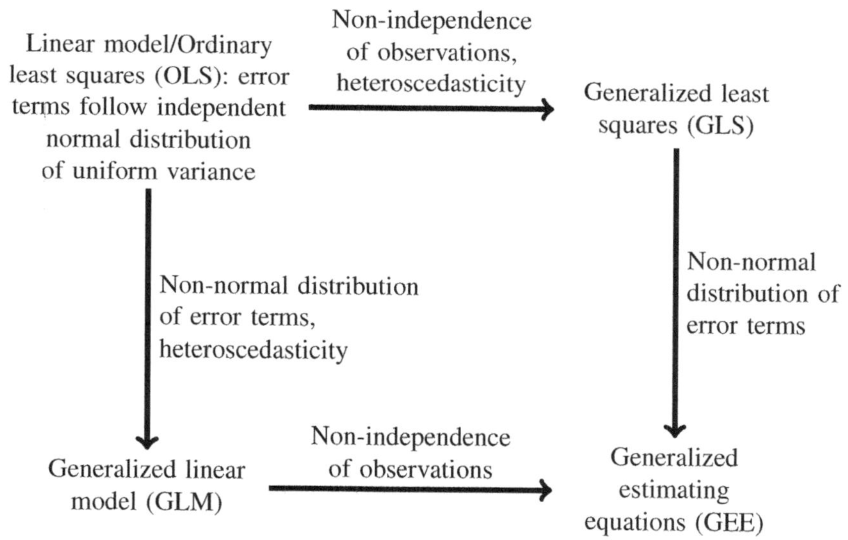

Figure 4.5. *The linear modeling framework and its extensions*

Regardless of the model, there is a need for a phylogeny with estimated branch lengths. It is typically obtained from a different source, for instance, molecular sequences. In many cases, the resulting branch lengths, which depict the evolutionary rates of particular genes, have a priori little to do with the evolution of the trait under study. Branch lengths, however, cannot be estimated from the trait data as this would result in too many free parameters. Grafen (1989) proposed a method to compute branch lengths for a given topology with a single parameter ρ. While the ρ parameter does not have a particular biological interpretation, it adjusts the amount of treelike evolution: while $\rho = 0$ results in all internal branches to be equal to 0, effectively turning the phylogeny into a star tree with null covariances, $\rho = 1$ reduces all terminal branches to 0 and the data to simply two points. Estimating ρ together will the regression parameters thus offers a ways to quantify the amount of phylogenetic correlation in the data, and to account for it when testing other effects. Finally, Martins and Hansen (1997) departed from the Brownian motion and considered a models where the covariance between two taxa depends on their phylogenetic distance t_{ij}:

$$V_{ij} = \gamma \exp(-\alpha \cdot t_{ij}). \qquad [4.11]$$

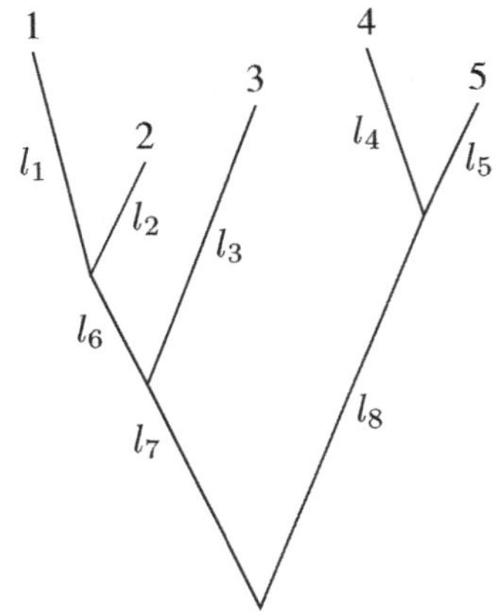

$$V_{brownian} = \sigma^2 \begin{pmatrix} l_1+l_6+l_7 & l_6+l_7 & l_7 & 0 & 0 \\ l_6+l_7 & l_2+l_6+l_7 & l_7 & 0 & 0 \\ l_7 & l_7 & l_3+l_7 & 0 & 0 \\ 0 & 0 & 0 & l_4+l_8 & l_8 \\ 0 & 0 & 0 & l_8 & l_5+l_8 \end{pmatrix}$$

Figure 4.6. *Variance-covariance matrix obtained from a rooted phylogeny with branch lengths, assuming a Brownian motion for the trait evolution. σ^2 is the intrinsic variance of the continuous trait*

The matrix V corresponding to the tree in Figure 4.6 is:

$$V_{Martins} = \gamma \exp \left[-\alpha \begin{pmatrix} 0 & l_{1,2} & l_{1,6,3} & l_{1,6,7,8,4} & l_{1,6,7,8,5} \\ l_{2,1} & 0 & l_{2,6,3} & l_{2,6,7,8,4} & l_{2,6,7,8,5} \\ l_{3,6,1} & l_{3,6,2} & 0 & l_{3,7,8,4} & l_{3,7,8,5} \\ l_{4,8,7,6,1} & l_{4,8,7,6,2} & l_{4,8,7,3} & 0 & l_{4,5} \\ l_{5,8,7,6,1} & l_{5,8,7,6,2} & l_{5,8,7,3} & l_{5,4} & 0 \end{pmatrix} \right] \qquad [4.12]$$

where $l_{x,y,z,\ldots} = l_x + l_y + l_z + l_{\ldots}$ and $\exp[M]$ denotes the element-wise exponential of a matrix M. The coefficients α and γ can be interpreted in terms of constraining force and interspecific variance of a model of stabilizing selection (Martins and Hansen 1997).

4.2.4. *Correlation between discrete traits*

Although methods for correlated evolution of discrete traits can be studied by the linear model framework, discrete traits coevolution has been the subject of specific developments. Building on the methodology of maximum parsimony, the transitions between trait states can be inferred and mapped on the branches of a phylogeny. In its simplest form, one first reconstructs the ancestral states at all internal nodes, and then counts one transition event on each branch where the states at the ancestral and descendant nodes are different (we will see a more advanced approach in section 4.3.2). For a single binary trait with two states 0 and 1, there are two possible transitions: $0 \to 1$, and $1 \to 0$. For a pair of binary traits, there are four possible pairs of states, 00, 01, 10, and 11, and 12 possible transitions. Ridley (1983) built a contingency table by counting the number of occurrences of each combination of traits at the end of branches where at least one trait had undergone a transition. As mutations on different branches of the tree can be considered independent, one can test independence using standard tests (χ^2 or Fisher's exact test).

The method of Ridley (1983) tests for covariation of traits, regardless the direction of mutation. It does not distinguish whether mutations in the first trait tend to cause mutations in the second. Such information can be inferred from branches where only one of the two traits undergoes a transition, noting the state of the other trait. Maddison (1990) introduced a method where one of the two traits is treated as a conditional variable, and assessed whether the transitions at the other trait are distributed randomly according to the conditioned trait along the phylogeny.

Harvey and Pagel (1991) extended the methods of Ridley (1983) and Maddison (1990) using a continuous time Markov model of trait evolution to compute the likelihood of the species distribution of states, given the underlying phylogeny. For binary traits, the generator of the Markov model is

$$Q = \begin{pmatrix} -q_{0\to1} & q_{0\to1} \\ q_{1\to0} & -q_{1\to0} \end{pmatrix}, \qquad [4.13]$$

where $q_{x\to y}$ denotes the rate of transition from state x to state y. This model is conceptually similar to the one introduced by Felsenstein (1981) for modeling sequence evolution, from which it borrows the so-called pruning algorithm (dynamic programming) to efficiently integrate over all possible ancestral states.

One can then properly compute a likelihood for the mutations in the tree. The likelihood function can be maximised to estimate the transition matrix. Interestingly, Harvey and Pagel (1991) used this likelihood framework to compute the mean number of mutations on each branch of the phylogeny for each trait independently. A correlation coefficient can subsequently be computed to assess the independence between the evolution of the two traits.

This approach was further extended by Pagel (1994), who considered pairs of trait states in the Markov chain: four combined states and at most 12 transitions. More specifically, he used the following generator to model the correlated evolution of both traits:

$$Q_{\text{pair}} = \begin{pmatrix} -q_{00\to01} - q_{00\to10} & q_{00\to01} & q_{00\to10} & 0 \\ q_{01\to00} & -q_{01\to00} - q_{01\to11} & 0 & q_{01\to11} \\ q_{10\to00} & 0 & -q_{10\to00} - q_{10\to11} & q_{10\to11} \\ 0 & q_{11\to01} & q_{11\to10} & -q_{11\to01} - q_{11\to10} \end{pmatrix} \quad [4.14]$$

further assuming that the two traits cannot mutate simultaneously and thus reducing the transition matrix to eight parameters. When the two traits are independent, the transition matrix can even further simplify to four parameters, two transitions for each trait. (Pagel 1994) then suggested to compare the likelihood maximized over the four or eight parameters (respectively, $\mathcal{L}_{4p}^{\max}$ and $\mathcal{L}_{8p}^{\max}$). As both models are nested – the four-parameter model is defined in a sub-space of the general one – we can test whether the likelihood increase due to the extra parameters is significant. A model with more parameters always fits better, and the use of likelihood ratio tests is standard to compare nested models: under independence, the quantity $2(\ln \mathcal{L}_{8p}^{\max} - \ln \mathcal{L}_{4p}^{\max})$ converges to a χ^2 distribution with 4 degrees of freedom. The generality of this framework permits to test various hypotheses about the evolutionary process, including the order of mutations along the phylogeny. The relatively large number of parameters involved, however, restricts its application to relatively large data sets.

4.2.5. *Examples of correlated traits*

Applications of phylogenetic comparative analysis fall within two categories: (i) inferring the factors that drive the evolution of one or several traits, and (ii) assessing the correlated evolution between several traits. One of the most studied trait is the body mass index, which correlates with many other physiological and environmental variables (Cooper and Purvis 2010). It was mostly studied in mammals, for which large data sets are available, as well as a well-studied phylogeny. These studied revealed an important historical component (Gittleman

1985; Cheverud et al. 1985), but also geographical and environmental factors such as temperature, altitude and species richness (Cooper and Purvis 2010). Paradis and Claude (2002) used GEEs to reassess the relationship between species dispersal and population synchrony (defined as the correlation of population size variations between localities; Lande et al. 1999). It accounts for dispersal, habitat and the interaction of the two factors, together with a long-term effect (time series) and the underlying phylogeny of the species. Their model confirms the existence of a habitat-specific effect of dispersal, but shows that the long-term effect vanishes when phylogeny is properly accounted for.

Other studies used phylogenetic comparative analysis to assess correlated evolution between distinct traits. An example of such correlated traits is the average versus sex dimorphism in species body mass. The "Rensch rule" (Rensch 1950) stipulates that sexual dimorphism increases with the average body mass in species where the male is the largest, while it decreases in species where the female is larger than the male (Fairbairn 1997). Abouheif and Fairbairn (1997) used PIC to put this rule to the test with 21 animal taxa. They found that, besides one single exception, allometric relationships follow the pattern of Rensch rule. Lenormand and Dutheil (2005) looked at another sexual dimorphism, the difference of recombination rate between females and males (heterochiasmy). They fitted a GEE model to a plant data set and assessed the factors driving the evolution of heterochiasmy. There the trait under study was the ratio of sex-specific average recombination rate, and the explanatory variables included, in addition to the phylogeny of the species, the presence of sex chromosomes and the selfing rate. This model allowed the authors to test the Haldane–Huxley hypothesis, which stipulates that in species where recombination is suppressed in one sex, this sex is always the heterogametic one. The pattern, however, does not hold for so-called heterochiasmate species, where both sexes recombine but at a distinct rate. Lenormand and Dutheil (2005) showed that the presence of sex chromosomes had no significant effect on the difference of recombination rate between sexes, but that the selfing rate did, as predicted by a model with differential selection during the haploid phase (Lenormand 2003).

4.2.6. *Jointly modeling traits and sequences*

An interesting complementary approach tested for correlated evolution between transition rates of molecular evolution and several morphological traits (Lartillot and Poujol 2011). Rates of molecular evolution, such as dN, the rate of amino acid changes (non-synonymous) or dS, the rate of synonymous mutation rate, are modeled as continuous traits. Although one can only hardly measure rates, even at the leaf of the tree, the sequences at the leaves of the trees are informative on the mutation processes. The mutational rates are modeled explicitly, as in classical phylogeny, but their rates vary as a Brownian motion, whereas they are held constant in standard phylogeny. The ambition of such models is to find morphological traits, or life history traits, that strongly correlate with molecular rates variation. Lartillot and Poujol (2011) found that body mass and longevity positively correlate with the intensity of selection, proxied by the ratio dN/dS. The same framework was also used to explore the interplay between biased gene conversion, selection and other traits (Lartillot 2013). Finally, it was also used to explore how the variation of mutation rates and their correlation to traits could bias the inference of the phylogenetic tree itself (Lartillot and Delsuc 2012).

4.3. Correlated evolution within genomes

Genetic sequences can be viewed as traits with simple, well-defined discrete states (four states for each site). Unlike morphological traits, the mechanisms of their evolution are well understood, with explicit models of transitions along a phylogeny. As a result, increasingly complex models of sequence evolution have been developed (Yang 2006). Correlated evolution of the molecular traits can thus be investigated using statistical methods that are similar to the ones described in section 4.1 for morphological traits. The current section presents different approaches that have been conducted, classified according to the selection level that they consider, from single nucleotides to complete genomes.

4.3.1. *Within genes, between nucleotides*

One can trace back the study of intramolecular coevolution to the research on RNA molecules. When Holley et al. (1965) sequenced the first transfer RNA (tRNA) molecule, they proposed three possible structures based on the occurrence of stretches of Watson–Crick pairs, forming putative double helices. When a second tRNA sequences was obtained, the only common possible structure was one with a "cloverleaf" shape, which was then proved to be common to all tRNAs (Westhof and Auffinger 2012). This conservation of tRNA structure opened the way to comparative sequence analysis as a way to extract information regarding evolutionary constraints and, ultimately, biological functionality – as envisioned by Crick (Cobb 2017).

The Watson–Crick base-pairing, that associates A-U and G-C, leads to an important evolutionary prediction: if two positions in a RNA molecule are interacting to form a Watson–Crick pair within a double helix, the paired nucleotides with states A-U, U-A, G-C and C-G have a similar fitness and are thus inter-exchangeable. Conversely, any of the other pairs leads to a structural instability and, therefore, a lower fitness. Consequently, when comparing multiple homologous paired positions in different genes, we expect a high frequency of A-U, U-A, G-C and C-G, and a low frequency, if any, of the other combinations. This results in a two-site fitness landscape where Watson–Crick pairs are "peaks" connected by two mutation events, one at each site, as illustrated in Figure 4.1(D) (Dutheil et al. 2010). A resulting property of this landscape is that the fitness effect of a mutation at one site (and consequently, its probability of fixation) depends on the state at the interacting site. Such epistatic interactions result in a scenario of coevolution by compensation: a deleterious mutation can be compensated by another mutation at an interacting site, and the evolution of the two positions become correlated. This reasoning served as a basis for many molecular coevolution studies. Identifying correlated evolution within a gene tells something about the structural and functional constraints of the encoded molecule.

This principle was first applied to the prediction of RNA structures, given the clear expectation of coevolution at sites located within double-stranded helices (see http://www.rna.icmb.utexas.edu/CAR/ for a historical perspective). The comparative analysis of RNA sequences starts with a multiple sequence alignment. The most simple approaches initially counted the number of Watson–Crick pair occurrences in all pairs of alignment columns, while subsequently developed approaches used more elaborated statistical methods. One of the earliest methods, but one that is still popular, uses the so-called mutual information (MI) measure, defined as the some of the Shannon entropy of the two positions, minus the joint entropy of the pair (Chiu and Kołodziejczak 1991):

$$MI(x,y) = -\sum_i f(x_i)\log(f(x_i)) - \sum_j f(y_j)\log(f(y_j)) + \sum_i \sum_j f(x_i y_j)\log(f(x_i y_j)) \qquad [4.15]$$

where x and y are two columns in the alignments. $f(x_i)$ and $f(y_j)$ denote the frequencies of each nucleotide i at site x and each nucleotide j at site y, respectively. Besides, $f(x_i y_j)$ denotes the frequency of each pair of nucleotides i and j at the two sites. Given that $\sum_i f(x_i) = 1$ and $\sum_j f(y_j) = 1$, we have $\sum_i f(x_i y_j) = f(y_j)$ and $\sum_j f(x_i y_j) = f(x_i)$. Equation 4.15 can then be factorized and reorganized in:

$$MI(x,y) = \sum_i \sum_j f(x_i y_i) \log\left(\frac{f(x_i y_j)}{f(x_i) \times f(y_j)}\right). \qquad [4.16]$$

If the two positions x and y are evolving independently, then $f(x_i y_i) = f(x_i) \times f(y_i)$, and $MI(x,y) = 0$. A MI value departing from 0 is therefore a signature of non-independent evolution.

A difficulty arises, however, when trying to evaluate the significance of $MI(x, y)$ as sequences partially share history (again represented by a phylogeny). Consequently, MI values will generally be positive for most pairs of sites (Dutheil 2012) and properly assessing the null distribution of MI values requires accounting for the phylogeny. This can be achieved by correcting the $MI(x, y)$ value using information from the full sequence alignment (assuming coevolving pairs are a minority among all possible pairs) or by simulations (see Dutheil (2012) for a review).

While it is possible to properly account for the evolutionary history of the species when computing the null-distribution of the MI statistics, this statistic is based on the sequence patterns only and is intrinsically agnostic of the underlying evolutionary paths. Several authors, therefore, designed models of coevolving positions, related to the model introduced by Pagel (1994) (see section 4.1), restricted to pairs of sites to avoid an excessively large number of parameters. For nucleotides, the model has 16 states and at most 240 transitions (but as for the original model of Pagel, the number of transitions can be reduced by banning double mutations and by imposing constraints on the reversibility of transitions). Tillier and Collins (1995, 1998) were the first to design a nucleotide model, which they applied to ribosomal sequences. A convenient aspect of this framework is that it is straightforward to derive the expected model when both sites evolve independently. This independence model can be used as a null-model and compared to the correlated model using likelihood ratio tests. It requires, however, to fit a model for all pairs of sites, which can be relatively resource demanding. Yeang et al. (2007) used a similar model, which they combined with a hidden Markov chain along the alignment in order to capture spatial dependencies, for instance within secondary structure motifs. Finally, Dib et al. (2014) used a general Pagel model, and employed a Bayesian inference procedure in order to jointly estimate model parameters, but also the coevolutionary "profile" of coevolving states among possible state pairs, as well as the underlying phylogeny (Meyer et al. 2019). Such approaches, combining sophisticated statistical inference and efficient computational implementations, are very powerful tools to study coevolutionary processes within sequences.

4.3.2. *Within proteins, between amino acids*

Characterizing correlated evolution within proteins is intrinsically more challenging, because of the larger number of states. While the 20 amino acids lead to a putative 400 possible pairs of states for two coevolving positions, all these combinations may not differ in their fitness. Because the 20 amino acids have redundant biochemical properties, some may be interchangeable at some positions in the protein without any notable fitness effect.

The year 1994 was particularly important for the study of protein coevolution. Shindyalov et al. (1994) introduced a new probabilistic method that "mapped" substitutions on the branches of a phylogenetic tree in order to assess how likely two positions are to undergo substitutions on the same branches of the tree, if they were evolving independently. Shindyalov et al. (1994) and Gbel et al. (1994) then asked how the three-dimensional structure of proteins impacts the occurrence of coevolution; in particular, whether residues in contact in the tertiary structure tend to have a correlated evolution. Neher (1994) developed a different statistical framework to address the same question. While his approach does not account for the underlying phylogeny, it proposes an attempt to account for the biochemical properties of proteins. By weighting each amino acid by a given biochemical index, he assessed how correlated biochemical properties are between sites in a protein.

Pollock et al. (1999) used an approach similar to that employed by Tillier and Collins (1998) to model the correlated evolution of RNA sites, but for proteins. In order to reduce the complexity of the model, which would contain a total of 400 states, they grouped several amino acids according to their biochemical properties. If two categories are considered (for instance, "large" and "small" residues), the model only has four states (large-large, large-small, small-large, small-small) to account for. It is similar to the general model introduced by Pagel (1994), yet the authors lowered the number of parameters (by making the model reversible).

Tufféry and Darlu (2000) introduced an intermediate approach. Building on Shindyalov et al.'s (1994) work, they first used a phylogeny to reconstruct the ancestral sequences at all inner nodes in the tree, and inferred on which branch and site substitutions occurred. They then evaluated which positions in the alignment tended to undergo substitutions on the same branches of the tree, introducing the term *co-substitution* to denote branches with mutations at both sites. Dutheil et al. (2005) and Dutheil and Galtier (2007) later extended this method to account for the branch lengths of the tree and the uncertainty of the ancestral states, a procedure called *substitution mapping*. They further introduced the possibility to include biochemical weights, following Neher (1994). A complementary development considers not only co-substitutions, but also sequentially ordered events (see Figure 4.7(d)), where substitutions at one site (the leading events) trigger substitutions at another site (the lagging events) (Kryazhimskiy et al. 2011; Behdenna et al. 2016). Once mutation events are placed on the tree, it is possible to compute analytically the probabilities under independence of any type of pairs (co-substitutions, ordered pairs, exclusive pairs, etc.) (Behdenna et al. 2016). Such semi-parametric approaches are more efficient than full-model based approaches as they do not explicitly model site dependence. Yet the underlying models account for many aspects of the evolutionary process and can be applied to both nucleic acids and proteins (Dutheil et al. 2005).

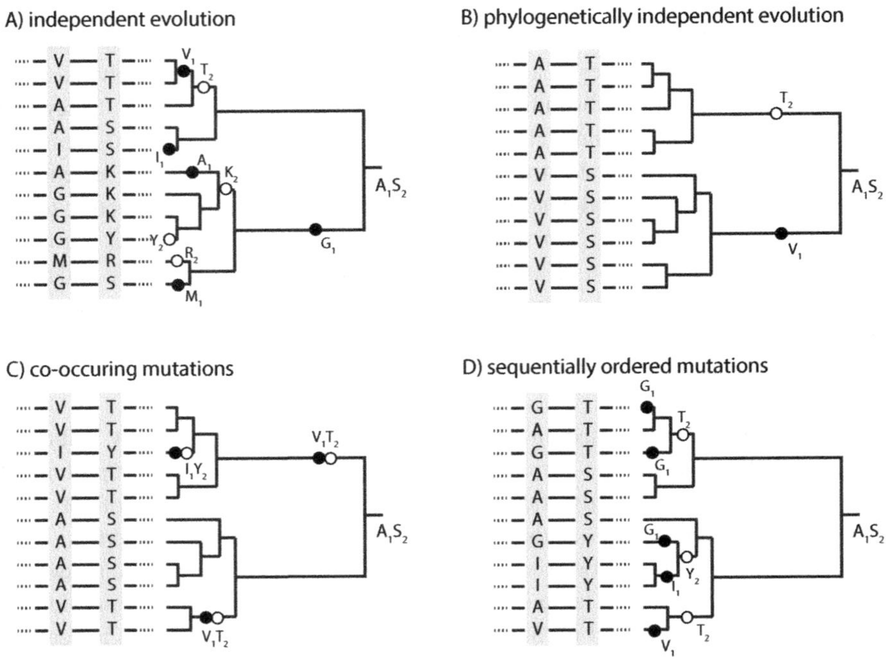

Figure 4.7. *Four possible outcomes of the comparison between two columns of an alignment. The gray area brings the focus to two columns of a larger alignment. The phylogenetic tree on the side illustrates the process of substitution mapping on the tree. Ancestral state is given at the root, mutations at the first site are displayed by black disks, while mutations at the second site by empty circles. (A) For most pairs of sites, no correlated variation is observed neither in the alignment nor in the tree. (B) Although the two pairs evolve independently, the phylogenetic inertia can mimic patterns of correlated variation in the alignment. The pattern would only be statistically significant for MI-like methods when phylogeny is ignored. (C) A pattern of co-variation corresponding to co-substitutions in the tree, suggestive of a strong reciprocal interaction between both sites. (D) A pattern of correlated evolution with a time lag between the first mutation and the following one, suggestive of an asymmetrical induction of mutations at site 2 on mutations at site 1*

Parallel to the development of phylogenetically grounded methods, methods solely based on the alignment have recently shown spectacular development. A popular and successful approach is to fit a generalized Ising model (Potts model) into the alignment of L amino acid sequences, disregarding the phylogeny. Ising models consist of discrete variables that represent atomic "spins" that can be in one of two states (+1 or -1). The spins are arranged in a graph, usually a lattice, allowing each spin to interact with its neighbors (pairwise interactions). Generally, neighbors spins are only compatible when of the same sign. One can then compute the probability of the whole graph to be in a configuration using marginal probabilities of each spin and pairwise interactions between them. The Potts model allows for more than two states (not only +1/-1). These models, developed in statistical physics, take into account local abundances and pairwise interactions, but disregard higher order interactions. DCA (direct coupling analysis) (Weigt et al. 2009; Morcos et al. 2011) is a successful implementation of a method based on this model.

The Potts model assumes that each sequence of the alignment has a probability of being sampled that is given by a Boltzmann-like distribution. The higher the probability, the more abundant is the sequence in nature. Low probability sequences correspond to deleterious combinations of the amino acids that one should not encounter in nature. A sequence $a_1 a_2 \cdots a_n$ has a probability $P(a_1 a_2 \cdots a_L) = \frac{1}{Z}\exp(-\mathcal{H}_{a_1 a_2 \cdots a_L})$ of being observed. Z is a normalization constant, equal to the integral of $\mathcal{H}$ over the whole sequence space. The "Hamiltonian" function $\mathcal{H}$ has both a local abundance component h and a pairwise epistatic component e. More precisely, it is computed as:

$$\mathcal{H}_{a_1 a_2 \cdots a_L} = \sum_{i=1}^{L} h_i(a_i) + \sum_{i=1}^{L} \sum_{j=i+1}^{L} e_{ij}(a_i, a_j)$$

The sums are taken over all positions of the alignment. In this formulation, the function h takes $21 \times L$ input variables (gap is a 21st symbol) and returns a quantity related to the marginal probability of observing one of the 21 symbols at position i. The e function takes $(21 \times L)^2$ input variables and returns a quantity related to the symmetrical interaction between amino acids (and gaps) for each pair of positions i, j, with $e_{ij}(a, b) = e_{ji}(b, a)$ and $e_{ii}(a, b) = 0$.

One reason proposed to justify the Potts model for protein sequences is that the correlation observed between two columns in an alignment is not a good indicator of a direct functional or structural interactions. In the case of structure for instance, a correlation can be the result of a contact between two residues (a true signal of correlated evolution), but can also reflect indirect interactions: for example, a_i and a_j are distant in the structure, but close to a third residue a_k, then resulting in indirect interactions. In the Potts model, the probability distribution only considers direct coupling effects. The hope is that those couplings reflect biologically relevant interactions. In this regard, only the e function is important.

In the end, the difficulty comes when the functions h and e must be inferred from the data, a challenge named the *inverse Potts problem*. Several techniques were proposed, from simple moment estimations to maximization of the $\mathcal{H}$ function in its gigantic parameter space (Weigt et al. 2009; Morcos et al. 2011; Ekeberg et al. 2013; Baldassi et al. 2014). The method performs especially well, when a very large number of "pseudo-independent" sequences are provided in the alignment. Indeed, it is noteworthy to mention that, to correct for the phylogenetic inertia, closely related sequences (by default with $> 70\%$ of identity) are grouped into clusters and each member of the cluster has a weight inversely proportional to the cluster size. This is also correct for bias sampling of the protein families in the database, as some clades are typically over-represented (e.g. vertebrates). DCA was originally developed to predict structural contacts between amino acids (Weigt et al. 2009), but as also used to find contacts in RNA (De Leonardis et al. 2015) or predict the effect of mutations on a query sequence (Figliuzzi et al. 2016).

While most methods typically exploit either evolutionary paths in phylogenies or counts in the alignment, others searched a compromise between the two, with the explicit goal to maximize the input information while keeping the computation time low. In CAPS (Fares and Travers 2006), the authors designed a score for two sites that uses all pairwise comparisons of the sequences; it thus takes into account the pairwise distances matrix between the sequences but does not reconstruct a tree. In the BIS method (Dib and Carbone 2012) and BIS2 (Champeimont et al. 2016), subsequent residues that show strong patterns of co-abundance are first merged into blocks. All pairs of blocks are then tested for co-evolution by computing a metric of co-abundance in the alignment (an MI-like score). The metric is finally recursively re-weighted by the co-abundances in all subtrees, starting with the root node (the root node encompasses all sequences). Consequently, the topology of the tree (not the branch lengths) is incorporated in the metric. Finally, more recently, persistence time in the phylogeny of co-conservations (not co-variations) for a pair of sites has been suggested to be a good sign of co-evolving positions (Laine et al. 2019).

Residues showing strong patterns of correlated evolution are typically in contact in the 3D structure of the folded protein (Pollock et al. 1999). It is noteworthy to mention that coevolving residues detected by methods geared to direct coupling such as DCA like are typically closer in space than methods solely based on correlation of abundances in the alignment, such as MI-like (Marks et al. 2011). Conversely, it has been proposed to use the graph of co-evolving residues to guide the folding of peptides, a major challenge of structural biology. Recently developed folding methods, such as in the EVfold suite, use the predictions of DCA to greatly improve the folding accuracy (Marks et al. 2011; Hopf et al. 2012).

The continuous development of innovative methods to study patterns of covariation in biological sequences has led to a better understanding of the relation between coevolution of sites and their functional linkage. Yet, several theoretical aspects of sequence coevolution remain poorly understood. What are the mechanisms by which co-substitutions occur? What is the role of compensatory mutations in sequence coevolution? What is the connection between coevolution and the rate of evolution at single sites? Answering these questions requires input from related fields of study, such as empirical fitness assessment (de Visser and Krug 2014), experimental evolution (Long et al. 2015), as well as modeling (Talavera et al. 2015).

4.3.3. *Within genomes, between genes*

Cellular functions are the results of multiple protein interactions. Many proteins interact with more than one partner, resulting in a network of protein interactions. As proteins involved in the same network are likely to share some selective constraints, they are expected to coevolve, similarly to sites within individual proteins. Inferring the interaction partners of a protein is of great interest to elucidate the functional role of a candidate gene. Therefore, several studies aimed at inferring correlated evolution to predict protein–protein interactions (de Juan et al. 2013).

A pioneering method, known as *mirror-tree*, detects genes with similar phylogenetic trees (Goh et al. 2000; Pazos and Valencia 2001). In its early implementation, the method simply looks at the correlation between distance matrices computed from two protein sequence alignments: because of common evolutionary constraints of interacting proteins, their rate of evolution will be correlated as they adapt simultaneously to the changing environment and, therefore, accumulate substitutions in the same branches of the tree. These assumptions are "co-departures from a clockwise evolution" (Galtier and Dutheil 2007). Yet, neutrally evolving proteins, which evolve under a molecular clock, will show a strong correlation of their genetic distances as they share a common history of speciation events. Accounting for this common phylogenetic context has been the focus of later developments (de Juan et al. 2013).

With the advent of genome sequencing, exhaustive information regarding the proteome content of species became available. This allowed the reconstruction of *phylogenetic profiles* (= *phyletic patterns*), defined as the patterns of presence/absence of homologous genes in a set of species. Barker and Pagel (2005) applied the method of Pagel (1994) (see section 4.2.4) to phyletic patterns from more than 10,000 protein pairs in order to detect linked genes. Using a set of known interacting proteins from *Saccharomyces cerevisiae*, they found that functionally interacting pairs indeed show a signal of coevolution, and that using phylogeny-aware coevolution detection methods significantly improves the prediction of functional interactions. The "discrete coevolution" model, however, requires several transition parameters to be estimated for each tested pair of proteins. Barker et al. (2007) later showed that fixing some of these parameters further improves the predictions.

An alternative approach originates from the studies of the rates of genes gains and losses. These studies summarize the patterns of presence-absence of many gene families into a large matrix of 0 and 1, with species as rows and genes as columns, similar to a sequence alignment. Using the model introduced by Harvey and Pagel (1991) for single binary characters, it is possible to infer rates of gains and losses by maximum likelihood. In this approach, the number of parameters to estimate is considerably reduced, as they are shared by all genes, which resolves some of the estimation issues when fitting the same model on each gene independently. Yet, it also makes unrealistic assumptions, as some gene families are known to undergo higher rates of gene gains and losses than others. To accommodate this issue, (Cohen et al. 2008) proposed a mixture model allowing for these rate parameters to follow an a priori distribution and, therefore, to vary between genes while introducing only a few additional parameters. Applying the substitution mapping approach introduced for mapping nucleotide and amino acid substitutions (Dutheil et al. 2005) to the presence-absence model, Cohen et al. (2012) inferred events of co-gain and co-losses along the phylogeny to infer co-evolving genes. Using simulations, they show that this approach largely outperforms simpler methods based on the correlation of patterns only. The authors also showed that it gave slightly better performances than the Barker and Pagel (2005) model, with less than 1% of the computational time, owing to the lower number of parameters to estimate (Cohen et al. 2013).

Finally, it is worth mentioning that the physical interaction between proteins involves biochemical mechanisms acting at the amino acid level. As a matter of fact, DCA was first developed to predict contact residues between two peptides (Weigt et al. 2009) and then improved to detect contact residues between two proteins (Szurmant and Weigt 2018). Conversely, DCA outcomes can be used to predict how two domains interact (Hopf et al. 2012). However, finding co-evolving sites between pairs of proteins greatly increase the computational load, $n_1 \times n_2$ comparisons have to be assessed between two genes encoding n_1 and n_2 amino acids, respectively. Therefore, few methods were subsequently developed to apply a DCA-like strategy at the scale of complete genomes, such as super DCA (Puranen et al. 2018). Obviously, MI-based methods can also be used to search for interacting partners of proteins (Bitbol 2018), overcoming the computational cost of the inverse Potts problem. Recently, an efficient implementation of MI was used to scan interactions between all pairs of nucleotides in alignments of complete bacterial genomes (Pensar et al. 2019).

Finally, it worth mentioning that the phylogenic approach was also applied to this problem. Yeang and Haussler (2007) tackled this challenge by introducing a model of coevolution with only one additional parameter compared to the independent model. This parameter penalizes simple substitutions while favoring cosubstitutions. Combining data filtering and using a computer grid, the authors could evaluate 0.1 trillion pairs of potentially interacting sites and found that most coevolving pairs are also functionally interacting. A general issue with the detection of inter-genic correlated evolution is the assumption of a unique phylogeny shared by the two interacting genes. While phenomena like horizontal gene transfers are implicitly incorporated in models of gene gains and losses, gene duplication events are usually not, and at best, duplicated genes are discarded from further analyses. This problem remains largely unaddressed (but see Yeang (2008)) and calls for a unification of methods to detect correlated evolution with methods of tree reconciliation.

4.4. Genetics is also correlated evolution

In this final section, we suggest that the whole history of gene mapping, since its very beginning, can be included in the large view of correlated evolution. Genetics can be considered as a specific case of correlated evolution where the first trait is discrete – presence/absence of a variant in the genome – and the second is a phenotype of interest. Geneticists search for positive correlations between a phenotypic trait that is observed and a genetic trait that is unknown. In genetics, there is obviously a very strong support that we should interpret these correlations as causalities, where the presence of the allele in the genome ultimately causes the phenotype. Most methodologies of genetic mapping rely on exhibiting the joint occurrence (i.e. a correlation) of one or several alleles with the phenotype. Our ambition is not to give a full and thorough review of the history and methods in genetics, but instead to suggest that ideas of correlated evolution may well have been growing with the rise of genetics in the early 20th century.

4.4.1. *In individuals*

After the rediscovery of Mendel work in 1900 (Moore 2001), started the rise of gene mapping, with research programs such as the one of TH Morgan group (Morgan 1926). The ambition was both to isolate the factors that segregated with phenotypes of drosophilas, and then to infer the linear genetic map of these factors. The phenotypes at stake were mostly natural mutants that were easily identifiable, such as mutants in eye-coloration or wing patterns. The experimental protocol consists mainly of two steps.

First testing whether a trait would segregate according to Mendel laws (each allele has a 0.5 chance of being transmitted to a progeny) by counting the number of descendants with or without the trait of interest in a controlled mating. It is noteworthy to mention that the existence of dominant and recessive alleles in diploids, described by Mendel (1901), add a modest complication for characterizing the transmissions. One must cross an individual with a dominant trait (A/-) with an individual without (-/-) to observe 50% of the progeny harboring the phenotype. The observation that a trait would be transmitted with known and explicit rules demonstrates that a yet unknown genetic factor perfectly correlates with the presence of the trait. Interestingly, the correlation between the allele and the trait is uniquely demonstrated by the observed segregating ratio.

The second part of Morgan's group work consists of controlled matings between males and females with two phenotypic traits. Using counts of co-occurrence between two traits in the progeny, they inferred the "linkage" between the corresponding genetic loci. A very simple estimation procedure of the linkage between two loci encoding two different dominant traits (say A and B) relies on counting in the progeny how many times a female[1] with both traits in linkage (an AB phenotype, corresponding to an AB/– genotype) mated with a male with no phenotype (genotype –/–) would produce progenies with 0, 1 or the 2 traits. When the original combination is kept, only descendants with 0 or 2 traits are observed. The mere observation of descendants with a single trait demonstrates the existence of new combinations of alleles that were named *recombinations*. The first estimator of the recombination probability is the observed fraction of descendants with a single trait. Consequently, recombination rate, computed as such, never exceed 0.5: traits mapped to independent loci have a 0.5 chance of being co-transmitted. The general method relied on observing correlation between traits to infer correlation between loci.

By combining the pairwise measures of recombination rate, the group showed that the loci underlying the traits were arranged linearly in what they called linkage groups. The order and the distances between the loci can be computed using pairwise recombination rates by the transformation: $d = -50\ln(1 - 2r)$, where r is the recombination rate. These distances, named centiMorgan (cM), transform the $[0, 0.5]$ range of recombination rate to proper $[0, \infty]$ distances.

Although controlled matings were further done with almost all living organisms (including bacteria), it was not possible for humans, for obvious ethical reasons. Therefore, to map traits to loci in humans, researchers had to rely on other methods such as pedigree analysis.

4.4.2. *In pedigrees*

Pedigree analysis turned out to be a major tool in human genetics until the rise of association mapping in the 21st century. In a pedigree analysis, geneticists follow the segregation of a phenotypic trait, a pathology in medical genetics, over two, three or ideally more generations. The trait is often binary: ill or not. Pedigree analysis

1 Males do not recombine in drosophilas.

is mostly suited to track dominant alleles. An example of a pedigree is presented in Figure 4.8.

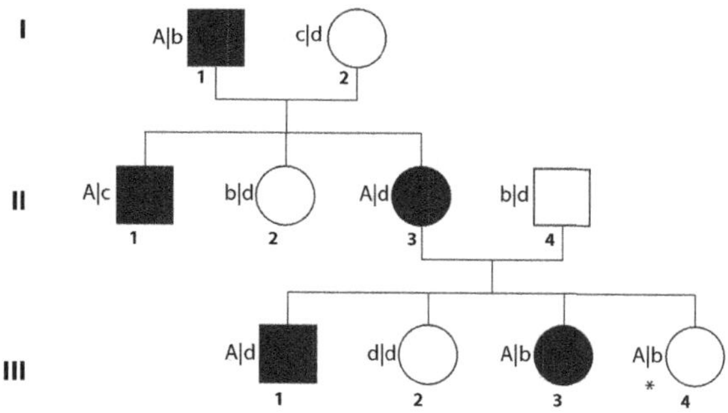

Figure 4.8. *This pedigree reports the transmission of a phenotypic trait segregating through 3 generations, labeled I to III. Males (squares) and females (circles) are colored in black when they show the phenotype. On the left of each individual, we report the observed genotype at a focal locus. Here, the suspicion is that the A allele at the locus co-segregates with the phenotype*

An implicit assumption of a pedigree analysis is that the phenotype is caused by an allele located at a single locus. By following the segregation of the trait, we infer the segregation of the causative allele. Using genotyping technologies, one can then genotype many loci throughout the genome for alleles that display a strong, possibly maximal, positive correlation with the phenotype. Geneticists leverage the correlated segregation between an observable phenotypic trait and a fully determined genotypic alleles to "map" a causative allele in a region of the genome. Until recently, only chosen markers distributed in genome were genotyped; consequently, the alleles co-segregating with the phenotype were not causal, but only genetically linked with an unseen causal variant located nearby. Nowadays, as complete genome sequences are available, the causal allele is among the set of alleles that co-segregate with the trait.

The most popular metrics of co-segregation is the *lod score* (Morton 1955), a log of an odd ratio. This odd ratio divides the likelihood of the observed co-segregation assuming linkage ($r < 0.5$) to the likelihood of the observed co-segregation assuming independence ($r = 0.5$):

$$S_r = \log_{10}\left(\frac{\mathcal{L}_{\mathbf{P}}(r < 0.5)}{\mathcal{L}_{\mathbf{P}}(r = 0.5)}\right)$$

To compute the likelihood, one follow the line of transmission of the allele with the phenotype. For example, in Figure 4.8, if one assumes that allele A is linked to a causal allele located at distance r, the probability of observing the genotype of individual II.1, knowing that he has the phenotype, is $(1-r).\frac{1}{2}$: A did not recombined from the causal variant $((1-r))$ and he received the allele c from his mother $(\frac{1}{2})$. The genotype of individual II.2 has probability $\frac{1}{4}$. In individual III.4, there must have been a recombination between the allele A and the causal variant; the probability is then $r.\frac{1}{2}$. Overall, we have:

$$\mathcal{L}_{\mathbf{P}}(r) = \frac{(1-r)^3 r}{2^{10}}$$

More generally, the lod score can be computed as

$$S_r = \log_{10}\left(2^{-(NR+R)} \times (1-r)^{NR} \times r^R\right)$$

where NR is the number of co-segregations of the focal allele with the phenotype (non-recombined) and R the number of recombinations. The value of r that maximizes the likelihood (and the lod-score) is the natural estimate $\hat{r}_{ML} = R/(R+NR)$. By scanning the genome for strong associations, geneticists usually identify one region containing several markers with maximal lod scores, named a haplotype. This region contains the unseen causal variant, which requires further sequencing and functional analyzes to be confirmed. It is usually recommended to reach a lod score of 3 to claim a strong co-segregation (Nyholt 2000): the odd of linkage is 1,000 times the odd of independence.

4.4.3. *In the population*

In rare cases, multiple haplotypes correlate with the phenotype. However, when the trait is not digenic but polygenic, another corpus of methodology is more suited: the association mapping. Fundamental hypotheses of association mapping stems from quantitative genetics, where traits of interest are no more binary but continuous. The apparent continuity emerges from the combination of many loci of small effects, an assumption known as the *infinitesimal model* (Barton et al. 2017).

In association mapping, instead of a family, one uses the population phenotypic diversity to find independent genetic variants that contribute to the phenotype. For a binary phenotype, the assumption is that detected variants modulate the probability of having the phenotype, or that the pathology becomes visible only once a certain quantity of an invisible continuous trait is reached.

Association mapping at the scale of the genome is known as a genome-wide association study (GWAS). In a GWAS, geneticists measure the independent contribution of all variants segregating in a cohort to the phenotype of interest, assuming additive independent contribution for each locus. In a simple case, for a given locus l hosting two alleles A and a, a genotype $\mathcal{G}$ is 0, 1 or 2 copies of the A allele. The statistical correlation between the number of A in the genotype and the trait is computed either by a contingency table or by linear regression.

Contingency table: for a binary trait $\mathcal{T} \in \{0, 1\}$, we can construct a contingency table, where $n_{i,j}$ represent the number of individuals with genotype $\mathcal{G} = i$ and phenotype $\mathcal{T} = j$, as illustrated in Table 4.1.

	$\mathcal{T} = 0$	$\mathcal{T} = 1$
$\mathcal{G} = 0$	$n_{0,0}$	$n_{0,1}$
$\mathcal{G} = 1$	$n_{1,0}$	$n_{1,1}$
$\mathcal{G} = 2$	$n_{2,0}$	$n_{2,1}$

Table 4.1. *Contingency table for a binary trait and a bi-allelic locus*

Assuming co-dominance, the contingency table can be reduced to a 2×2 table by simply counting the number of alleles of type A and a associated with both traits: $n_{A,0}, n_{A,1}, n_{a,0}, n_{a,1}$: it is deduced from the general table with $n_{A,j} = 2n_{2,j} + n_{1,j}$ and $n_{a,j} = n_{1,j} + 2n_{0,j}$. When one allele is dominant, the contingency table also shrinks to a 2×2 table by pooling the counts of genotype $\mathcal{G} = 1$ with the genotype $\mathcal{G} = 2$ when A is dominant ($n_{D,j} = n_{2,j} + n_{1,j}$; $n_{R,j} = n_{0,j}$). Association is then simply tested by a χ^2 homogeneity test with 1 or 2 degrees of freedom depending on the table size. Larger contingency tables can be used to test for association for more than two alleles. Although the contingency table could, in theory, be extended to cases of non-binary phenotypes, it is in practice more convenient to re-express the problem in terms of linear regression.

Linear regression: additive genetic effect can be estimated by the following linear model:

$$T_i = a_0 + a_1\, G_i + \varepsilon_i$$

This model can be used estimate both (i) the significance and strength of the correlation of a given allele with the phenotype and (ii) a_i, the amount of additive effect of the allele to the trait. Techniques to estimate the a_i, such as least squares, are described in section 1. Although the basic version reported here assumes co-dominance, dominance can be explicitly addressed by setting the values of G_i

adequately. Linear models handle elegantly cases of continuous traits, but are not well adapted when more than two alleles exist at a given locus.

Both contingency tables and linear models can be used to estimate the association of alleles at a given locus to the trait. In GWAS, one assesses correlations with as many as possible polymorphic loci of the genome. The hope is to find a haplotype that is significantly associated with the trait. As there is a very large number of loci and thus of statistical tests, the power to detect a single association even in large samples is weak. Thus, GWASs need by construction very large sample sizes to potentially detect significant associations. In brief, basic GWAS is based on counting co-occurrences of alleles and traits in individuals, ignoring their shared history.

One inherent challenge of measuring the independent contribution of the loci to the phenotype is the presence of epistasis between the loci. A phenotype can be the results of the synergistic interaction of multiple alleles (see Figure 4.1(b)). Correlations can hide correlations. Attempts to include epistasis in the detection of association include the use of MI, generalized to three-partners (Ignac et al. 2014), two genetic partners with symmetrical roles and one phenotypic trait that is the value to be predicted. In this model, both loci contribute together to the trait: they have an additive component and an interaction term, that is reminiscent of the DCA assumptions (see section 4.3.2).

Although erroneous associations could arise from multiple issues, one major annoyance is the stratification problem, that is the spurious association caused by population structuration (Lander and Schork 1994; Pritchard et al. 2000; Price et al. 2006). When a phenotype is mostly restricted to a subpopulation, all variants that characterize this subpopulation can be significantly associated with it. More generally, any correlation that is based on groups with and without a phenotype will suffer from the same bias, if there is some level of stratification. Therefore, like other correlated evolution methods based on counts of individuals, association mapping is biased by phylogenetic inertia that can create non-relevant associations. Although there are several ways to handle this problem of stratification, one interesting method is to incorporate the genetic kinship between individuals in the model.

More specifically, individuals can be assigned to a genotype space, whose number of dimensions equals the number of genotyped loci. Their proximity in the genotype space relates to their kinship and their common ancestors (McVean 2009), as individuals with recent ancestors are more similar genetically. A convenient way to summarize the genotype space is to transform the original space using PCA (Novembre and Stephens 2008). After the transformation, the first components of the new space are the ones explaining most of the variance, hence their name principal components (PC). Therefore, genetical proximity between individuals can be summarized (and often visualized) by only considering the first PCs that explain most of the genetic variance. One method (Price et al. 2006) correcting for stratification explicitly incorporates the positions of individuals, in the PCA space, within the linear model:

$$T_i = a_0 + a_1 \, G_i + \sum_{j}^{n} a_{j+1} \mathrm{PC}_{i,j} + \varepsilon_i$$

where $\mathrm{PC}_{i,j}$ is the value of the jth PC of individual i. Interestingly, PCA is also used to visually inspect the homogeneity between the individuals and eventually remove outliers from the analysis (Purcell et al. 2007). It should be mentioned that other multidimensional transformations, such as MDS (multidimensional scaling), are sometimes preferred over PCA.

Although GWAS were first developed for humans (Ozaki et al. 2002), they rapidly were applied to other organisms such as plants (Atwell et al. 2010), cattle (Mai et al. 2010) or even, more recently, bacteria (Chen and Shapiro 2015). Interestingly, in bacteria the impact of stratification is even stronger as recombination in bacterial genomes differs strongly from the meiotic recombination in eukaryotes. In bacteria, recombination is assimilated to gene conversion, as it replaces only locally a piece of the genome, without affecting its backbone. Thus, the kinship of bacterial genomes within species can be described by a tree, especially for species with a low level of recombination (the tree paradigm suffers major drawback when recombination is high or when divergence between lineages is large). Therefore, GWAS methods in bacteria were redesigned to correct for the genealogy of genomes, using methods almost identical to substitution mapping reported in section 4.3.2. However, there, all sites of the core genome are mapped on the tree together with the trait of interest. And, instead of looking at correlated evolution between all pairs of sites, the correlation is computed between each site and the trait (Collins and Didelot 2018). The observation that treeWAS, the recent GWAS version that is tree-aware, uses standard metric of correlated evolution that are precisely the same as the ones used to detect correlated evolution among sites in a genome, highlights again the theoretical and methodological proximity between classical genetics and correlated evolution.

4.5. Conclusion

In this chapter, we have reviewed several models and methods used to characterize correlated evolution. We have divided the chapter into three sections, covering coevolution within morphological traits, coevolution within genetic traits and finally coevolution between genetic alleles and phenotypic traits. As we have seen, most methods we have described are highly similar in these three fields of applications. Even more, several methods can be directly applied to any types of traits, provided they are re-coded in the same abstraction. This unity in the methodologies fosters the common underlying framework of all these applications. We hope that this review will help readers to connect fields that could at first sight be considered as unrelated. Models and methods evolve, similar to living organisms. Consequently, they also share common ancestries, explaining their relatedness.

We have described two main schools of modeling correlated evolution that were developed in parallel. The first one simply considers sampled species, individuals or sequences, disregarding their shared history, and compute various metrics of correlation among the traits. A second one explicitly uses the shared history to compute correlation among evolutionary or segregation paths of the traits. We shall emphasize that, without any doubt, both frameworks have been successfully applied to many different questions, demonstrating their respective inner strengths. We would also like to stress that there may be good reasons to use one or the other. Practically, it can be hard to assess with confidence a phylogeny. Indeed, many problems can arise like (a) individuals within species have not a single genealogy for their genome, making difficult to infer the tree sequence, (b) inference can be entailed with large errors, especially for very divergent species, (c) computation time can be too high in very large samples to infer a tree. Thus, in many cases, simply counting co-occurrences without looking at the phylogeny is an excellent choice. Conversely, when the genealogy can be assessed with confidence, computing correlations between evolutionary paths is powerful, as it follows the evolutionary processes rather solely their final outcomes. Typically on datasets of small size, the phylogenetic methods often perform better.

Interestingly, we have also seen that despite the change in traits one is willing to study (morphological, genetics, etc.), the fundamental principles and models of the methods remained invariant. However, as the data grow in number and in quality, the fine implementation of the methods has been largely improved.

We have show that the method of correlating traits in biology to deduce some causality using prior knowledge between a predictor trait (e.g. a gene) and a induced one (e.g. a phenotype) is an old practice; this, despite the rampant ambiguity between correlation and causation. It is tempting to remind that formal statistical correlations were popularized under the impulse of biometricians (Galton 1968; Pearson and Galton 1895), suggesting that correlated evolution in biology has more ancient roots than the rise of genetics.

4.6. References

Abouheif, E. and Fairbairn, D.J. (1997). A comparative analysis of allometry for sexual size dimorphism: Assessing Rensch's rule. *The American Naturalist*, 149(3), 540–562.

Achaz, G., Rodriguez-Verdugo, A., Gaut, B.S., Tenaillon, O. (2014). The reproducibility of adaptation in the light of experimental evolution with whole genome sequencing. In *Ecological Genomics: Ecology and the Evolution of Genes and Genomes, Advances in Experimental Medicine and Biology*, Landry, C.R. and Aubin-Horth, N. (eds). Springer, Dordrecht.

Atwell, S., Huang, Y.S., Vilhjálmsson, B.J., Willems, G., Horton, M., Li, Y., Meng, D., Platt, A., Tarone, A.M., Hu, T.T. et al. (2010). Genome-wide association study of 107 phenotypes in Arabidopsis thaliana inbred lines. *Nature*, 465(7298), 627–631.

Baldassi, C., Zamparo, M., Feinauer, C., Procaccini, A., Zecchina, R., Weigt, M., Pagnani, A. (2014). Fast and accurate multivariate Gaussian modeling of protein families: Predicting residue contacts and protein-interaction partners. *PLOS ONE*, 9(3), e92721.

Barker, D. and Pagel, M. (2005). Predicting functional gene links from phylogenetic-statistical analyses of whole genomes. *PLOS Comput. Biol.*, 1(1), e3.

Barker, D., Meade, A., Pagel, M. (2007). Constrained models of evolution lead to improved prediction of functional linkage from correlated gain and loss of genes. *Bioinformatics*, 23(1), 14–20.

Barton, N.H., Etheridge, A.M., Véber, A. (2017). The infinitesimal model: Definition, derivation, and implications. *Theoretical Population Biology*, 118, 50–73.

Behdenna, A., Pothier, J., Abby, S.S., Lambert, A., Achaz, G. (2016). Testing for independence between evolutionary processes. *Systematic Biology*, 65(5), 812–823.

Bitbol, A.-F. (2018). Inferring interaction partners from protein sequences using mutual information. *PLOS Computational Biology*, 14(11), e1006401.

Blomberg, S.P., Garland, T., Ives, A.R. (2003). Testing for phylogenetic signal in comparative data: Behavioral traits are more labile. *Evolution*, 57(4), 717–745.

Champeimont, R., Laine, E., Hu, S.-W., Penin, F., Carbone, A. (2016). Coevolution analysis of Hepatitis C virus genome to identify the structural and functional dependency network of viral proteins. *Scientific Reports*, 6(1), 1–20.

Chen, P.E. and Shapiro, B.J. (2015). The advent of genome-wide association studies for bacteria. *Current Opinion in Microbiology*, 25, 17–24.

Cheverud, J.M., Dow, M.M., Leutenegger, W. (1985). The quantitative assessment of phylogenetic constraints in comparative analyses: Sexual dimorphism in body weight among primates. *Evolution*, 39(6), 1335–1351.

Chiu, D.K. and Kołodziejczak, T. (1991). Inferring consensus structure from nucleic acid sequences. *Computer Applications in the Biosciences*, 7(3), 347–352.

Cobb, M. (2017). 60 years ago, Francis Crick changed the logic of biology. *PLOS Biology*, 15(9), e2003243.

Cohen, O., Rubinstein, N.D., Stern, A., Gophna, U., Pupko, T. (2008). A likelihood framework to analyse phyletic patterns. *Philosophical Transactions of the Royal Society of London B: Biological Sciences*, 363(1512), 3903–3911.

Cohen, O., Ashkenazy, H., Burstein, D., Pupko, T. (2012). Uncovering the co-evolutionary network among prokaryotic genes. *Bioinformatics*, 28(18), i389–i394.

Cohen, O., Ashkenazy, H., Levy Karin, E., Burstein, D., Pupko, T. (2013). CoPAP: Coevolution of presence-absence patterns. *Nucleic Acids Research*, 41(Web server issue), W232–W237.

Collins, C. and Didelot, X. (2018). A phylogenetic method to perform genome-wide association studies in microbes that accounts for population structure and recombination. *PLOS Computational Biology*, 14(2), e1005958.

Cooper, N. and Purvis, A. (2010). Body size evolution in mammals: Complexity in tempo and mode. *American Naturalist*, 175(6), 727–738.

De Leonardis, E., Lutz, B., Ratz, S., Cocco, S., Monasson, R., Schug, A., Weigt, M. (2015). Direct-coupling analysis of nucleotide coevolution facilitates RNA secondary and tertiary structure prediction. *Nucleic Acids Research*, 43(21), 10444–10455.

Dib, L. and Carbone, A. (2012). Protein fragments: Functional and structural roles of their coevolution networks. *PLOS ONE*, 7(11), e48124.

Dib, L., Silvestro, D., Salamin, N. (2014). Evolutionary footprint of coevolving positions in genes. *Bioinformatics*, 30(9), 1241–1249.

Dutheil, J. (2012). Detecting coevolving positions in a molecule: Why and how to account for phylogeny. *Briefings in Bioinformatics*, 13(2), 228–243.

Dutheil, J. and Galtier, N. (2007). Detecting groups of coevolving positions in a molecule: A clustering approach. *BMC Evolutionary Biology*, 7, 242.

Dutheil, J., Pupko, T., Jean-Marie, A., Galtier, N. (2005). A model-based approach for detecting coevolving positions in a molecule. *Molecular Biology and Evolution*, 22(9), 1919–1928.

Dutheil, J.Y., Jossinet, F., Westhof, E. (2010). Base pairing constraints drive structural epistasis in ribosomal RNA sequences. *Mol. Biol. Evol.*, 27(8), 1868–1876.

Ekeberg, M., Lövkvist, C., Lan, Y., Weigt, M., Aurell, E. (2013). Improved contact prediction in proteins: Using pseudolikelihoods to infer Potts models. *Physical Review*, 87(1), 012707.

Fairbairn, D.J. (1997). Allometry for sexual size dimorphism: Pattern and process in the coevolution of body size in males and females. *Annual Review of Ecology and Systematics*, 28(1), 659–687.

Fares, M.A. and Travers, S.A.A. (2006). A novel method for detecting intramolecular coevolution: Adding a further dimension to selective constraints analyses. *Genetics*, 173(1), 9–23.

Felsenstein, J. (1973). Maximum-likelihood estimation of evolutionary trees from continuous characters. *American Journal of Human Genetics*, 25(5), 471–492.

Felsenstein, J. (1981). Evolutionary trees from DNA sequences: A maximum likelihood approach. *Jounal of Molecular Evolution*, 17(6), 368–376.

Felsenstein, J. (1985). Phylogenies and the comparative method. *The American Naturalist*, 125(1), 1–15.

Felsenstein, J. (2003). *Inferring Phylogenies*, 2nd edition. Sinauer Associates, Sunderland, MA.

Figliuzzi, M., Jacquier, H., Schug, A., Tenaillon, O., Weigt, M. (2016). Coevolutionary landscape inference and the context-dependence of mutations in beta-lactamase TEM-1. *Molecular Biology and Evolution*, 33(1), 268–280.

Galtier, N. and Dutheil, J. (2007). Coevolution within and between genes. *Genome Dynamics*, 3, 1–12.

Galton, F. (1886). Regression towards mediocrity in hereditary stature. *The Journal of the Anthropological Institute of Great Britain and Ireland*, 15, 246–263.

Gittleman, J.L. (1985). Carnivore body size: Ecological and taxonomic correlates. *Oecologia*, 67(4), 540–554.

Gbel, U., Sander, C., Schneider, R., Valencia, A. (1994). Correlated mutations and residue contacts in proteins. *Proteins*, 18(4), 309–317.

Goh, C.S., Bogan, A.A., Joachimiak, M., Walther, D., Cohen, F.E. (2000). Co-evolution of proteins with their interaction partners. *Journal of Molecular Biology*, 299(2), 283–293.

Grafen, A. (1989). The phylogenetic regression. *Philosophical Transactions of the Royal Society London B: Biological Sciences*, 326(1233), 119–157.

Harvey, P.H. and Pagel, M.D. (1991). *The Comparative Method in Evolutionary Biology*. Oxford University Press, Oxford.

Holley, R.W., Apgar, J., Everett, G.A., Madison, J.T., Marquisee, M., Merrill, S.H., Penswick, J.R., Zamir, A. (1965). Structure of a ribonucleic acid. *Science*, 147(3664), 1462–1465.

Hopf, T.A., Colwell, L.J., Sheridan, R., Rost, B., Sander, C., Marks, D.S. (2012). Three-dimensional structures of membrane proteins from genomic sequencing. *Cell*, 149(7), 1607–1621.

Ignac, T.M., Skupin, A., Sakhanenko, N.A., Galas, D.J. (2014). Discovering pair-wise genetic interactions: An information theory-based approach. *PLOS ONE*, 9(3), e92310.

de Juan, D., Pazos, F., Valencia, A. (2013). Emerging methods in protein co-evolution. *Nature Reviews Genetics*, 14(4), 249–261.

Kay, K.M., Reeves, P.A., Olmstead, R.G., Schemske, D.W. (2005). Rapid speciation and the evolution of hummingbird pollination in neotropical Costus subgenus Costus (Costaceae): Evidence from nrDNA ITS and ETS sequences. *American Journal of Botany*, 92(11), 1899–1910.

Kryazhimskiy, S., Dushoff, J., Bazykin, G.A., Plotkin, J.B. (2011). Prevalence of epistasis in the evolution of influenza A surface proteins. *PLOS Genetics*, 7(2), e1001301.

Laine, E., Karami, Y., Carbone, A. (2019). GEMME: A simple and fast global epistatic model predicting mutational effects. *Molecular Biology and Evolution*, 36(11), 2604–2619.

Lande, R., Engen, S., Sæther, B. (1999). Spatial scale of population synchrony: Environmental correlation versus dispersal and density regulation. *The American Naturalist*, 154(3), 271–281.

Lander, E.S. and Schork, N.J. (1994). Genetic dissection of complex traits. *Science*, 265(5181), 2037–2048.

Lartillot, N. (2013). Interaction between selection and biased gene conversion in mammalian protein-coding sequence evolution revealed by a phylogenetic covariance analysis. *Molecular Biology and Evolution*, 30(2), 356–368.

Lartillot, N. and Delsuc, F. (2012). Joint reconstruction of divergence times and life-history evolution in placental mammals using a phylogenetic covariance model. *Evolution*, 66(6), 1773–1787.

Lartillot, N. and Poujol, R. (2011). A phylogenetic model for investigating correlated evolution of substitution rates and continuous phenotypic characters. *Molecular Biology and Evolution*, 28(1), 729–744.

Lenormand, T. (2003). The evolution of sex dimorphism in recombination. *Genetics*, 163(2), 811–822.

Lenormand, T. and Dutheil, J. (2005). Recombination difference between sexes: A role for haploid selection. *PLOS Biology*, 3(3), e63.

Lewontin, R. (2004). Building a science of population biology. In *The Evolution of Population Biology*, Singh, R.S. and Uyenoyama, M.K. (eds). Cambridge University Press, Cambridge.

Long, A., Liti, G., Luptak, A., Tenaillon, O. (2015). Elucidating the molecular architecture of adaptation via evolve and resequence experiments. *Nature Reviews Genetics*, 16(10), 567–582.

Maddison, W.P. (1990). A method for testing the correlated evolution of two binary characters: Are gains or losses concentrated on certain branches of a phylogenetic tree? *Evolution*, 44(3), 539–557.

Mai, M.D., Sahana, G., Christiansen, F.B., Guldbrandtsen, B. (2010). A genome-wide association study for milk production traits in Danish Jersey cattle using a 50K single nucleotide polymorphism chip. *Journal of Animal Sciences*, 88(11), 3522–3528.

Marks, D.S., Colwell, L.J., Sheridan, R., Hopf, T.A., Pagnani, A., Zecchina, R., Sander, C. (2011). Protein 3D structure computed from evolutionary sequence variation. *PLOS ONE*, 6(12), e28766.

Martins, E.P. (1994). Estimating the rate of phenotypic evolution from comparative data. *The American Naturalist*, 144(2), 193–209.

Martins, E.P. and Hansen, T.F. (1997). Phylogenies and the comparative method: A general approach to incorporating phylogenetic information into the analysis of interspecific data. *The American Naturalist*, 149(4), 646–667.

McVean, G. (2009). A genealogical interpretation of principal components analysis. *PLOS Genetics*, 5(10), e1000686.

Mendel, J.G. (1901). Experiments in plant hybridization. *Journal of the Royal Horticultural Society*, 26, 1–32.

Meyer, X., Dib, L., Silvestro, D., Salamin, N. (2019). Simultaneous Bayesian inference of phylogeny and molecular coevolution. *Proceedings Natlonal Academy of Sciences USA*, 116(11), 5027–5036.

Moore, R. (2001). The “rediscovery” of Mendel’s work. *Bioscene*, 27(2), 13–24.

Morcos, F., Pagnani, A., Lunt, B., Bertolino, A., Marks, D.S., Sander, C., Zecchina, R., Onuchic, J.N., Hwa, T., Weigt, M. (2011). Direct-coupling analysis of residue coevolution captures native contacts across many protein families. *Proceedings of the National Academy of Sciences USA*, 108(49), E1293–E1301.

Morgan, T.H. (1926). *The Theory of the Gene: 2nd Edition*. Creative Media Partners, LLC, Sacramento.

Morton, N.E. (1955). Sequential tests for the detection of linkage. *American Journal of Human Genetics*, 7(3), 277–318.

Neher, E. (1994). How frequent are correlated changes in families of protein sequences? *Proceedings National Academy of Sciences USA*, 91(1), 98–102.

Novembre, J. and Stephens, M. (2008). Interpreting principal component analyses of spatial population genetic variation. *Nature Genetics*, 40(5), 646–649.

Nyholt, D.R. (2000). All LODs are not created equal. *American Journal of Human Genetics*, 67(2), 282–288.

Ozaki, K., Ohnishi, Y., Iida, A., Sekine, A., Yamada, R., Tsunoda, T., Sato, H., Sato, H., Hori, M., Nakamura, Y., Tanaka, T. (2002). Functional SNPs in the lymphotoxin-α gene that are associated with susceptibility to myocardial infarction. *Nature Genetics*, 32(4), 650–654.

Pagel, M. (1994). Detecting correlated evolution on phylogenies: A general method for the comparative analysis of discrete characters. *Proceedings of the Royal Society of London B: Biological Sciences*, 255(1342), 37–45.

Paradis, E. and Claude, J. (2002). Analysis of comparative data using generalized estimating equations. *Journal of Theoretical Biology*, 218(2), 175–185.

Pazos, F. and Valencia, A. (2001). Similarity of phylogenetic trees as indicator of protein-protein interaction. *Protein Engineering*, 14(9), 609–614.

Pearson, K. and Galton, F. (1895). VII. Note on regression and inheritance in the case of two parents. *Proceedings of the Royal Society of London*, 58(347–352), 240–242.

Pensar, J., Puranen, S., Arnold, B., MacAlasdair, N., Kuronen, J., Tonkin-Hill, G., Pesonen, M., Xu, Y., Sipola, A., Sánchez-Busâ, L. et al. (2019). Genome-wide epistasis and co-selection study using mutual information. *Nucleic Acids Research*, 47(18), e112–e112.

Phillips, P.C. (2008). Epistasis – The essential role of gene interactions in the structure and evolution of genetic systems. *Nature Review Genetics*, 9(11), 855–867.

Pollock, D.D., Taylor, W.R., Goldman, N. (1999). Coevolving protein residues: Maximum likelihood identification and relationship to structure. *Journal of Molecular Biology*, 287(1), 187–198.

Price, A.L., Patterson, N.J., Plenge, R.M., Weinblatt, M.E., Shadick, N.A., Reich, D. (2006). Principal components analysis corrects for stratification in genome-wide association studies. *Nature Genetics*, 38(8), 904–909.

Pritchard, J.K., Stephens, M., Rosenberg, N.A., Donnelly, P. (2000). Association mapping in structured populations. *American Journal of Human Genetics*, 67(1), 170–181.

Puranen, S., Pesonen, M., Pensar, J., Xu, Y.Y., Lees, J.A., Bentley, S.D., Croucher, N.J., Corander, J. (2018). SuperDCA for genome-wide epistasis analysis. *Microbial Genomics*, 4(6), e000184.

Purcell, S., Neale, B., Todd-Brown, K., Thomas, L., Ferreira, M.A.R., Bender, D., Maller, J., Sklar, P., de Bakker, P.I.W., Daly, M.J., Sham, P.C. (2007). PLINK: A tool set for whole-genome association and population-based linkage analyses. *American Journal of Human Genetics*, 81(3), 559–575.

Rensch, B. (1950). Die Abhängigkeit der relativen Sexualdifferenz von der Körpergröße. *Bonner Zoologische Beiträge*, 1, 58–69.

Ridley, M. (1983). *The Explanation of Organic Diversity: The Comparative Method and Adaptations for Mating*. Oxford University Press, Oxford/New York.

Shindyalov, I.N., Kolchanov, N.A., Sander, C. (1994). Can three-dimensional contacts in protein structures be predicted by analysis of correlated mutations? *Protein Engineering*, 7(3), 349–358.

Szurmant, H. and Weigt, M. (2018). Inter-residue, inter-protein and inter-family coevolution: Bridging the scales. *Current Opinion in Structural Biology*, 50, 26–32.

Talavera, D., Lovell, S.C., Whelan, S. (2015). Covariation is a poor measure of molecular coevolution. *Molecular Biology and Evolution*, 32(9), 2456–2468.

Tillier, E.R.M. and Collins, R.A. (1995). Neighbor joining and maximum likelihood with RNA sequences: Addressing the interdependence of sites. *Molecular Biology and Evolution*, 12(1), 7–7.

Tillier, E.R.M. and Collins, R.A. (1998). High apparent rate of simultaneous compensatory base-pair substitutions in ribosomal RNA. *Genetics*, 148(4), 1993–2002.

Tufféry, P. and Darlu, P. (2000). Exploring a phylogenetic approach for the detection of correlated substitutions in proteins. *Molecular Biology and Evolution*, 17(11), 1753–1759.

Van Valen, L. (1973). A new evolutionary law. *Evolutionary Theory*, 1(1), 1–30.

de Visser, J.A.G.M. and Krug, J. (2014). Empirical fitness landscapes and the predictability of evolution. *Nature Reviews Genetics*, 15(7), 480–490.

Weigt, M., White, R.A., Szurmant, H., Hoch, J.A., Hwa, T. (2009). Identification of direct residue contacts in protein–protein interaction by message passing. *Proceedings of the National Academy of Sciences*, 106(1), 67–72.

Westhof, E. and Auffinger, P. (2012). Transfer RNA structure [Online]. Available at: http://www.els.net/WileyCDA/ElsArticle/refId-a0000527.html.

Woolhouse, M.E.J., Webster, J.P., Domingo, E., Charlesworth, B., Levin, B.R. (2002). Biological and biomedical implications of the co-evolution of pathogens and their hosts. *Nature Genetics*, 32(4), 569–577.

Wright, S. (1932). The roles of mutation, inbreeding, crossbreeding and selection in evolution. *Proceedings of the Sixth International Congress on Genetics*, 1, 356–366.

Yang, Z. (2006). *Computational Molecular Evolution*. Oxford University Press, Oxford.

Yeang, C.-H. (2008). Identifying coevolving partners from paralogous gene families. *Evolutionary Bioinformatics Online*, 4, 97–107.

Yeang, C.-H., Darot, J.F.J., Noller, H.F., Haussler, D. (2007). Detecting the coevolution of biosequences – An example of RNA interaction prediction. *Mol. Biol. Evol.*, 24(9), 2119–2131.

Yeang, C.-H. and Haussler, D. (2007). Detecting coevolution in and among protein domains. *PLOS Computational Biology*, 3(11), e211.

Yi, X. and Dean, A.M. (2019). Adaptive landscapes in the age of synthetic biology. *Molecular Biology and Evolution*, 36(5), 890–907.

5

A Century of Genomic Rearrangements

Anne BERGERON[1] and Krister M. SWENSON[2]
[1] *LACIM, Université du Québec à Montréal, Canada*
[2] *LIRMM, CNRS, Université de Montpellier, France*

5.1. Introduction

In February 1920, the prestigious PNAS journal published a short letter entitled *The Evidence for the Linear Order of the Genes* (Morgan et al. 1920) adding to the debate between two opposing theories on the organization of genetic markers on a chromosome. On the one hand, Morgan and his colleagues argued that, in a species, the genes were arranged one after the other in a linear structure (see also Chapter 4). On the other hand, William Castle, proposed a three-dimensional organization which, according to him, better explained the observed *distances* between the different markers (Castle 1919).

Over the course of the 20th century, subsequent discoveries finally proved that Morgan's team was right. The first genetic maps were developed, the identification of inversions in the order of markers between different species were discovered and trees that rebuild the history of these events were constructed.

It took a few more decades, with the work of Crick and Watson (Crick et al. 1954), to establish that DNA molecules are the physical framework for this linear organization. In the 21st century, advances in sequencing techniques have facilitated the comparison of the genomes of an increasing number of species, bringing to light a great diversity of rearrangement types: explaining the nature of these rearrangements has shifted from being a biological problem to an algorithmic one.

At the same time, Castle's hypothesis has been partially justified by highlighting the three-dimensional structure adopted by the chromosomes within the cell nucleus (Lieberman-Aiden et al. 2009). Indeed, rearrangements occur between linear molecules, but the positions affected by these rearrangements are often determined by their relative location in space (Nikiforova et al. 2000; Yaffe et al. 2010; Véron et al. 2011; Swenson and Blanchette 2019).

This chapter develops the mathematical and algorithmic techniques utilized to address *gene* rearrangement problems, both in their original sense as markers of inherited traits, and in their modern interpretation as families of transcripts.

5.2. Orderings of genes and the rearrangements that act on them

In order to study rearrangements, we need abstract and simplified definitions for genes and chromosomes. These definitions are abstract in the sense that every gene will correspond to a chromosome segment identified by a number and a relative orientation. They are simplified because the discoveries of the last decades have significantly revised and corrected the initial idea that considered a gene as a discrete marker for hereditary traits (see, for example, Pearson 2006).

When comparing the gene order of two species, several problems arise, the primary one being the determination of whether two genes are *equal*. This biological question concerns *homologous* genes, which are genes that share a common ancestor. When a cell splits, it transmits a copy of its DNA to each of its daughter cells, and so forth for every new cell; theoretically all descendants of the ancestral cell share identical genes.

In practice, several events can alter the DNA of certain daughters, such as errors in the copying mechanism or exposure to environmental factors. If the common ancestor existed relatively recently – for instance, a few million years ago – homologous genes can be easily recognized, because they are still almost identical and many bioinformatics tools have automated this process. We will therefore assume in this chapter that the problem of deciding whether two genes are equal is solved.

When comparing the gene content between two organisms, several problems emerge: some genes can be repeated in the same organism, referred to as *duplications*; some genes may be present in one organism but not in another, referred to as *insertions* and *deletions*; finally, even in the absence of duplications, insertions or deletions, the composition and order of genes in chromosomes may vary, which is the product of *rearrangements*.

This chapter focuses on the rearrangement problem between representative genomes of different species, and we will assume that these genomes have the same gene content, without duplications. Moreover, we consider only one copy of each chromosome, ignoring possible heterogeneity in polyploid species.

5.2.1. *Basic representations and definitions*

Modern genome sequencing and annotation techniques identify the position and orientation (positive or negative) of genes on the chromosomes. This information is often represented by diagrams, as shown in Figure 5.1, where genes are represented by oriented arrows and labeled with an identifier that enables recognition of homologous genes.

When studying rearrangement problems, the representation of genomes is further simplified. The gene identifiers are integers, with a positive or negative sign representing their orientations relative to one of the genomes, chosen as reference. This arbitrarily chosen reference often has genes labeled in ascending order. Figure 5.2 shows an example of two genomes each having two chromosomes, with nine genes.

In this example, the genome B is represented by:

$$B = (\circ\ 1\ 2\ 3\ 4\ 5\ \circ)\ (\circ\ 6\ 7\ 8\ 9\ \circ)$$

where the symbol $\circ$ indicates the end of a chromosome, and the genome A is represented by:

$$A = (\circ\ \text{-}1\ 4\ 5\ \text{-}3\ \circ)\ (\circ\ \text{-}8\ 6\ 9\ 7\ \text{-}2\ \circ)$$

This notation indicates, for example, that genes 3 and 4 are found in different positions and opposite relative orientations in the genomes A and B.

Since genes are oriented, the two *extremities* of a gene can be differentiated as the *tail* or the *head*, named after the tail and head of the arrows in Figure 5.2. The extremities of a gene g will be represented by symbols g_t and g_h, respectively, noting that the tail appears to the left of the head for a gene with a positive orientation in the figure. An *adjacency* of a genome is defined as a pair $\{u, v\}$ of extremities (note that adjacencies $\{u, v\}$ and $\{v, u\}$ are the same). When a gene extremity e also corresponds to the end of a chromosome, called a telomere, we include the adjacency $\{e, \circ\}$.

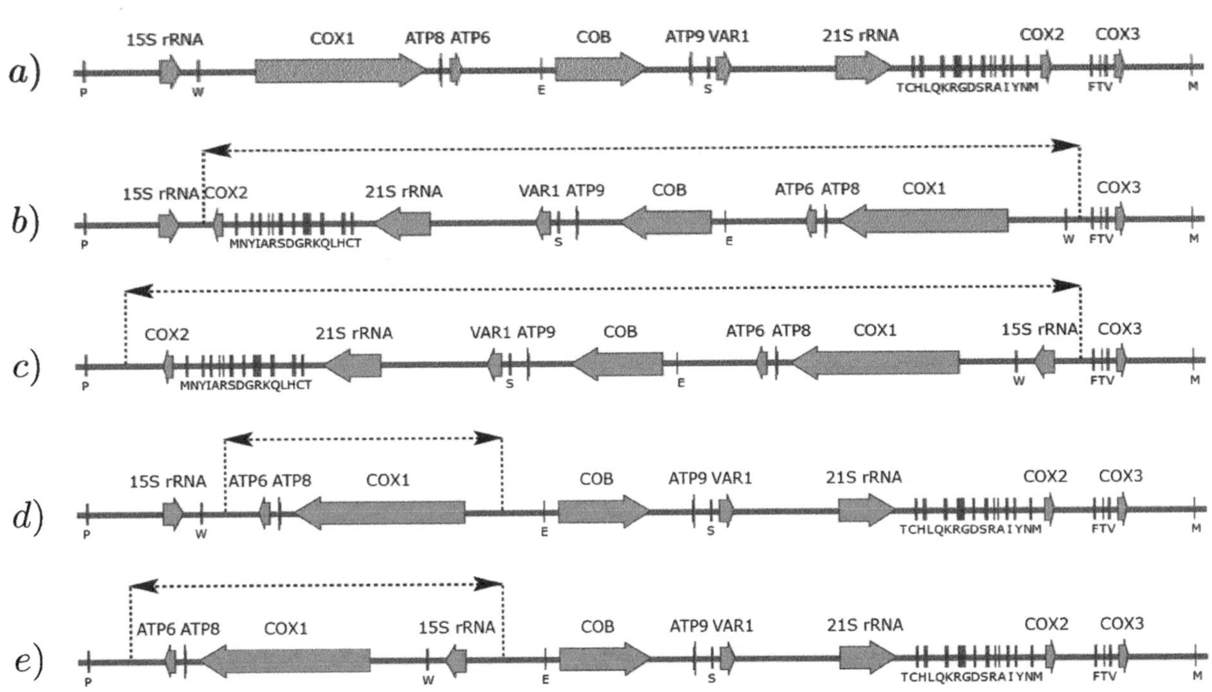

Figure 5.1. *Typical representation of chromosome segments showing rearrangements, in this case, in the mitochondria of different yeast species. COX1 and 21sRNA genes, for example, appear in different relative orientations when comparing species* a *and* d. *Adapted from De Chiara et al. (2020). For a color version of this figure, www.iste.co.uk/didier/models.zip*

$$\{\circ,1_h\}\ \{1_t,4_t\}\ \{4_h,5_t\}\ \{5_h,3_h\}\ \{3_t,\circ\} \quad \{\circ,8_h\}\ \{8_t,6_t\}\ \{6_h,9_t\}\ \{9_h,7_t\}\ \{7_h,2_h\}\ \{2_t,\circ\}$$

$$A = (\ \circ\ \overleftarrow{-1}\ \ \overrightarrow{4}\ \ \overrightarrow{5}\ \ \overleftarrow{-3}\ \circ\ )\ (\ \circ\ \overleftarrow{-8}\ \ \overrightarrow{6}\ \ \overrightarrow{9}\ \ \overrightarrow{7}\ \ \overleftarrow{-2}\ \circ\ )$$

$$\{\circ,1_t\}\ \{1_h,2_t\}\ \{2_h,3_t\}\ \{3_h,4_t\}\ \{4_h,5_t\}\ \{5_h,\circ\} \quad \{\circ,6_t\}\ \{6_h,7_t\}\ \{7_h,8_t\}\ \{8_h,9_t\}\ \{9_h,\circ\}$$

$$B = (\ \circ\ \overrightarrow{1}\ \ \overrightarrow{2}\ \ \overrightarrow{3}\ \ \overrightarrow{4}\ \ \overrightarrow{5}\ \circ\ )\ (\ \circ\ \overrightarrow{6}\ \ \overrightarrow{7}\ \ \overrightarrow{8}\ \ \overrightarrow{9}\ \circ\ )$$

Figure 5.2. *Genome representations $A =$ (∘ -1 4 5 -3 ∘) (∘ -8 6 9 7 -2 ∘) and $B =$ (∘1 2 3 4 5 ∘) (∘6 7 8 9 ∘), with their adjacency sets. Observe that the conserved adjacency $\{4_h, 5_t\}$ is the only one common to both sets*

For example, each genome in Figure 5.2 is composed of 11 adjacencies, of which seven are internal, and four are located at the end of the chromosomes. These adjacencies all differ among genomes except $\{4_h, 5_t\}$, which is due to the fact that in both genomes genes 4 and 5 follow each other in the same orientation. In this case, we say that the adjacency is *conserved*.

The notion of conserved adjacencies leads to the formalization of *equivalent genomes*. In reality, chromosomes are not ordered and do not have a fixed orientation. With the following definition, we can test when two genomes are identical.

DEFINITION 5.1.– *Two duplication-free genomes with the same gene content are* equivalent *if and only if their adjacency sets are equal.*

For example, the genomes $A =$ (∘1 2 ∘) (∘ 3 4 ∘) and $B =$ (∘ -4 -3 ∘) (∘1 2 ∘) are equivalent since their adjacencies are, in each case, $\{\{\circ, 1_t\}, \{1_h, 2_t\}, \{2_h, \circ\}, \{\circ, 3_t\}, \{3_h, 4_t\}, \{4_h, \circ\}\}$.

5.2.2. DCJ operations and the breakpoint graph

Most of the rearrangements that preserve the gene content of a genome can be modeled by a simple operation introduced by Yancopoulos et al. (2005). The operation acts on one or two DNA strands and is based on natural biological mechanisms that are capable of repairing chromosome breakage following a chemical, radiological or mechanical accident.

When a simple breakage occurs, these mechanisms usually perform their work correctly. Similar to fixing burst plumbing pipes, after the break is located, the faulty section is cut and both ends are joined. If two strands of DNA that are close to each other suffer the same accident, the previous strategy can give rise to aberrations, such as connecting the hot water outlet to the cold water inlet. With the formalism introduced in the previous section, if two adjacencies $\{u, v\}$ and $\{r, s\}$ are cut, there are three ways to join them:

1) $\{u, s\}, \{r, v\}$;
2) $\{u, r\}, \{v, s\}$;
3) $\{u, v\}, \{r, s\}$.

Only the first two joins change the structure of a genome. We refer to either of these operations by the acronym *DCJ* (double-cut-and-join).

For example, if a DCJ operation is performed on the adjacencies $\{1_h, 2_t\}$ and $\{3_h, 4_t\}$ of the genome $A =$ (∘ 1 2 ∘) (∘ 3 4 ∘) , it yields either:

$A' =$ (∘ 1 4 ∘) (∘ 3 2 ∘)

or alternatively:

$A'' =$ (∘ 1 -3 ∘) (∘ -4 2 ∘)

It is possible that the two cut adjacencies belong to the same chromosome. For example, a DCJ operation on the adjacencies $\{1_h, 2_t\}$ and $\{3_h, 4_t\}$ of the genome $A =$ (∘ 1 2 3 4 ∘) can produce the genome:

$A' =$ (∘ 1 -3 -2 4 ∘)

EXERCISE 5.1.– What would be the genome obtained with the alternative DCJ operation acting on $\{1_h, 2_t\}$ and $\{3_h, 4_t\}$ of genome A?[1]

Given two genomes, our immediate goal is to determine the minimum number and nature of the DCJ operations that result in transforming one into the other. To this end, we define a classical structure, the *breakpoint graph*, which is the primary tool for the comparison of adjacency sets of genomes.

DEFINITION 5.2 (Breakpoint graph).– *Let A and B be two genomes with n genes. A* breakpoint graph *is an undirected graph whose edges are colored in gray or black. The graph's vertices are the extremities of the n genes, as well as the telomere symbol* ∘*: the* black *edges are the adjacencies of the genome A, and the* gray *edges are the adjacencies of genome B.*

---

1 Hint: Yes, it is possible for a chromosome to have no telomeres. This is called a *circular* chromosome.

The breakpoint graph for genomes $A = (\circ$ -1 4 5 -3 $\circ)$ $(\circ$ -8 6 9 7 -2 $\circ)$ and $B = (\circ$ 1 2 3 4 5 $\circ)$ $(\circ$ 6 7 8 9 $\circ)$ is shown in Figure 5.3. Note that the two dashed edges can be ignored for now, as they will later be used to enumerate rearrangement scenarios. This graph can be used to easily compute the *distance* between two genomes, denoted by d_{DCJ}, which corresponds the minimum number of DCJ operations required to transform one genome into another.

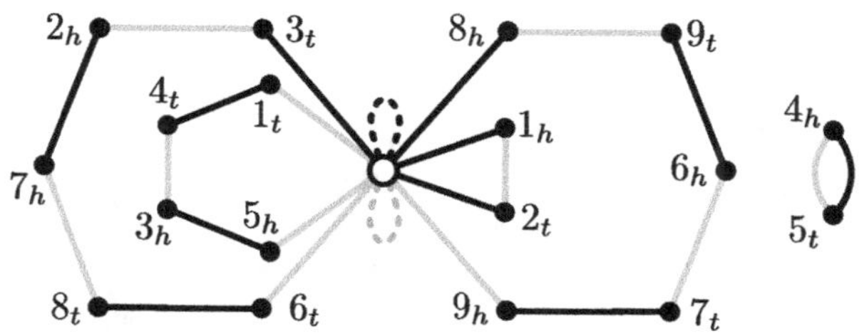

Figure 5.3. *Breakpoint graph for genomes* $A = (\circ$ *-1 4 5 -3* $\circ)$ $(\circ$ *-8 6 9 7 -2* $\circ)$ *and* $B = (\circ$ *1 2 3 4 5* $\circ)$ $(\circ$ *6 7 8 9* $\circ)$

An *elementary* cycle in a graph is a cycle that does not contain the same vertex twice. In a breakpoint graph, a cycle is *balanced* if it has the same number of gray and black edges. The graph in Figure 5.3 has three balanced elementary cycles, two of which contain the vertex $\circ$.

A balanced elementary cycle is said to be *complete* if all of its vertices are gene extremities, conversely it is said *incomplete* if it contains the vertex $\circ$. The distance DCJ is given by the formula:

$$d_{DCJ} = n - (c + i/2)$$

where c is the number of complete cycles and i is the number of incomplete cycles. This formula does not appear to depend on the number of unbalanced elementary cycles, such as the two cycles shown in Figure 5.3: $(5_h, 3_h, 4_t, 1_t, \circ)$ and $(1_h, 2_t, \circ)$. This is an illusion, however, because there are dependencies between the number of unbalanced cycles and the values of n, c and i, which make several different formulas equivalent (Bergeron and Stoye 2013).

EXERCISE 5.2.– It is instructive to build the breakpoint graph of equivalent genomes. What is the graph for genomes $A = (\circ$ 1 2 $\circ)$ $(\circ$ 3 4 $\circ)$ and $B = (\circ$ -4 -3 $\circ)$ $(\circ$ 1 2 $\circ)$? Using the graph, deduce that $d_{DCJ}(A, B) = 0$.

In the next section, we will need to compute all of the sequences of DCJ operations of minimum length. We therefore conclude this section by showing, with a simple example, how at least **one** sequence of minimum length can be computed. Such a sequence will be called an *optimal scenario.*

Consider the genomes $A = (\circ\ 1\ \text{-}3\ 5\ 2\ \text{-}4\ 6\ \circ)$ and $B = (\circ\ 1\ 2\ 3\ 4\ 5\ 6\ \circ)$. The corresponding breakpoint graph is shown in Figure 5.4. It contains two complete cycles and two incomplete cycles (with no unbalanced cycles), which implies a distance of $6 - (2 + 2/2) = 3$ DCJ operations.

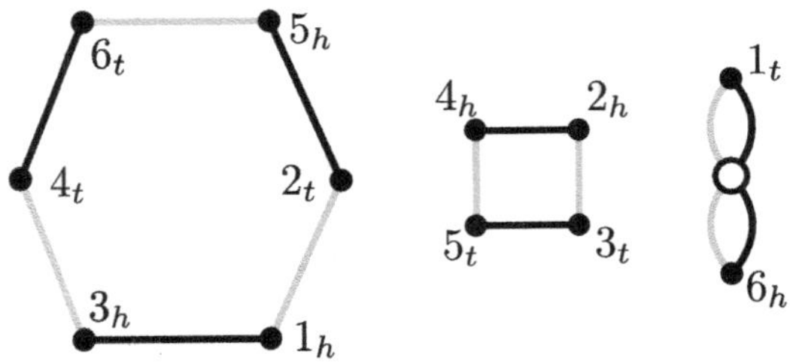

Figure 5.4. *Breakpoint graph for genomes $A = (\circ 1\ \text{-}3\ 5\ 2\ \text{-}4\ 6\ \circ)$ and $B = (\circ 1\ 2\ 3\ 4\ 5\ 6\ \circ)$*

To reduce the distance $n-(c+i/2)$ between two genomes, the number of complete or incomplete cycles has to be increased. Our goal for this example will be to increase the number of complete cycles.

The cycle of length 4 is easy to split it into two cycles of length 2 with a single DCJ operation, as shown in Figure 5.5. This rearrangement corresponds to cutting the black adjacencies $\{2_h, 4_h\}$ and $\{3_t, 5_t\}$ of genome A to construct the gray adjacencies $\{2_h, 3_t\}$ and $\{4_h, 5_t\}$ of genome B. As illustrated in Figure 5.5, this operation transforms the genome $A = (\circ\ 1\ \text{-}3\ 5\ 2\ \text{-}4\ 6\ \circ)$ into the genome $A' = (\circ\ 1\ \text{-}3\ \text{-}2\ \text{-}5\ \text{-}4\ 6\ \circ)$.

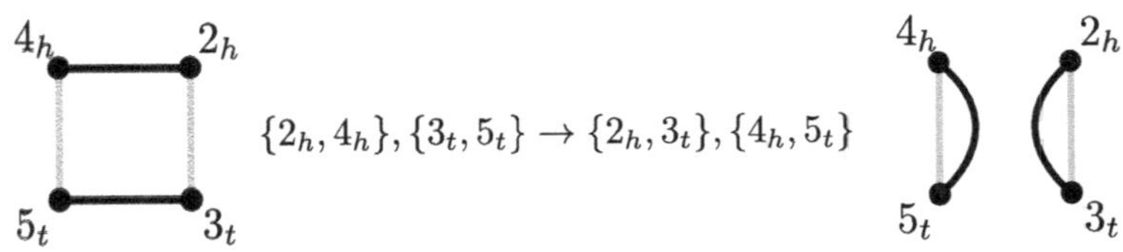

Figure 5.5. *Cycle splitting with a DCJ operation*

For longer cycles, as in the case of the cycle of length 6, several strategies are possible. Three DCJ operations split this cycle into one of length 2 and one of length 4. These are as follows:

$$\{1_h, 3_h\}, \{2_t, 5_h\} \rightarrow \{1_h, 2_t\}, \{3_h, 5_h\}$$

$$\{1_h, 3_h\}, \{4_t, 6_t\} \rightarrow \{3_h, 4_t\}, \{1_h, 6_t\}$$

$$\{2_t, 5_h\}, \{4_t, 6_t\} \rightarrow \{2_t, 4_t\}, \{5_h, 6_t\}$$

To complete a scenario, each of these operations is followed by another one that splits the resulting square in two. There are thus three different optimal scenarios to solve the cycle of length 6, which gives us a total of nine different scenarios since the operations on the two initial cycles can be performed independently.

5.3. Counting DCJ scenarios

In this section, we will address the problem of enumerating the set of all optimal scenarios that transform one genome into another. The basic principle can be described in terms of balanced cycles. When the breakpoint graph is solely composed of cycles of length 2 and complete cycles, the problem is relatively simple. This was the case for the genomes in Figure 5.4. To increase the number of balanced cycles, a cycle must be divided into two; no DCJ operation cutting adjacencies from two balanced cycles can increase the number of elementary cycles. By enumerating all the decompositions of the breakpoint graph into balanced cycles, we reduce the problem of counting all scenarios to that of counting the number of scenarios that solve a balanced cycle.

5.3.1. *Scenarios for a balanced cycle of length 2m*

If two black edges $\{u, v\}$ and $\{r, s\}$ in a cycle of length $2m$ are chosen, only one of the two possible DCJ operations will split the cycle in two. We encourage the reader to apply the alternative operation to the small cycle shown in Figure 5.5. In an optimal scenario, a DCJ operation is thus characterized by a pair of black edges, and the four implicated extremities form a quadrilateral when the cycle is drawn without any crossings. The effect of the operation will be to split the cycle into two using the black edges.

Since each of these DCJ operations splits a cycle in two, we have the following.

PROPOSITION 5.1.– The number of DCJ operations to solve a cycle of length $2m$ is $m - 1$.

If the quadrilaterals corresponding to all the operations of an optimal scenario for a cycle are drawn, we obtain a *quadrangulation* of the $2m$-vertex polygon, which is a partition of a polygon into non-overlapping rectangles. Two examples of quadrangulations are shown in Figure 5.6 for a cycle of length 2×8. Conversely, a given quadrangulation characterizes all possible optimal scenarios whose DCJ operations share the same edge pairs.

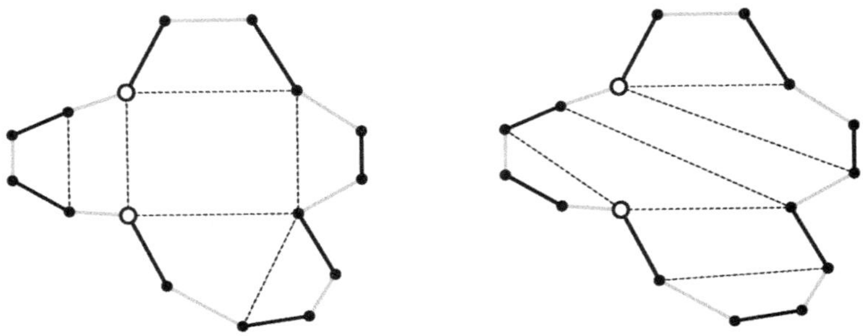

Figure 5.6. *Two of the 7752 possible quadrangulations of a 16-vertex polygon*

Fortunately, we know how to enumerate the quadrangulations of a $2m$-vertex polygon, their number given by the formula $\frac{1}{2m-1}\binom{3m-3}{m-1}$, (Baryshnikov 2001).

5.3.2. *The (many) cycle decompositions of a breakpoint graph*

Now that we know all the ways to resolve a balanced cycle, our next goal is to compute all the ways to decompose the breakpoint graph into balanced cycles while preserving the distance $d_{DCJ} = n - (c + i/2)$.

Any decomposition of the breakpoint graph into balanced cycles will contain all of the balanced elementary cycles. Matters are less clear, however, for unbalanced elementary cycles such as $(5_h, 3_h, 4_t, 1_t, \circ)$ and $(1_h, 2_t, \circ)$ in Figure 5.3. An unbalanced cycle is characterized by the color, gray or black, of its two incident edges at the vertex $\circ$: we will therefore distinguish the set G of *gray* cycles and the set B of *black* cycles, according to this color.

One way to build balanced cycles uses the dashed edges to balance an unbalanced cycle. These dashed edges make it possible to model the fact that a DCJ operation can change the number of chromosomes, and as such, the number of telomeres in a genome, while maintaining the same gene content. (One of the most famous of these rearrangements involves chromosome 2 in human beings, which was a result of the fusion of two chromosomes in a common ancestor of human and Neanderthal (Ijdo et al. 1991)).

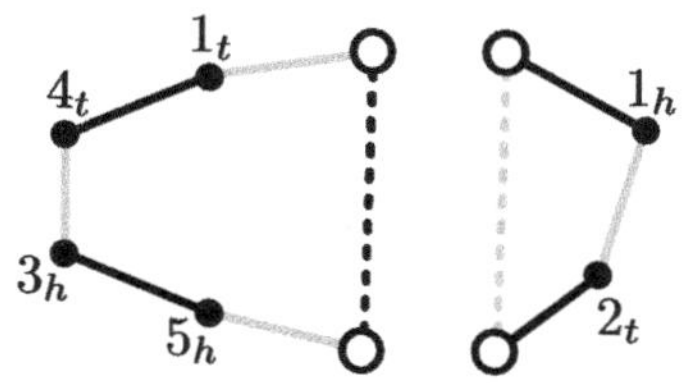

Figure 5.7. *An unbalanced gray or black cycle can be merged with a dashed cycle of the opposite color, black or gray to create a balanced cycle*

Consider an unbalanced cycle augmented with a dashed edge of the appropriate color, creating a balanced cycle. Figure 5.7 shows an example of these fusions for both cycles $(5_h, 3_h, 4_t, 1_t, \circ)$ and $(1_h, 2_t, \circ)$. Viewing the unbalanced cycle in this way does not affect the distance, since each DCJ operation that splits one of these augmented cycles into two gives either one complete cycle and one unbalanced cycle, or two incomplete cycles. For example, the cycle $(5_h, 3_h, 4_t, 1_t, \circ)$ will be split according to the three possible quadrangulations of the augmented cycle, into:

- $(3_h, 4_t,)(5_h, 1_t, \circ)$;
- $(5_h, \circ)(3_h, 4_t, 1_t, \circ)$;
- $(1_t, \circ)(5_h, 3_h, 4_t, \circ)$.

A second way to create balanced cycles is to merge a gray cycle $g \in G$ with a black cycle $b \in B$. The result is a balanced cycle that passes through the vertex $\circ$ twice. In the example of Figure 5.3, the sets G and B contain a single cycle. Figure 5.8 shows the two possible fusions (depending on the choice of the edges adjacent to the telomere) for these two cycles. For clarity, we choose to draw these cycles as *circles* where the vertex $\circ$ is repeated twice. Once again, the distance is preserved since each DCJ operation that splits one of these cycles into two produces either one complete cycle and one unbalanced cycle, or two incomplete cycles.

5.3.2.1. *Computing the number of decompositions*

We now estimate the number of different decompositions in the breakpoint graph. Since the balanced cycles belong to any decomposition, one simply has to count the number of ways to merge the elements of G and B.

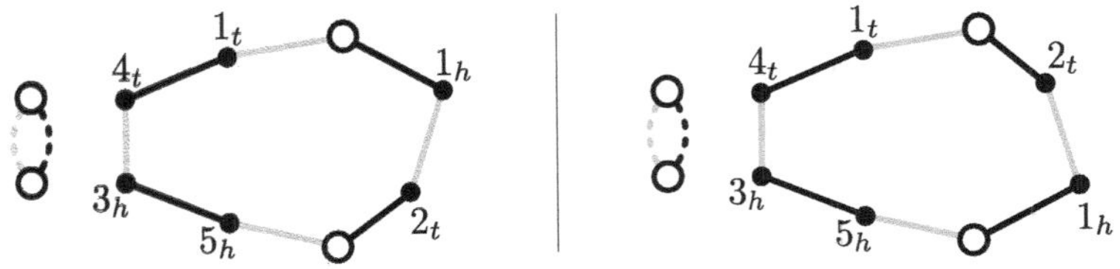

Figure 5.8. *Two ways of merging the gray cycle* $(5_h, 3_h, 4_t, 1_t, \circ)$ *and the black cycle* $(1_h, 2_t, \circ)$ *to create a balanced cycle (dashed cycles are not used)*

Let j be the number of elements of G and ℓ the number of elements of B. We assume that $j \leq \ell$, because the final result will be symmetrical with respect to j and ℓ. A breakpoint graph balanced cycle decomposition is characterized by the following choices:

1) The choice of k cycles of G that will be merged with k cycles of B. This choice can be achieved in $\binom{j}{k}$ ways, where $0 \leq k \leq j$.

2) The choice of an orientation for each fusion, there being 2^k possibilities.

3) For each of the k cycles of G, the choice of fusion partner in B. There are $(\ell)(\ell - 1) \cdots (\ell - (k-1)) = k!\binom{\ell}{k}$ such choices.

This gives us the formula:

$$\sum_{k=0}^{j} 2^k k! \binom{j}{k} \binom{\ell}{k}$$

The sum indeed counts the three decompositions of Figure 5.8, since $j = \ell = 1$ in this example. Unfortunately, this number grows very rapidly, which excludes the possibility of examining each optimal scenario in a reasonable amount of time, even for small values of j. For example when $j = 16$, which is a plausible value knowing that j is of the same order of magnitude as the number of chromosomes, the value of this sum is strictly greater than $\sum_{k=0}^{16} 2^k k! = 1\ 415\ 527\ 220\ 320\ 869\ 563$.

The good news is that the number of *different* cycles that can appear in a scenario remains limited. Indeed, the number of different cycles obtained from the fusion of a cycle $g \in G$ with a cycle $b \in B$ is $2j\ell$, and the number of different cycles obtained by augmenting a cycle b or g with a dashed edge is $j + \ell$. For $j = \ell = 16$, we have 480 cycles, which remains very reasonable.

We will see in section 5.4 that computations on all possible minimum length scenarios depend only on the number of different cycles, which makes them efficient enough to be applicable in practice.

5.4. Chromosomal contact data and weighted scenarios

So far we have addressed the problem of rearrangements in a very abstract manner, in the sense that all DCJ operations are equally likely to occur. In reality, chromosomes inhabit a three-dimensional space, and the probability of a rearrangement occurring between two loci can vary greatly, depending on how a chromosome is twisted upon itself or how it co-inhabits the cellular space with another chromosome.

Using again the plumbing analogy, mistakenly connecting the hot water outlet of Mrs. Tremblay from Verdun with the cold water inlet of her upstairs neighbor makes much more sense than connecting Mrs. Tremblay's hot water outlet with the cold water inlet of Mr. Segura from Montpellier.

In this section, we introduce a notion of weight that will allow us to measure the plausibility of an optimal scenario. As in many optimization problems, a greedy strategy that would consist of choosing the best operation at each stage does not guarantee that the resulting scenario is the best overall. In fact, we know that a DCJ operation on a cycle considerably reduces the choice of subsequent operations since all subsequent moves will imperatively have to occur inside one of the two newly created cycles.

We present a polynomial-time algorithm capable of choosing a minimum weight scenario among all possible optimal scenarios.

5.4.1. *A model incorporating chromosomal contacts*

Figure 5.9 illustrates a fictitious three-dimensional configuration of the genomes $A = (\circ\ \text{-}1\ 4\ 5\ \text{-}3\ \circ)\ (\circ\ \text{-}8\ 6\ 9\ 7\ \text{-}2\ \circ)$ and $B = (\circ\ 1\ 2\ 3\ 4\ 5\ \circ)\ (\circ\ 6\ 7\ 8\ 9\ \circ)$ from Figure 5.3. The adjacencies are represented in this figure by dashed chromosome segments. Note that some pairs of adjacencies are much closer to one another than others.

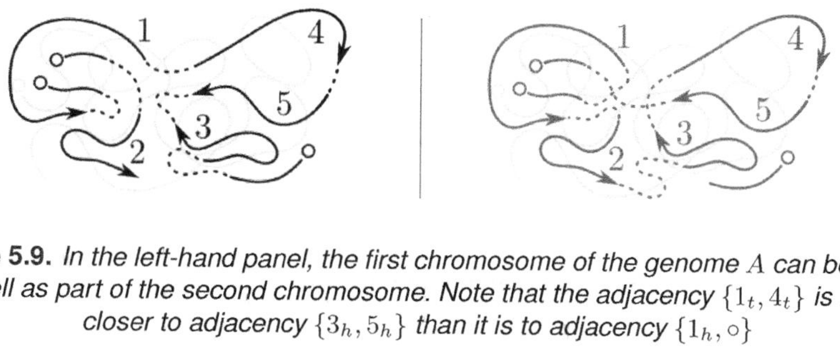

Figure 5.9. *In the left-hand panel, the first chromosome of the genome A can be seen as well as part of the second chromosome. Note that the adjacency $\{1_t, 4_t\}$ is much closer to adjacency $\{3_h, 5_h\}$ than it is to adjacency $\{1_h, \circ\}$*

One approach for measuring the relative proximity of DNA segments is based on the *number of contacts* observed between the segments in a collection of cells. The high-throughput version of this laboratory experiment is called Hi-C (Lieberman-Aiden et al. 2009). These data are then transformed into real values, under the assumption that the physical distance between two segments is inversely proportional to their number of contacts.

To integrate this information into our model, we associate each adjacency $\{u, v\}$ with the corresponding DNA segment x. A DCJ operation will be modeled, as in previous sections, by adding information regarding the segments involved. For example, the operation $\{u, v\}, \{r, s\} \rightarrow \{u, r\}, \{v, s\}$ will become:

$$(\{u, v\}, x), (\{r, s\}, y) \rightarrow (\{u, r\}, x), (\{v, s\}, y)$$

$$\text{or} \quad (\{u, v\}, x), (\{r, s\}, y) \rightarrow (\{u, r\}, y), (\{v, s\}, x)$$

depending on whether the new adjacencies inherit from segment x or segment y. Figure 5.10 gives an example of such an operation.

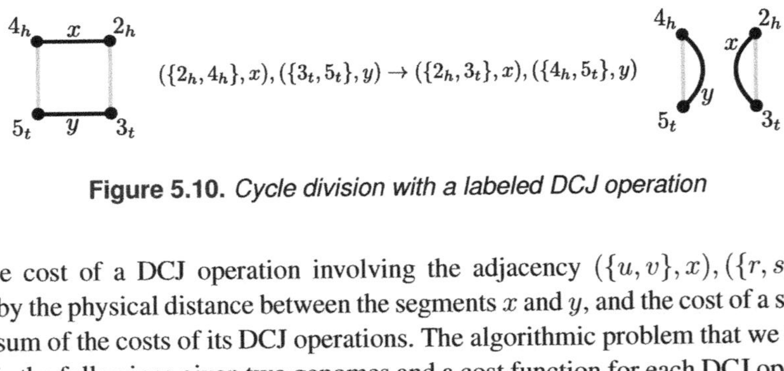

Figure 5.10. *Cycle division with a labeled DCJ operation*

The cost of a DCJ operation involving the adjacency $(\{u, v\}, x), (\{r, s\}, y)$ is given by the physical distance between the segments x and y, and the cost of a scenario is the sum of the costs of its DCJ operations. The algorithmic problem that we want to solve is the following: given two genomes and a cost function for each DCJ operation, compute the minimum cost of a scenario among all optimal scenarios.

Suppose that we know the minimum cost solution for each possible cycle in the various decompositions of the breakpoint graph (of which there are $c+i+2j\ell+j+\ell$). A minimum cost scenario can then be found with the following algorithm:

1) compute the costs of the c complete cycles and i incomplete cycles;

2) compute the costs of the $2j\ell + j + \ell$ possible merged cycles;

3) find a valid cycle merging with minimum cost;

4) compute the total cost of the retained cycles.

The problem of finding a valid merging of gray and black cycles turns out to be a classical graph theory problem: *"find a minimum cost perfect matching on a bipartite graph"*. Martello (2010) retraces the history of this "assignment problem", which goes back to the 19th century. Several polynomial time algorithms have since been proposed, including those based on the *Hungarian method* (Kuhn 1955).

We present a sketch of the bipartite graph construction, whose vertices are said to be on the *left* or the *right*. Those on the left are the j cycles of G, as well as the ℓ gray dashed cycles; the vertices on the right are the ℓ cycles of B, as well as the j black dashed cycles. Let $g \in G$, g_p be a gray dashed cycle, $b \in B$ and b_p a black dashed cycle. The costs associated with the different edges connecting these vertices are as follows:

– $c(g, b_p)$ and $c(g_p, b)$ are the costs associated with cycles augmented by dashed edges (as shown in Figure 5.7);

– $c(g, b)$ is the minimum cost of the two possible merges of g and b (as shown in Figure 5.8);

– $c(g_p, b_p) = 0$ since these cycles contain no genes.

A *perfect* matching is a set of edges that joins each vertex on the left to exactly one vertex on the right. There always exists a perfect matching in our construction, since it is a complete bipartite graph with $\ell + j$ vertices on each side. A solution to the minimum cost perfect matching problem implies a way to merge the unbalanced cycles, so that the resulting set of balanced cycles has a minimum cost.

In the next section, we solve the problem of calculating the minimum cost for each balanced cycle.

5.4.2. *Planar trees and an algorithm for exploring them*

As we have seen, all optimal scenarios solving a cycle of length $2m$ have $m - 1$ DCJ operations. What differentiates them in terms of cost is the set of the $m - 1$ pairs of labels involved in each operation, the cost of a cycle being the sum of the costs of these pairs.

In this section, we establish a bijection between the optimal labeled scenarios for solving a cycle of length $2m$ and the *planar trees* with $m - 1$ branches, representing a set of label pairs compatible with a scenario of DCJ operations (Simonaitis et al. 2019). We will then show how to build a minimum cost tree.

We start by recalling some properties of trees and planar trees.

5.4.3. *Planar trees*

A *tree* is a connected acyclic graph. A binary tree with m vertices has $m - 1$ edges, called *branches*. A tree with at least one branch always has at least two vertices of degree 1, which we call *leaves*.

Consider a tree with vertices ordered by the integers on the interval $[0..m-1]$. We say that a tree is *planar* if its branches do not intersect when the vertices appear in increasing order, on equidistant points on a circle. Note that the branches of a planar tree are chords of the circle, and can thereby be associated with, and ordered by, their *lengths*. Figure 5.11 shows a planar tree with eight vertices. The shortest branches of this tree are $\{1,2\}$, $\{3,4\}$ and $\{6,7\}$. The longest ones are $\{0,3\}$ and $\{0,5\}$.

An interval of consecutive vertices on the m-vertex circle will be denoted by $[i..j]$. When speaking of a circle, this notation must be interpreted as the set of vertices going from i to j by traveling the circle in a clockwise fashion. For example, in Figure 5.11, the interval $[1..3]$ designates the set $\{1,2,3\}$, whereas the interval $[3..1]$ denotes the set $\{3,4,5,6,7,0,1\}$. Each branch $\{i,j\}$ of a planar tree P determines two subtrees of P, denoted by $P[i..j]$ and $P[j..i]$, which contain the vertices i and j as well as the vertices between i and j on one side of the branch or the other.

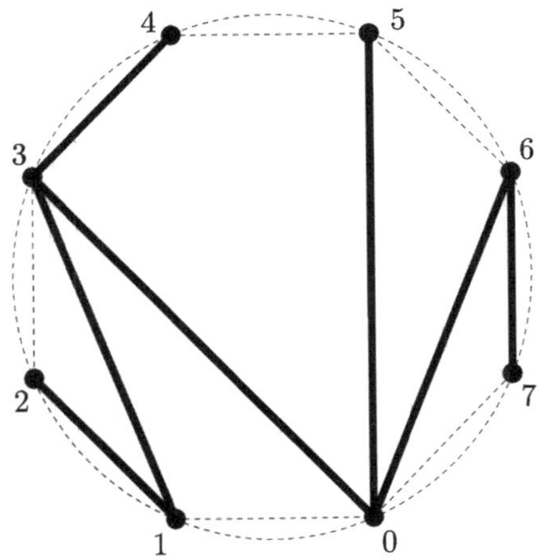

Figure 5.11. *Planar tree with eight vertices inscribed on a circle*

Finally, the following result will be necessary.

PROPOSITION 5.2.– In a planar tree, there is at least one leaf adjacent to one of its neighbors on the circle.

PROOF.– Let i be a leaf of the tree P and j the unique vertex adjacent to i such that the branch $\{i, j\}$ is of minimum length. Since $\{i, j\}$ is a branch, $P[i..j]$ and $P[j..i]$ will both be trees. If one of these trees contains only one branch, then the leaf i is adjacent to one of its two neighbors on the circle.

Otherwise, assume that the number of vertices in $P[i..j]$ is less than or equal to that in $P[j..i]$. Note that all branches of tree $P[i..j]$ are shorter than the branch $\{i, j\}$. Since $P[i..j]$ has at least two leaves, one of them belongs to the tree P, which contradicts the hypothesis that the branch $\{i, j\}$ is of minimum length. □

5.4.3.1. *A bijection between labeled scenarios and planar trees*

Our objective is to find an optimal scenario with minimum cost for a balanced cycle of length $2m$. As the cost of a scenario is equal to the sum of the costs of the label pairs involved in a DCJ operation, we consider as equivalent two scenarios that involve the same set of label pairs. In this section, we give a bijection between the sets of label pairs and the planar trees.

Consider a balanced cycle of length $2m$, labeled by DNA segments $(x_0, x_1, \ldots, x_{m-1})$. The choice of the name x_0 is arbitrary, but the others must respect the order induced by the cycle.

Given an optimal labeled scenario of length $m - 1$, we associate it to a planar tree on m vertices in the following way: if a DCJ operation involves the pair of labels $\{x_i, x_j\}$, the branch $\{i, j\}$ is in the tree. This tree only depends on the set of label pairs, and it is planar because two intersecting branches would imply that one of the scenario's DCJ operations would have acted on two different cycles.

Conversely, given a planar tree with $m - 1$ edges, a scenario of that length can be recursively constructed. If a leaf i is adjacent to one of its neighbors $j = (i + 1) \mod m$ or $j = (i - 1) \mod m$, a labeled DCJ operation can always be performed to create a new cycle of length 2 with label x_i. The remaining cycle is of length $2(m - 1)$, and none of the DCJ operations required to solve it implicates the label x_i, since i was a leaf. Proposition 5.2 guarantees the existence of such a leaf.

5.4.3.2. *Constructing a minimum cost tree*

Chromosome contact data supplies us with a cost $c(x_i, x_j)$ that we can apply to a DCJ operation involving labels x_i and x_j. Similarly, the same cost $c(i, j)$ can be associated with a branch $\{i, j\}$ of a planar tree. The fictitious genome in Figure 5.9 shows relative distances between the pairs of adjacencies corresponding to the cycle in the left-hand panel in Figure 5.8.

In the previous section, we reduced the problem of finding the minimum cost for a cycle of length $2m$ to that of finding a planar tree P of minimum cost c_{min} in the set

$\mathcal{P}_{[0..m-1]}$ of all planar trees on vertices in the interval $[0..m-1]$. In general, if $[k..\ell]$ is a sub-interval of $[0..m-1]$, we will denote by $c_{min}(\mathcal{P}_{[k..\ell]})$ the minimum cost of a planar tree whose vertices are $[k..\ell]$ in the circle of size m.

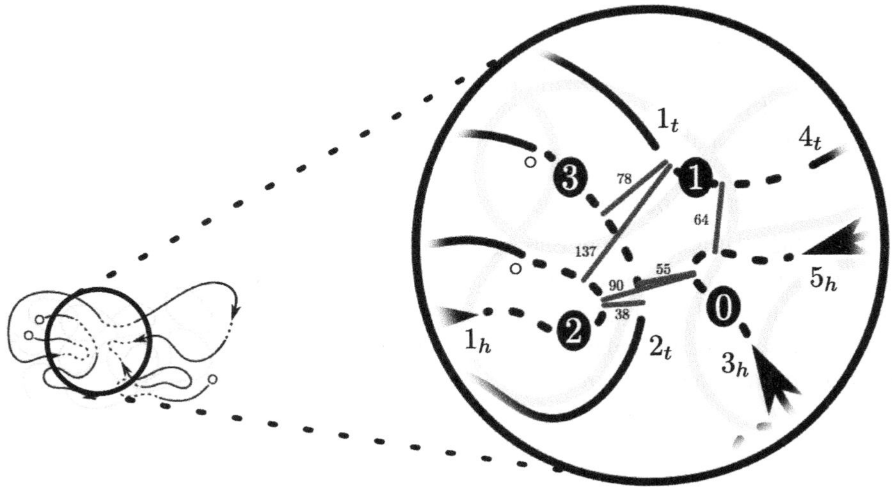

Figure 5.12. *A close-up of the adjacencies of the fictitious genomes. Notice, for example, that the distance between the adjacency $\{3_h, 5_h\}$, labeled by $i = 0$, and the adjacency $\{1_t, 4_t\}$, labeled by $j = 1$, is equal to 64. Thus, we set the cost $c(0, 1) = 64$. For a color version of this figure, www.iste.co.uk/didier/models.zip*

Let $\{i, j\}$ be a branch of a planar tree P. The cost $c(P)$ can be calculated by means of the following formula, which will be useful for the construction of a recurrence relation:

$$c(P) = c(P[i..j]) + c(P[j..i]) - c(i, j)$$

By splitting the circle in two along the branch $\{i, j\}$, the computation of the cost of a tree whose set of vertices is $[0..m-1]$ can be reduced to the cost of two trees on smaller circles. These two trees are elements of the sets $\mathcal{P}_{[i..j]}$ and $\mathcal{P}_{[j..i]}$, both of which contain the branch $\{i, j\}$. The problem is thus reduced to computing the minimum cost of a tree P in $\mathcal{P}_{[i..j]}$ assuming that the branch $\{i, j\}$ belongs to P, which we will denote by $c_{min}(\mathcal{P}_{[i..j]}|\{i, j\})$.

By testing each of the possible chords $\{i, j\}$ of the circle whose vertices are $[k..\ell]$, the following equation can be deduced:

$$c_{min}(\mathcal{P}_{\mathcal{I}=[k..\ell]}) = \min_{i \neq j \in \mathcal{I}} \{c_{min}(\mathcal{P}_{[i..j] \cap \mathcal{I}}|\{i, j\}) + c_{min}(\mathcal{P}_{[j..i] \cap \mathcal{I}}|\{i, j\}) - c(i, j)\}$$

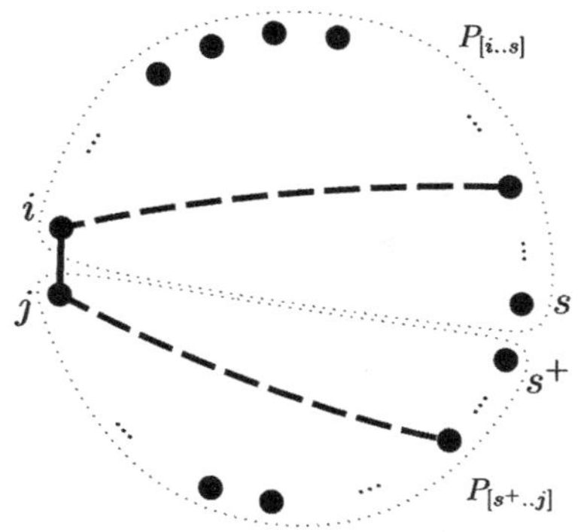

Figure 5.13. *Decomposition of a tree in $P[i..j]$, containing the branch $\{i,j\}$, into two sub-trees $P[i..s]$ and $P[s^+..j]$. One of the two subtrees could potentially be empty*

Since the vertices i and j are neighbors in $\mathcal{P}_{[i..j]\cap[k..\ell]}$, a planar tree containing this branch can be decomposed into two subtrees $P[i..s]$ and $P[s^+..j]$, where s^+ is the vertex after s in the circle $[k..\ell]$ (s^+ is $s+1$, unless $s=\ell$). Thus, we calculate the minimum cost over all values of s:

$$c_{min}(\mathcal{P}_{[i..j]\cap\mathcal{I}}|\{i,j\}) = \min_{s\in[i..j-1]\cap\mathcal{I}} \{c(i,j) + c_{min}(\mathcal{P}_{[i..s]}) + c_{min}(\mathcal{P}_{[s^+..j]})\}$$

knowing that $c_{min}(\mathcal{P}_{[i..i]}) = 0$.

An efficient way to calculate $c_{min}(\mathcal{P}_{[0..m-1]})$ is to calculate the values of $c_{min}(\mathcal{P}_{[i..j]\cap[k..\ell]}|\{i,j\})$ and $c_{min}(\mathcal{P}_{[i..j]})$ for all the chords $\{i,j\}$ of the circle, from shortest to longest. We show these elementary calculations for part of the cycle $(5_h, 3_h, 4_t, 1_t, \circ, 1_h, 2_t, \circ)$ using the cost matrix $c(i,j)$ given by Figure 5.12. The value of $c_{min}(\mathcal{P}_{\mathcal{I}=[0..3]})$ is given by:

$$c_{min}(\mathcal{P}_{\mathcal{I}=[0..3]}) = \min \begin{cases} c_{min}(\mathcal{P}_{[0..1]\cap\mathcal{I}}|\{0,1\}) + c_{min}(\mathcal{P}_{[1..0]\cap\mathcal{I}}|\{0,1\}) - c(0,1) \\ c_{min}(\mathcal{P}_{[0..2]\cap\mathcal{I}}|\{0,2\}) + c_{min}(\mathcal{P}_{[2..0]\cap\mathcal{I}}|\{0,2\}) - c(0,2) \\ c_{min}(\mathcal{P}_{[0..3]\cap\mathcal{I}}|\{0,3\}) + c_{min}(\mathcal{P}_{[3..0]\cap\mathcal{I}}|\{0,3\}) - c(0,3)* \\ c_{min}(\mathcal{P}_{[1..2]\cap\mathcal{I}}|\{1,2\}) + c_{min}(\mathcal{P}_{[2..1]\cap\mathcal{I}}|\{1,2\}) - c(1,2) \\ c_{min}(\mathcal{P}_{[1..3]\cap\mathcal{I}}|\{1,3\}) + c_{min}(\mathcal{P}_{[3..1]\cap\mathcal{I}}|\{1,3\}) - c(1,3) \\ c_{min}(\mathcal{P}_{[2..3]\cap\mathcal{I}}|\{2,3\}) + c_{min}(\mathcal{P}_{[3..2]\cap\mathcal{I}}|\{2,3\}) - c(2,3) \end{cases}$$

Since vertices 0 and 3 are neighbors, $c_{min}(\mathcal{P}_{[3..0]\cap\mathcal{I}}|\{0,3\}) = c(0,3)$, the starred expression is thus equal to $c_{min}(\mathcal{P}_{[0..3]\cap[0..3]}|\{0,3\})$, and as such to:

$$c_{min}(\mathcal{P}_{[0..3]\cap[0..3]}|\{0,3\}) = \min \begin{cases} c(0,3)+c_{min}(\mathcal{P}_{[0..0]})+c_{min}(\mathcal{P}_{[1..3]}) \\ c(0,3)+c_{min}(\mathcal{P}_{[0..1]})+c_{min}(\mathcal{P}_{[2..3]})* \\ c(0,3)+c_{min}(\mathcal{P}_{[0..2]})+c_{min}(\mathcal{P}_{[3..3]}) \end{cases}$$

Once again, the starred expression is easily calculated, giving the minimum cost $c(0,3) + c(0,1) + c(2,3) = 55 + 64 + 38 = 157$. The corresponding planar tree, containing the branches $\{0, 3\}$, $\{0, 1\}$ and $\{2, 3\}$, is depicted in Figure 5.14, along with the corresponding minimum weight scenario.

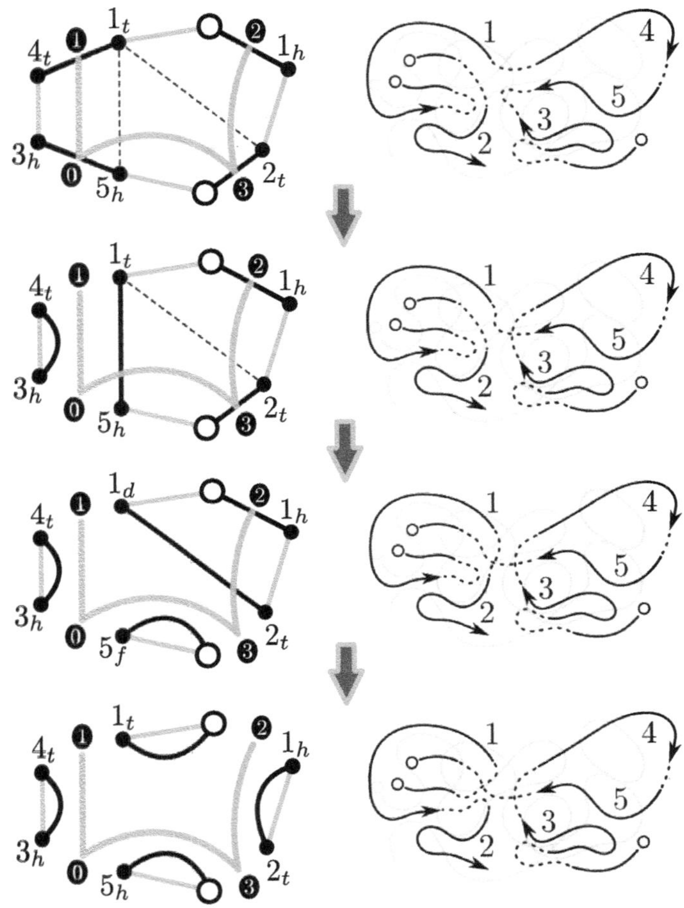

Figure 5.14. *The minimum weight scenario for the cycle* $(5_h, 3_h, 4_t, 1_t, \circ, 1_h, 2_t, \circ)$. *The associated planar tree is composed of edges* $\{0, 1\}$, $\{0, 3\}$ *and* $\{2, 3\}$. *The right-hand column shows the effect of DCJ operations on genomes* A *and* B

5.5. Conclusion

This chapter has covered various techniques for the comparison of the positions of the common genes shared by two species, in both their linear and three-dimensional organization. The comparison of two genomes provides information such as the minimum number of DCJ operations that separate two species, but methods to discover the nature and order of these operations remain mostly underdeveloped. To answer such questions, we may need to simultaneously consider large sets of genomes, or we may need to introduce more biological information.

For most of the 20th century, data on gene orders was virtually non-existent. Morgan's group was able to experimentally identify possible relationships between gene organizations of a few species of *drosophila* fruit-flies from the west coast of North America. Since the number of rearrangements was small, they were able to reconstruct the phylogenetic relationships between these species using elementary techniques (Sturtevant and Dobzhansky 1936).

One of the first systematic studies involving multiple species, published in the PNAS journal in 1992, compared the order of genes in 16 insect mitochondria (Sankoff et al. 1992). The study served as a catalyst for the computer science community, and the formal framework that was introduced precipitated a race to the first efficient algorithm solving the inversion distance problem between two genomes (Hannenhalli and Pevzner 1995).

Pair-wise distance calculations were subsequently used as an alternative phylogenetic signal, providing input to fast techniques for inferring trees (see Chapter 6). Since then, genomic rearrangement models have multiplied and been refined over the last 30 years: in addition to inversions, a wide range of biological operations have been defined to fit the different branches of the tree of life. We advise interested readers to refer to Fertin et al. (2009) for an in-depth study on the subject.

Today genome sequencing has almost become routine, and biological data accumulate much faster than the analysis tools are developed. There is still significant work to be done so that the techniques developed in this chapter can be applied on a large scale. One hurdle is the fact that genomes are often reconstructed from short sequences using a close species as reference, which tends to hide certain recent rearrangements. Also, we have assumed throughout this chapter that the problem of deciding whether two genes are homologous is solved, and that genomes do not contain duplicated or missing genes. Unfortunately, these assumptions are rarely met in the biological world, requiring further approximations or methodological development.

The tools necessary for tackling these problems are both technological and computational. Genome sequencing is becoming more and more efficient, and the difficult step of assembling very short sequences will gradually be replaced by the direct reading of much longer ones (Wenger et al. 2019). The identification of homologous genes is being informed by more than just the simple notion of sequence similarity, such as the genomic neighborhood of a gene or its individual genealogy with respect to that of its species (Chauve et al. 2013). When the number of duplicated genes is modest, linear programming techniques make it possible to master the combinatorial explosion of necessary for the computations (Shao et al. 2015; Simonaitis et al. 2019).

The long-term goal is to understand when, how and why rearrangements have contributed to the biological diversity that can be observed. We have at least another century to achieve it.

5.6. References

Baryshnikov, Y. (2001). On Stokes sets. In *New Developments in Singularity Theory*, Siersma, D., Wall, C.T.C., Zakalyukin, V. (eds). Kluwer Academic Publishers, Dordrecht.

Bergeron, A. and Stoye, J. (2013). *The Genesis of the DCJ Formula*. Springer, London.

Castle, W.E. (1919). Is the arrangement of the genes in the chromosome linear? *Proceedings of the National Academy of Sciences*, 5(2), 25–32.

Chauve, C., El-Mabrouk, N., Guéguen, L., Semeria, M., Tannier, E. (2013). Duplication, rearrangement and reconciliation: A follow-up 13 years later. In *Models and Algorithms for Genome Evolution*, Chauve, C., El-Mabrouk, N., Tannier, E. (eds). Springer, London.

Crick, F.C., Watson, J.D., Bragg, W.L. (1954). The complementary structure of deoxyribonucleic acid. *Proceedings of the Royal Society of London. Series A: Mathematical and Physical Sciences*, 223(1152), 80–96.

De Chiara, M., Friedrich, A., Barré, B., Breitenbach, M., Schacherer, J., Liti, G. (2020). Discordant evolution of mitochondrial and nuclear yeast genomes at population level. *BMC Biology*, 18(1), 49.

Fertin, G., Labarre, A., Rusu, I., Vialette, S., Tannier, E. (2009). *Combinatorics of Genome Rearrangements*. MIT Press, Cambridge.

Hannenhalli, S. and Pevzner, P.A. (1995). Transforming men into mice (polynomial algorithm for genomic distance problem). In *36th Annual Symposium on Foundations of Computer Science*. Milwaukee, WI, 23–25 October [Online]. Available at: https://doi.org/10.1109/SFCS.1995.492588.

Ijdo, J.W., Baldini, A., Ward, D.C., Reeders, S.T., Wells, R.A. (1991). Origin of human chromosome 2: An ancestral telomere-telomere fusion. *Proceedings of the National Academy of Sciences*, 88(20), 9051–9055.

Kuhn, H.W. (1955). The Hungarian method for the assignment problem. *Naval Research Logistics Quarterly*, 2(1/2), 83–97.

Lieberman-Aiden, E., Van Berkum, N.L., Williams, L., Imakaev, M., Ragoczy, T., Telling, A., Amit, I., Lajoie, B.R., Sabo, P.J., Dorschner, M.O. et al. (2009). Comprehensive mapping of long-range interactions reveals folding principles of the human genome. *Science*, 326(5950), 289–293.

Martello, S. (2010). Jenő Egerváry: From the origins of the Hungarian algorithm to satellite communication. *Central European Journal of Operations Research*, 18(1), 47–58.

Morgan, T.H., Sturtevant, A.H., Bridge, C.B. (1920). The evidence for the linear order of the genes. *Proceedings of the National Academy of Sciences*, 6(4), 162–164.

Nikiforova, M.N., Stringer, J.R., Blough, R., Medvedovic, M., Fagin, J.A., Nikiforov, Y.E. (2000). Proximity of chromosomal loci that participate in radiation-induced rearrangements in human cells. *Science*, 290(5489), 138–141.

Pearson, H. (2006). What is a gene? *Nature*, 441(7092), 398–401.

Sankoff, D., Leduc, G., Antoine, N., Paquin, B., Lang, B.F., Cedergren, R. (1992). Gene order comparisons for phylogenetic inference: Evolution of the mitochondrial genome. *Proceedings of the National Academy of Sciences*, 89(14), 6575–6579.

Shao, M., Lin, Y., Moret, B.M. (2015). An exact algorithm to compute the double-cut-and-join distance for genomes with duplicate genes. *Journal of Computational Biology*, 22(5), 425–435.

Simonaitis, P., Chateau, A., Swenson, K. (2019). A general framework for genome rearrangement with biological constraints. *Algorithms for Molecular Biology*, 14(15).

Sturtevant, A. and Dobzhansky, T. (1936). Inversions in the third chromosome of wild races of drosophila pseudoobscura and their use in the study of the history of the species. *Proceedings of the National Academy of Sciences*, 22, 448–450.

Swenson, K.M. and Blanchette, M. (2019). Large-scale mammalian genome rearrangements coincide with chromatin interactions. *Bioinformatics*, 35(14), i117–i126.

Véron, A.S., Lemaitre, C., Gautier, C., Lacroix, V., Sagot, M.-F. (2011). Close 3D proximity of evolutionary breakpoints argues for the notion of spatial synteny. *BMC Genomics*, 12, 303 [Online]. Available at: http://dx.doi.org/10.1186/1471-2164-12-303.

Wenger, A.M., Peluso, P., Rowell, W.J., Chang, P.-C., Hall, R.J., Concepcion, G.T., Ebler, J., Fungtammasan, A., Kolesnikov, A., Olson, N.D. et al. (2019). Accurate circular consensus long-read sequencing improves variant detection and assembly of a human genome. *Nature Biotechnology*, 37(10), 1155–1162.

Yaffe, E., Farkash-Amar, S., Polten, A., Yakhini, Z., Tanay, A., Simon, I. (2010). Comparative analysis of DNA replication timing reveals conserved large-scale chromosomal architecture. *PLOS Genetics*, 6(7), e1001011.

Yancopoulos, S., Attie, O., Friedberg, R. (2005). Efficient sorting of genomic permutations by translocation, inversion and block interchange. *Bioinformatics*, 21(16), 3340–3346.

6

Phylogenetic Inference: Distance-Based Methods

Fabio PARDI

LIRMM, CNRS, Université de Montpellier, France

When evolutionary distances between pairs of taxa can be estimated, they can be used to rapidly infer a phylogenetic tree. In this chapter, we first explain how distances should be defined and estimated, and then focus on the task of constructing a phylogenetic tree that corresponds to those distances. More specifically, we highlight the properties that distances must satisfy in principle, and the classic approaches to tree inference, *least squares* and *minimum evolution*. Finally, we concentrate on the most commonly known methods, in particular *neighbor joining* (NJ) and a wide range of algorithms inspired therefrom.

6.1. Introduction

Distance-based phylogenetic inference is motivated by a simple but key observation, illustrated in Figure 6.1. Let us assume that we are able to know the precise number of changes that occurred on the evolutionary path between any two taxa i and j in a collection of n taxa ($1 \leq i \leq j \leq n$). These "changes" can be defined in many ways, but the reasoning can be applied to anything that can be counted, such as the number of substitutions that occurred in a gene or the number of times there was a change in gene order between the genome of i and j. It turns out that, as long as these events are frequent enough, this information is all we need to reconstruct the topology of the unrooted tree describing the evolution of our n taxa.

Models and Methods for Biological Evolution,
coordinated by Gilles DIDIER and Stéphane GUINDON.

This notion (formalized in a classic and well-known result that we will later state in theorem 6.2) constitutes the theoretical foundation of distance-based approaches to phylogenetic inference. In practice, unfortunately, the number of changes that have occurred between taxa, their *evolutionary distance*, cannot be observed, but can only be estimated. It is then based on the *estimated* distances that tree reconstruction is carried out, the goal being to find the tree that best accounts for the estimated distances, sometimes including the uncertainty in these estimates.

The use of distances or similarities between taxa is one of the oldest approaches in systematic biology (Sneath 1957; Sokal and Michener 1958). In early studies, a very clear distinction was made between, on the one hand, hierarchical clustering techniques for taxonomic classification (Sokal and Sneath 1963), and, on the other hand, optimization methods aimed directly at phylogenetic inference (Cavalli-Sforza and Edwards 1967; Fitch and Margoliash 1967). These two lines of work finally converged in the late 1980s, leading to the very popular NJ method, (Saitou and Nei 1987) which combined algorithmic ideas from classification (for instance, UPGMA (Sokal and Michener 1958); ADDTREE (Sattath and Tversky 1977)) to phylogenetic optimization principles (for instance, ME (Kidd and Sgaramella-Zonta 1971); OLS (Cavalli-Sforza and Edwards 1967) see below). Since NJ, phylogenetics has seen a revival of distance-based methods, often related to or inspired by NJ. We will cover several of them in section 6.4.3.

These methods are widely used for their computational efficiency, an advantage that makes them particularly suitable to inferring very large phylogenies or large numbers of phylogenies (for bootstrapping, for example). They are also used to provide a guiding tree for progressive alignment (Larkin et al. 2007) or a initial tree for more sophisticated inference approaches such as those based on maximum likelihood (ML) (Guindon and Gascuel 2003). More generally, the use of estimated distances between biological sequences is an obvious way to cope with the huge datasets generated by modern, ever faster and cheaper sequencing techniques. This is evidenced by the continued success of NJ, which to this day remains the most cited algorithm in phylogenetics (about 3,000 citations per year since 2010 for the original article by Saitou and Nei (1987)).

In this chapter, we present the major ideas underlying distance-based methodology. We begin by explaining the importance of estimating distances that reflect the number of changes that actually occurred between two taxa i and j, rather than the most parsimonious number of changes that are necessary to explain the differences between i and j. After a concise description of the task of estimating evolutionary distances, the focus here will be on the second crucial step: inferring the tree that corresponds to the estimated distances.

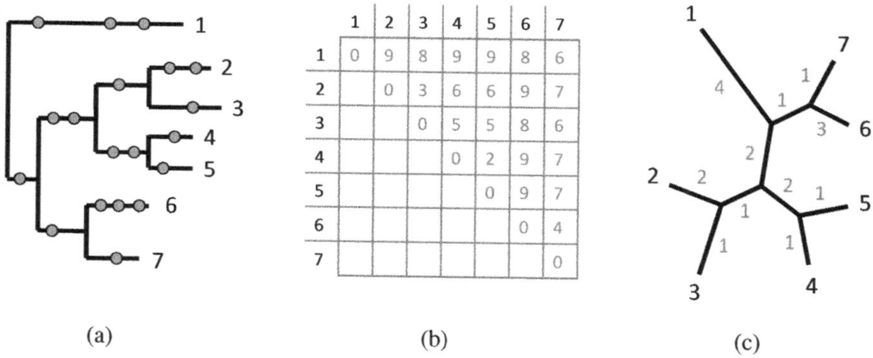

	1	2	3	4	5	6	7
1	0	9	8	9	9	8	6
2		0	3	6	6	9	7
3			0	5	5	8	6
4				0	2	9	7
5					0	9	7
6						0	4
7							0

Figure 6.1. *The fundamental idea of distance-based tree reconstruction. A number of changes (gray circles) occur along the branches of an unknown phylogeny (a). Suppose that we have a matrix reporting the number of changes between each pair of taxa (b). Then, as long as at least one change has occurred on each internal branch, we are able to accurately reconstruct the topology of the unrooted tree underlying the true phylogeny, including the number of changes for each branch (c)*

6.2. Mathematical basis

In order for us to expose the classical results on the use of distances for phylogenetic inference, a few basic concepts have to be defined. Throughout this chapter, we identify the taxa of interest with indices $1, 2, \ldots, n$. These "taxa" can be anything that is the end product of a tree-shaped evolutionary process, from genetic sequences to cooking recipes passed down from generation to generation.

A *dissimilarity* on taxa $1, 2, \ldots, n$ is a collection (a matrix, a vector, a function, it does not matter) $\mathrm{d} = (d_{ij})$ such that $d_{ij} \geq 0$, $d_{ij} = d_{ji}$ and $d_{ii} = 0$ for all $i, j \in \{1, 2, \ldots n\}$. A *distance metric* is a dissimilarity for which we also have $d_{ij} \leq d_{ik} + d_{kj}$[1]. We note that in phylogenetics the word "distance" is often used to designate a dissimilarity, for example, when speaking of "distance matrix". The possible confusion is easily avoided by realizing that what is usually meant by "distance" is an *estimated* distance (in fact something that is close to a tree distance, defined below).

A *phylogenetic tree* T on $1, 2, \ldots, n$ is composed of (1) a tree topology τ (an unrooted tree with no degree-2 nodes), whose leaves are uniquely labeled with $1, 2, \ldots, n$; and (2) the lengths $\lambda_e \geq 0$ assigned to each branch e of τ. The branch

1 The standard distance definitions also require that $d_{ij} = 0$ implies $i = j$, but we will not make that assumption here.

lengths are used to model evolutionary change along each branch (in section 6.3.1, we will see that for molecular sequences, λ_e is generally defined as the mathematical expectation of the number of substitutions per site occurring on e).

Because its branches have lengths, a phylogenetic tree T naturally determines a distance metric $\mathbf{d}^T = (d_{ij}^T)$ on its taxa $1, 2, \ldots, n$, where d_{ij}^T is simply defined as the length of the path connecting i and j in T. In other words, if τ_{ij} represents the set of branches between i and j in τ:

$$d_{ij}^T = \sum_{e \in \tau_{ij}} \lambda_e \qquad [6.1]$$

When $\mathbf{d} = \mathbf{d}^T$ for a phylogenetic tree T, it is said that $\mathbf{d}$ is a *tree distance* (on T). Tree distances are sometimes called *patristic* distances (because they follow the ancestral path between i and j) or still *additive* distances (which refers to the sum in [6.1]).

Not all dissimilarities are tree distances. An interesting observation is that if we take any four taxa i, j, h, k and consider the sums:

$$d_{ij} + d_{hk}, \quad d_{ih} + d_{jk}, \quad d_{ik} + d_{jh} \qquad [6.2]$$

then, when $\mathbf{d}$ is a tree distance, two of these sums must be equal and the third sum must be less than or equal to the other two sums. This can be understood by considering a practical example: if we take the taxa $1, 2, 3$ and 6 in the tree of Figure 6.1c, we then have $d_{12} + d_{36} = d_{13} + d_{26} = 17$, which is greater than $d_{16} + d_{23} = 11$. The first two sums are equal because they are sums on the same set of branches, while $d_{16} + d_{23}$ is only on a subset of these branches. The relations between the sums in equation [6.2] can be succinctly expressed in what is known as the *four-point condition*: for all i, j, h, k:

$$d_{ij} + d_{hk} \leq \max\{d_{ih} + d_{jk}, d_{ik} + d_{jh}\}$$

Surprisingly, the four-point condition is not only necessary, but also sufficient for $\mathbf{d}$ to be a tree distance:

THEOREM 6.1.– $\mathbf{d}$ *is a tree distance if and only if* $\mathbf{d}$ *satisfies the four-point condition.*

The study of tree distances is an old subject and is one of the foundations of mathematical phylogenetics (Zaretskii 1965; Pereira 1969; Buneman 1971). Perhaps the most important results for distance-based phylogenetic inference is the following.

THEOREM 6.2.– *Let T be a phylogenetic tree with strictly positive branch lengths. Then T is the only phylogenetic tree with strictly positive branch lengths having* $\mathbf{d}^T$ *as the tree distance.*

There cannot be another phylogenetic tree T' with a different topology than that of T (i.e. not isomorphic to it), and having strictly positive and different branch lengths from those of T, such that $\mathbf{d}^T = \mathbf{d}^{T'}$.

To prove this theorem, the main idea is that the four-point condition for a quartet $\{i, j, h, k\}$ induces a tree topology on this quartet and also imposes some constraints on a subset of branch lengths. For example, in Figure 6.1, the fact that $d_{16} + d_{23} < d_{12} + d_{36} = d_{13} + d_{26}$ implies that the path between 1 and 6 does not share any branches with the path between 2 and 3, and this determines the topology of the subtree linking taxa 1, 2, 3 and 6. On the other hand, the path between 1 and 2 and the path between 3 and 6 have branches in common, and the total length of these shared branches is determined by the expression $1/2 \cdot (d_{12} + d_{36} - d_{16} - d_{23})$ (equal to 3 in Figure 6.1). Collectively, this information for all quartets $\{i, j, h, k\}$ is largely sufficient to determine T, including its branch lengths. We refer the reader to Barthélémy and Guénoche (1991) or Semple and Steel (2003) for a full proof of this result.

The first natural question following theorem 6.2 is as follows: Are all the elements of $\mathbf{d}^T$ necessary for determining T, or does this uniqueness result also hold when only some of the distances are known? It turns out that much of the content of $\mathbf{d}^T$ is in effect redundant. For example, for a fully resolved (namely binary) phylogenetic tree T with strictly positive branch lengths, the knowledge of a carefully chosen subset of $2n - 3$ distances from $\mathbf{d}^T$ is sufficient to uniquely determine T (again topology *and* branch lengths) (Dress et al. 2012). The intuition behind this idea (for branch lengths) will appear later in Figure 6.3. See also (Huber and Kettleborough 2015; Kettleborough et al. 2015) for recent studies on this topic.

However, the question that interests us most in this chapter is: how can the unique tree T be found from $\mathbf{d}^T$? In fact, this is an easy problem and many algorithms can solve it efficiently with runtime complexities as low as $O(n \log n)$ (Hein 1989). Although this may seem like very good news for phylogenetic inference, it should be kept in mind that in practice $\mathbf{d}^T$ is not directly observable from biological data. The best we can do is to produce an estimate $\boldsymbol{\delta}$ of $\mathbf{d}^T$. In general, $\boldsymbol{\delta}$ will not be a tree distance (sometimes not even a distance metric) but just a dissimilarity that, at best, comes close to $\mathbf{d}^T$. Designing methods to infer T from an *approximation* of a tree distance is far from being a trivial exercise.

A distance-based phylogenetic inference method can then be described as a function that takes a dissimilarity as input and returns a phylogenetic tree. A good distance-based method is one that is capable of constructing a tree T', which is close to T when the input dissimilarity $\boldsymbol{\delta}$ is close to $\mathbf{d}^T$. For example, a minimal prerequisite for any distance-based method is topological consistency:

DEFINITION 6.1.– *A distance-based tree inference method $\mathcal{M}$ is* topologically consistent *if, for any binary tree T with strictly positive branch lengths, for $\boldsymbol{\delta}$ sufficiently close to d^T, using $\mathcal{M}$ on $\boldsymbol{\delta}$ gives a tree T' with the same topology as T.*

Advanced studies on the theoretical properties of these methods seek to characterize the robustness of inferring the correct tree topology to deviations of $\boldsymbol{\delta}$ from d^T. Following a seminal article by Atteson (1999), interesting results are still obtained on this topic (Mihaescu et al. 2009; Gascuel and Steel 2016). A precise description of this line of work is beyond the scope of this chapter.

We will come back to tree inference itself in section 6.4. Before that, we are giving a brief overview of the most common approaches to producing distance estimates $\boldsymbol{\delta} = (\delta_{ij})$.

6.3. Distance estimation

The first component of a distance-based approach to tree inference is distance estimation. The most important point to remember from the previous sections is that not all dissimilarities can be adopted in phylogenetics. Only those dissimilarities that can be considered as estimates of a tree distance have a rigorous justification.

Ideally, as increasingly more data are collected, the estimated distances will become closer and closer to a tree distance d^T for a tree T that correctly represents the evolutionary history of the taxa under consideration. This, combined with the use of any topologically consistent method (see definition 6.1), guarantees in principle the convergence of the inferred tree to the correct tree topology.

6.3.1. *Estimating distances from aligned sequences*

The best studied method for estimating evolutionary distances is from aligned sequences (nucleotides or proteins). What we need is a collection of n homologous sequences (one for each taxon $1, 2, \ldots, n$) and, for each pair of taxa i and j, a pairwise alignment of the sequences of i and j (we do not need a multiple alignment containing all the sequences).

How can we then compute the distance estimates $\boldsymbol{\delta}$? It should be intuitive that counting the number of differences, or *mismatches*, between every pair of sequences (or even taking a weighted version of such a count) is not a good idea. Although this naive approach does produce a distance metric (also known as the *Hamming distance*) it often results in a distance that significantly deviates from a tree distance. In other words, since what we should be trying to estimate is the number of substitutions that actually occurred between i and j (see again Figure 6.1), the number of mismatches m_{ij} between i and j is in fact a very poor estimate: as several substitutions can occur at the same site, m_{ij} often underestimates the actual number of substitutions (see Figure 6.2 for an example).

```
i GAATACTCAAA
  ||.|.|.|.||
k GACTGCCCGAA
  ||.|||.||.|
j GATTGCTCGGA
```

Figure 6.2. *The number of mismatches underestimates the number of substitutions. Let us assume that the sequence k exists at an intermediate stage on the evolutionary path between i and j. While the number of mismatches m_{ij} between i and j is 4 (letters in bold), the number of substitutions that occurred between i and j is at least $m_{ik} + m_{kj} = 7$*

How can we estimate the number of substitutions that occurred between two aligned sequences? We now describe, in very general terms, the ML approach to solve this estimation problem, where the number of substitutions is commonly normalized by the length of the sequences. Interested readers are referred to other works (Felsenstein 2004; Yang 2006) for a detailed treatment of distance estimation from aligned sequences.

Substitution models (see Chapter 2) determine a matrix of substitution probabilities $\mathbf{P}(\lambda) = (p_{xy}(\lambda))$, where x and y denote nucleotides, amino acids or other biological characters, and $p_{xy}(\lambda)$ denotes the probability that x becomes y after evolving along a branch of length λ. We emphasize that λ is not expressed in units of time. Instead, the rate of substitution models is usually scaled so that λ equals the expected number of substitutions per site along a branch of this length.

Let π_x be the stationary probability of x. This can be defined as the limit of $p_{yx}(\lambda)$ when $\lambda \to \infty$, and it is sometimes estimated by simply taking the frequency of x in the data set. In a time-reversible model, the probability of observing two characters x and y at both ends of a branch of length λ is given by $\pi_x p_{xy}(\lambda) = \pi_y p_{yx}(\lambda)$.

Now let i_k and j_k be the kth aligned characters of sequences i and j, respectively. The likelihood of d_{ij} as the path length between i and j can then be expressed as follows:

$$L(d_{ij}) = \prod_{k=1}^{m} \pi_{i_k} p_{i_k j_k}(d_{ij}) \qquad [6.3]$$

where m is the number of characters aligned in the alignement of i and j (sites with gaps in the alignment are ignored). The ML estimate of the distance d_{ij} is the value that maximizes $L(d_{ij})$, and it can be obtained numerically or analytically, depending on the model.

As an example, we consider the simplest substitution model for DNA, the Jukes–Cantor model (Jukes and Cantor 1969) (see Chapter 2). In this model, we have $\pi_A = \pi_C = \pi_G = \pi_T = {}^1/_4$ and, if $x \neq y$:

$$p_{xy}(\lambda) = \frac{1}{4}\left(1 - e^{-\frac{4}{3}\lambda}\right), \qquad p_{xx}(\lambda) = \frac{1}{4}\left(1 + 3e^{-\frac{4}{3}\lambda}\right)$$

We now recall that m_{ij} denotes the number of mismatches between i and j among aligned sites. The likelihood [6.3] can then be written as:

$$L(d_{ij}) = \frac{1}{4^m}\left(1 - e^{-\frac{4}{3}d_{ij}}\right)^{m_{ij}}\left(1 + 3e^{-\frac{4}{3}d_{ij}}\right)^{m-m_{ij}} \qquad [6.4]$$

The value of d_{ij} that maximizes [6.4] is given by:

$$\delta_{ij} = -\frac{3}{4}\ln\left(1 - \frac{4}{3}\frac{m_{ij}}{m}\right)$$

In other words, the estimate δ_{ij} of the distance between the two sequences is a strictly increasing function of the proportion of mismatches ${}^{m_{ij}}/_m$. Moreover, it is easy to see that $\delta_{ij} \geq {}^{m_{ij}}/_m$, which corresponds to the fact that the number of substitutions is always greater than or equal to the number of mismatches. In a way, δ_{ij} "corrects" ${}^{m_{ij}}/_m$ by taking into account the unobserved substitutions between i and j.

Other models result in distance estimates that are functions of several characteristics of pairwise alignment (e.g. functions that operate with transitions and transversions differently); therefore, in general, distance estimates are not simply corrected versions of the proportion ${}^{m_{ij}}/_m$. Since estimating a pairwise distance using ML is equivalent to ML-based inference of an unrooted 2-taxon tree (with a single branch length to optimize), numerical techniques for distance estimation are largely the same as those for branch length optimization.

6.3.2. *Other approaches to estimate distances*

Historically, the use of aligned sequences was not the first way to estimate evolutionary distances. For example, some of the oldest distance-based methods used data such as morphological characters (Sokal and Michener 1958), allele frequencies (Cavalli-Sforza and Edwards 1967), the strength of immunological cross-reactions

(Sarich and Wilson 1967) and DNA-DNA hybridization data (Sibley and Ahlquist 1984). New ways of computing evolutionary distances continue to appear in the literature. A thorough review of the proposed approaches falls beyond the scope of this chapter; nonetheless, we briefly outline here an overview of two broad families of ideas.

One line of work is motivated by the observation that sequence alignment is a complex task requiring the use of algorithms that are resource intensive, potentially inaccurate, sensitive to sequencing errors and difficult to apply to genomes that have undergone structural rearrangements. To solve these problems, *alignment-free* methods, while still based on nucleotide or protein sequences, aim to completely avoid the alignment step. A large body of literature on these ideas has emerged over the past 20 years, with applications that go well beyond the calculation of dissimilarities in phylogenetics. Alignment-free methods quantify the similarity/dissimilarity between sequences based on characteristics such as (1) the frequencies of certain word classes (e.g. k-mers which are contiguous subsequences of length k) (Sims et al. 2009; Sims and Kim 2011), (2) the length of shared substrings (Ulitsky et al. 2006; Leimeister and Morgenstern 2014) and (3) micro-alignments (Haubold et al. 2015; Leimeister et al. 2017). Far more approaches than those cited here have been proposed based on concepts often related to data compression and information theory (Vinga 2014). Indeed, the similarity between two sequences is related to the degree to which knowledge about one sequence provides information about the other (e.g. by allowing the description of the other one to be shortened). See Haubold (2014), Zielezinski et al. (2017), Zielezinski et al. (2019) and Leimeister (2018) for overviews of this timely research topic.

A second line of work is based on the observation that genomes do not evolve solely through nucleotide substitutions and small insertions and deletions, acting "microscopically" at the sequence level. Genomes also evolve in terms of their macroscopic structure through rearrangements that change the position and orientation of large genomic regions (e.g. inversions, duplications, translocations, etc.). Comparing the large-scale structure of the genomes of several species can be very informative about their evolutionary history but, to date, the standard phylogenetic methodology does not make use of this information. A line of work on modeling large-scale genome evolution has emerged over the past 30 years, pioneered by David Sankoff and Pavel Pevzner. Some of these works focus on calculating the smallest number of rearrangement events needed to convert one genome into another. This number can be used in a distance-based context (but this is not ideal, see below). We refer readers to Chapter 5 for a more detailed overview of this area.

We conclude with a word of caution. Given that phylogenetic inference is often not their primary intended application, many of the approaches discussed in this section lead to dissimilarities or distance metrics that cannot be interpreted as evolutionary distance estimates (in other words, they cannot be expected to converge to a tree distance $\mathbf{d}^T$). When this is the case, the construction of a tree on the basis of these dissimilarities must be considered as a form of clustering, rather than phylogenetic inference. No theoretical guarantee such as topological consistency holds in this case.

6.4. Tree inference

We organize our overview of tree inference methods around three specific questions: Any combination of answers to these questions defines a different distance-based method.

– Q1 – How to fit branch lengths? A method has to be described to fit the branch lengths of any tree topology τ, such that the resulting tree distances d^T_{ij} are as close as possible to the estimated distances δ_{ij}. This is usually done using least squares techniques borrowed from regression (Cavalli-Sforza and Edwards 1967; Fitch and Margoliash 1967).

– Q2 – How to choose among tree topologies? A criterion needs to be defined to assign a score to any tree topology. This is usually achieved by first fitting its branch lengths (see Q1), and then by scoring the plausibility of the resulting tree T. An obvious choice to evaluate T is the same least squares criterion used to fit its branch lengths. However, as we describe below, many recent methods are based on a different criterion, minimum evolution (Kidd and Sgaramella-Zonta 1971).

– Q3 – How to optimize? An algorithm has to be designed to seek the optimal tree topology relatively to the criterion in Q2. Since this is usually a computationally hard optimization problem, the algorithms for this task are heuristic and based on simple but effective ideas such as sequential addition, agglomerative clustering (described in section 6.4.3) or hill-climbing (Swofford et al. 1990; Felsenstein 2004).

In the following, we will begin by studying the methodology to answer Q1 (section 6.4.1). We will next introduce one of the most popular criteria for Q2, *minimum evolution* (section 6.4.2), which is the basis for what is still the best known distance-based methods, *neighbor joining*. We will describe this method together with other algorithms based on the same general strategy (section 6.4.3). Finally, we will briefly describe recent approaches that, strictly speaking, are not distance-based methods, but share with them several ideas and the same emphasis on computational efficiency (section 6.4.4). Since the heuristic techniques used to answer Q3 are often the same as those used in other approaches for tree inference such as ML and parsimony, we will not address these techniques here.

6.4.1. *Fitting branch lengths with least squares*

Given the estimated distances (δ_{ij}), the goal of least-squares phylogenetic inference (Cavalli-Sforza and Edwards 1967; Fitch and Margoliash 1967) is to find a phylogenetic tree T that minimizes a quadratic function of the differences between the estimates δ_{ij} and the tree distances d^T_{ij}. A few choices are possible for this quadratic function (see equations [6.5] and [6.6]). Although many versions of this problem have been proven computationally hard (Day 1987), here we focus on the simpler problem of assigning branch lengths to a tree of fixed topology τ. Exact and polynomially computable solutions are available for this task.

The least squares method was introduced in phylogenetics in the 1960s. The proposed objective function was:

$$\sum_{i<j} w_{ij}(\delta_{ij} - d^T_{ij})^2 \qquad [6.5]$$

with $w_{ij} = 1$ (Cavalli-Sforza and Edwards 1967) or $w_{ij} = {}^1/_{\delta^2_{ij}}$ (Fitch and Margoliash 1967). The choice $w_{ij} = 1$ defines *ordinary least squares* (OLS), and both choices are special cases of *weighted least squares* (WLS). In WLS, the weight w_{ij} represents the degree of confidence that can be attached to the distance estimate δ_{ij}. Ideally, w_{ij} should be inversely proportional to the variance of δ_{ij}, but in practice the choice of weights can be tricky, because these variances are difficult to evaluate. One possible choice that we will revisit several times is that of *balanced* weights, where w_{ij} is proportional to $2^{-|\tau_{ij}|}$ for a binary tree (we recall that τ_{ij} denotes the set of branches between i and j in τ). Balanced weights assign lower confidence to distances between taxa that are separated by several branches in τ.

WLS approaches do not take into account the correlations between distance estimates δ_{ij}, δ_{kl} for different pairs of taxa. These correlations can be significant when the paths τ_{ij}, τ_{kl} do share several branches. To account for these correlations, generalized least square (GLS) (Chakraborty 1977; Bulmer 1991) seeks to minimize:

$$\sum_{i<j}\sum_{k<l} w_{ij,kl}(\delta_{ij} - d^T_{ij})(\delta_{kl} - d^T_{kl}) \qquad [6.6]$$

where $w_{ij,kl}$ should (ideally) be the elements of the inverse of the variance–covariance matrix for the distance estimates δ_{ij}. Just as OLS is a special case of WLS, WLS is the special case of GLS obtained by setting $w_{ij,kl} = 0$ if $\{i,j\} \neq \{k,l\}$. In practice,

GLS is rarely used for phylogenetic inference because of the difficulty of evaluating covariances, and due to its high computational costs compared to those of OLS and WLS.

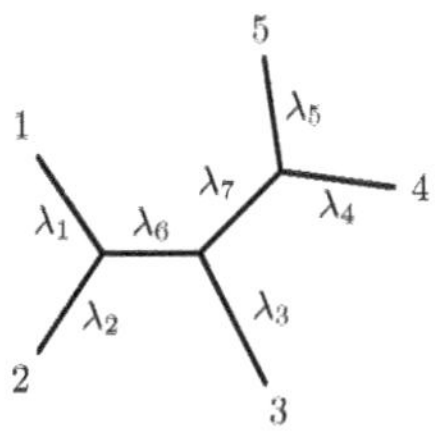

$$\begin{bmatrix} 1 & 1 & 0 & 0 & 0 & 0 & 0 \\ 1 & 0 & 1 & 0 & 0 & 1 & 0 \\ 1 & 0 & 0 & 1 & 0 & 1 & 1 \\ 1 & 0 & 0 & 0 & 1 & 1 & 1 \\ 0 & 1 & 1 & 0 & 0 & 1 & 0 \\ 0 & 1 & 0 & 1 & 0 & 1 & 1 \\ 0 & 1 & 0 & 0 & 1 & 1 & 1 \\ 0 & 0 & 1 & 1 & 0 & 0 & 1 \\ 0 & 0 & 1 & 0 & 1 & 0 & 1 \\ 0 & 0 & 0 & 1 & 1 & 0 & 0 \end{bmatrix} \times \begin{bmatrix} \lambda_1 \\ \lambda_2 \\ \lambda_3 \\ \lambda_4 \\ \lambda_5 \\ \lambda_6 \\ \lambda_7 \end{bmatrix} = \begin{bmatrix} \lambda_1 + \lambda_2 \\ \lambda_1 + \lambda_3 + \lambda_6 \\ \lambda_1 + \lambda_4 + \lambda_6 + \lambda_7 \\ \lambda_1 + \lambda_5 + \lambda_6 + \lambda_7 \\ \lambda_2 + \lambda_3 + \lambda_6 \\ \lambda_2 + \lambda_4 + \lambda_6 + \lambda_7 \\ \lambda_2 + \lambda_5 + \lambda_6 + \lambda_7 \\ \lambda_3 + \lambda_4 + \lambda_7 \\ \lambda_3 + \lambda_5 + \lambda_7 \\ \lambda_4 + \lambda_5 \end{bmatrix} = \begin{bmatrix} d^T_{12} \\ d^T_{13} \\ d^T_{14} \\ d^T_{15} \\ d^T_{23} \\ d^T_{24} \\ d^T_{25} \\ d^T_{34} \\ d^T_{35} \\ d^T_{45} \end{bmatrix}$$

Figure 6.3. *Tree distances are linear functions of branch lengths. The matrix A_τ can be used to express $\mathbf{d}^T$ as $A_\tau \boldsymbol{\lambda} = \mathbf{d}^T$. Note that this is an overdetermined system of 10 linear equations. Any subset of seven linearly independent equations out of the original 10 suffices to determine the seven unknowns*

The optimal branch lengths for the WLS and GLS criteria (equations [6.5] and [6.6]) can be concisely expressed in matrix notation. To this end, we will consider $\boldsymbol{\delta}$ and $\mathbf{d}^T$ as column vectors whose elements correspond to subsets of two taxa and are ordered in the usual lexicographic order (see, for example, the vector on the right-hand side of Figure 6.3). Moreover, given any topology τ, let $A_\tau = (a_{ij,e})$ be a matrix whose rows correspond to subsets of two taxa (in lexicographic order) and whose columns correspond to the branches of τ. Its elements are such that $a_{ij,e} = 1$ if e is on the path between i and j in τ, and $a_{ij,e} = 0$ otherwise. Given these notations,

$$A_\tau \boldsymbol{\lambda} = \mathbf{d}^T$$

where $\boldsymbol{\lambda} = (\lambda_e)$ denotes the branch lengths of T in vector form (see Figure 6.3 for an example illustrating these notations). The objective functions of OLS, WLS and GLS can then be written in the matrix form:

$$(\boldsymbol{\delta} - A_\tau \boldsymbol{\lambda})^t \, W \, (\boldsymbol{\delta} - A_\tau \boldsymbol{\lambda}) \qquad [6.7]$$

where $W = (w_{ij,kl})$ contains the weights and t denotes matrix transposition. WLS and OLS are obtained when W is a diagonal matrix and the identity matrix, respectively. The branch lengths that minimize [6.7] can then be expressed as:

$$\widehat{\boldsymbol{\lambda}}_\tau = \left(A_\tau^t W A_\tau\right)^{-1} A_\tau^t W \boldsymbol{\delta} \qquad [6.8]$$

The matrix calculations in equation [6.8] are expensive, but several properties of the matrices in it can be exploited to speed up the computations (Gascuel 1997b;

Bryant and Waddell 1998). In some cases, asymptotically optimal algorithms have been found because of the existence of simpler analytical formulas than [6.8]. This was first remarked for OLS (Vach 1989; Rzhetsky and Nei 1993), then for WLS with balanced weights (Desper and Gascuel 2004), and later extended to a larger subclass of WLS, namely the cases where the weights w_{ij} are *multiplicative* (Mihaescu and Pachter 2008; Pardi and Gascuel 2012). The practical importance of these algorithmic results lies in the fact that the computation of $\widehat{\boldsymbol{\lambda}}_\tau$ is a frequently called subroutine of many distance-based tree inference methods.

We conclude by noting that none of the approaches described above guarantee that the fitted branch lengths $\widehat{\boldsymbol{\lambda}}_\tau$ are positive. Least squares approaches sometimes lead to negative branch lengths, which does not correspond to any biological reality. Branch lengths can be constrained to be non-negative in a least squares criterion (NNLS [*non-negative least squares*]; Lawson and Hanson 1974). This approach is available in programs such as FITCH (Felsenstein 1997) and PAUP* (Swofford 2003), but it further increases computational costs.

6.4.2. *Scoring trees: from least squares to minimum evolution*

Once a means of adjusting branch lengths has been determined, the next question (Q2) is how to measure the plausibility of the resulting trees. One possibility is to adopt the same least squares criterion that was used for fitting branch lengths. In this case, a topology is considered plausible if its tree distances can be closely matched to the estimated distances. Since these approaches involve least squares as the answer to both Questions Q1 and Q2, we will call them âĂIJLS+LSâĂİ.

It is interesting to note that any method providing the optimal tree with respect to an LS+LS criterion is topologically consistent (see Definition 6.1). This follows from Theorem 6.2, which implies that if $\boldsymbol{\delta}$ converges to $\mathbf{d}^T$, the least squares criterion [6.7] can only become arbitrarily close to 0 when τ is the topology of T (see, for example, Pardi and Gascuel (2012) for more details on this argument). In practice, LS+LS approaches are known to be accurate when negative branch lengths are not allowed (Kuhner and Felsenstein 1994).

An alternative approach, *minimum evolution* (ME), uses as a score for a tree a function $S(\widehat{\boldsymbol{\lambda}}_\tau)$ of the fitted branch lengths. In ME, this function is a sum expressing the total length of the tree. The intuition behind ME is the same as in maximum parsimony (MP): just as simple explanations are preferable to complex ones, the shortest trees are here considered the most plausible.

Different versions of ME differ in how negative branch lengths are counted in $S(\widehat{\boldsymbol{\lambda}}_\tau)$. For example, they can simply be excluded from the sum (Swofford et al. 1990), or $S(\widehat{\boldsymbol{\lambda}}_T)$ can be defined as the sum of absolute values of branch lengths (Kidd and Sgaramella-Zonta 1971). Currently, the most common approach is to define $S(\widehat{\boldsymbol{\lambda}}_T)$ as the sum of all the branch lengths, regardless of their sign (Saitou and Imanishi 1989; Rzhetsky and Nei 1993). Although this seems to favor negative branch lengths, in practice negative branch lengths are often more than compensated by large positive branch lengths that increase the total length of the tree.

From a theoretical point of view, why ME works is not as clear as one might think. In the 1990s, it was known that ME tree inference with OLS-fitted branch lengths ("OLS + ME") is topologically consistent (Rzhetsky and Nei 1993). Nevertheless, it was later shown that this does not hold more generally for WLS + ME (Gascuel et al. 2001). It turns out that topological consistency is valid only for a subclass of WLS + ME. To date, this important prerequisite has been demonstrated for all cases of WLS + ME with multiplicative weights, and for the three ways of treating negative branch lengths in $S(\widehat{\boldsymbol{\lambda}}_T)$ (Pardi and Gascuel 2012).

An important special case thereof is WLS + ME with balanced weights, which is also known as BME (*balanced minimum evolution*) (Pauplin 2000; Desper and Gascuel 2002, 2004; Lefort et al. 2015). This optimization criterion has several interesting mathematical characteristics (Semple and Steel 2004), in particular the fact that for fully resolved (binary) trees, the length of the fitted tree can be elegantly expressed as a simple function of the estimated distances:

$$S(\widehat{\boldsymbol{\lambda}}_T) = \sum_{i<j} 2^{1-|\tau_{ij}|}\delta_{ij} \qquad [6.9]$$

BME is important to shed some light on a number of central methods in distance-based phylogenetics, including NJ, as we explain below.

6.4.3. *NJ and other agglomerative algorithms*

NJ is an agglomerative clustering algorithm, that is, it constructs a tree in a bottom-up fashion, alternating the following two steps until the tree is complete:

– *Selection step*: based on the estimated distances, two *active* taxa i and j are chosen to be "joined". That is, they are linked to a new taxon (ij) representing their immediate common ancestor (initially, all taxa are active).

– *Reduction step*: i and j are removed from the list of active taxa, and new distances are defined between the new active taxon (ij) and the remaining active taxa.

These two steps are common to all agglomerative algorithms. For NJ, there are several equivalent ways to specify these steps (Saitou and Nei 1987; Studier and

Keppler 1988; Gascuel 1994). Here, we describe the fastest algorithm (Studier and Keppler 1988). In the selection step, NJ agglomerates the taxa i and j that minimize:

$$(|A| - 2)\delta_{ij} - \sum_{k \in A} (\delta_{ik} + \delta_{jk}), \quad [6.10]$$

where A is the set of active taxa. As for the reduction step, the new distance between (ij) and any other active taxon k is defined by:

$$\delta_{(ij)k} = \frac{1}{2}(\delta_{ik} + \delta_{jk} - \delta_{ij}) \quad [6.11]$$

The criterion [6.10] may seem somewhat mysterious at first sight, but it is possible to show that if $\boldsymbol{\delta} = \mathbf{d}^T$ for a phylogenetic tree T, then the taxa i and j that minimize [6.10] must form a "cherry" in T (i.e. i and j are connected by a path of exactly 2 branches). See also Bryant (2005) for a thorough study of [6.10]. Moreover, it is easy to see that if $\boldsymbol{\delta} = \mathbf{d}^T$, the value of $\delta_{(ij)k}$ obtained with [6.11] is equal to the length of the path between k and the only node separating i and j. These observations on [6.10] and [6.11] imply the topological consistency of NJ.

Different choices for the selection and reduction steps lead to other well-known clustering algorithms: single linkage clustering (Sneath 1957), average linkage clustering, commonly referred to as UPGMA (Sokal and Michener 1958), WPGMA (Sokal and Sneath 1963), ADDTREE (Sattath and Tversky 1977), UNJ (Gascuel 1997b) and BIONJ (Gascuel 1997a). The last method is a special case of the MVR approach (Gascuel 2000a), which adapts the reduction step to take into account the variances and covariances of the distance estimates δ_{ij}. In spirit, it is similar to another agglomerative clustering algorithm, Weighbor (Bruno et al. 2000), which also modifies the selection criterion and uses a different formula (Bulmer 1991) to evaluate the variances of the distances.

Among these methods, the fastest are the classical clustering algorithms (single linkage, UPGMA and WPGMA), which construct a tree in $O(n^2)$ time (Sibson 1973; Murtagh 1984; Gronau and Moran 2007). However, these clustering approaches are only accurate (in fact, topologically consistent) when the estimated distances are close to a tree distance for a tree whose leaves are all at the same distance from a central point, to be interpreted as their last common ancestor. This occurs when evolutionary rates are constant and branch lengths are directly proportional to time, a hypothesis known as the *strict molecular clock*. For this reason, these fast algorithms are not a good choice when suspecting that the strict molecular clock hypothesis does not hold.

Besides these fast clustering algorithms and ADDTREE and MVR (which run in $O(n^4)$ time), all the clustering algorithms presented here build a tree in $O(n^3)$ time.

This includes NJ where each of the $O(n)$ selection steps is performed in time $O(n^2)$ via the pre-computation of the sums $\sum_{k \in A} \delta_{ik}$ (Studier and Keppler 1988). One of the reasons for the ongoing success of NJ is the fact that it has long been considered as achieving a very good compromise between accuracy and speed.

Subsequently to NJ's widespread use, a lot of work has been devoted to developing efficient implementations for NJ, requiring less memory and execution time than the standard implementation. Among these works, we mention QuickTree (Howe et al. 2002), QuickJoin (Mailund and Pedersen 2004), the bucket-based method of (Zaslavsky and Tatusova 2008), NINJA (Wheeler 2009), RapidNJ and ERapidNJ (Simonsen et al. 2010). A common idea shared by these approaches is to speed up the identification of the optimal taxon pair with respect to the criterion [6.10] by restricting the search to a subset of pairs guaranteed to contain the optimal pair. Although the time complexity remains $O(n^3)$ in the worst case, these approaches lead to a spectacular improvement in efficiency, allowing tree inference with more than 50,000 taxa in a few hours on a normal PC (Wheeler 2009; Simonsen et al. 2010).

Even faster algorithms can be obtained by employing heuristics instead of exact algorithms to find a good pair of taxa with respect to the criterion [6.10]. This is the case of FastNJ (Elias and Lagergren 2009) and RelaxedNJ (Evans et al. 2006), implemented in Clearcut (Sheneman et al. 2006). In the case of FastNJ, the taxa to be clustered are selected from a list of $O(n)$ pairs, which implies an execution time of $O(n^2)$. Some loss of accuracy is to be expected for FastNJ and RelaxedNJ, as these approaches are not guaranteed to build the same tree as NJ. Despite this, these methods are still topologically consistent (Elias and Lagergren 2009).

A question that puzzled phylogeneticists for some time is whether NJ is related to known optimization criteria (see section 6.4.2), and more generally if it can be described as a combination of answers to Questions Q1 and Q3 (section 6.4). For example, Felsenstein's textbook speculated that "NJ has some relation to OLS and some to minimum evolution, without being definable as an approximate algorithm for either" (Felsenstein 2004, p. 170). Similar connections were suggested by other authors (Saitou and Nei 1987; Gascuel 1994), and it was long assumed that the closest optimization criterion to NJ was OLS + ME. However, if the good performance of NJ was due to its relationship with OLS + ME, we would then expect that better (shorter) trees with respect to this criterion be also more accurate than the trees inferred with NJ. This is in fact contradicted by several experiments based on simulated data (Saitou and Imanishi 1989; Kumar 1996; Gascuel 2000b; Desper and Gascuel 2002).

A solution to this conundrum came from realizing that the real optimization criterion behind NJ is BME (Desper and Gascuel 2005; Gascuel and Steel 2006). To understand this result, it should be first pointed out that BME also assigns a length $S(\widehat{\boldsymbol{\lambda}}_\tau)$ to the trees that are not fully resolved (non-binary). For these tree topologies, there exists a slightly more complex formula than that in equation [6.9] (Semple and Steel 2004; Gascuel and Steel 2006). Now consider a agglomerative algorithm such that, at any time, all active nodes are connected to a central node (the algorithm starts with the star tree on $\{1, 2, \dots, n\}$ when all taxa are active). At each selection step, this algorithm chooses the pair of active taxa (i, j) whose clustering produces the tree with the smallest BME length $S(\widehat{\boldsymbol{\lambda}}_\tau)$. It turns out that this greedy algorithm makes *exactly* the same sequence of choices as NJ (Gascuel and Steel 2006).

Motivated by the observation that NJ can be seen as a greedy algorithm for BME, other methods guided by BME have been proposed (Desper and Gascuel 2002; Catanzaro et al. 2012). These methods differ from NJ in their response to Question Q3, namely by the algorithm for determining the optimal tree with respect to the BME criterion. For example, FastME (Desper and Gascuel 2002; Lefort et al. 2015) implements heuristics to build an initial tree and to perform a local search in the space of tree topologies. While the runtime of FastME is comparable to NJ, it offers greater reconstruction accuracy (Desper and Gascuel 2002, 2004; Vinh and von Haeseler 2005), thus confirming the attractiveness of BME as optimization criterion for distance-based phylogenetic inference.

6.4.4. *Beyond distances*

A major computational bottleneck of all the methods that we have presented so far is that they rely on the initial estimation and storage of distances for all taxon pairs. This requires $O(n^2)$ space in memory, and $O(m \cdot n^2)$ time, assuming that distances are estimated from aligned sequences of length m. Nevertheless, as we mentioned in section 6.2, not all distances are necessary to reconstruct a phylogeny, which means that, in principle, it should be possible to improve these bounds. Some distances, those with high variances, may even be misleading for tree inference. In addition, for large datasets with hundreds of thousands of taxa, reducing memory usage may be a necessity: simply storing the entire distance matrix for 100,000 taxa usually requires about 20 GB of memory, which can be problematic for many users.

In recent years, a few approaches have been proposed for circumventing this bottleneck. One of the most widely used is FastTree (Price et al. 2009) whose strategy for building an initial tree is inspired by NJ, but with two fundamental differences, the first aiming at improving inference accuracy and the second aiming at improving computational efficiency. We briefly describe the main distinctive points of FastTree, because they explain its solid performance and its popularity.

The first difference with NJ and other distance-based methods is that FastTree stores sequence profiles for active taxa (instead of a distance matrix) and only computes the distance between two profiles if the corresponding pair is a candidate for joining. As reduction step, a sequence profile for the new node (ij) is calculated as the mean of the profiles for i and j. The taxon pairs to join are selected based on a criterion similar to [6.10], but where distances are not corrected and are defined on the basis of the profiles stored for the active taxa.

The second difference resides in maintaining a list of $O(\sqrt{n})$ "top-hits" for each taxon (i.e. putative neighbors), which is combined with the strategies of FastNJ (Elias and Lagergren 2009) and RelaxedNJ (Evans et al. 2006) to reduce the number of taxon pairs to consider for joining. Over the entirety of its execution, FastTree considers only $O(n^{1.5} \log n)$ pairs, against $O(n^3)$ for NJ and $O(n^2)$ for FastNJ. The construction of an initial tree, with a claimed complexity of $O(m \cdot n^{1.5} \log n)$ time and $O(mn + n^{1.5})$ space, is then followed by a local search based on topology rearrangements (NNI, SPR) employing techniques similar to those implemented in FastME (Desper and Gascuel 2002).

FastTree is faster and requires less memory than the distance-based methods discussed so far, and it can easily handle datasets containing hundreds of thousands of sequences. Given the central role of sequence profiles, it is difficult to consider FastTree as a purely distance-based method. FastTree shares ideas with character-based methods such as MP, benefiting in particular from information on ancestral sequences.

6.5. Conclusion

Despite the simplicity of their approach and the information that is lost by replacing a set of sequences with their distances, distance-based methods are not only efficient but also remarkably accurate. The reconstruction accuracy of distance-based methods has been the subject of numerous simulation studies (Saitou and Imanishi 1989; Kuhner and Felsenstein 1994; Kumar 1996; Gascuel 2000b; Nakhleh et al. 2002; Desper and Gascuel 2004). The general idea of these works is to generate sequences using known substitution models and model trees, and then to compare the trees constructed by the tested methods to the model trees.

These studies show that, although not comparable to likelihood-based methods, the accuracy of distance-based methods is competitive with MP, MP being superior for model trees with short branches, and inferior when the impact of multiple substitutions at the same site becomes important. The reason for this is largely intuitive: although distance-based methods naturally take into account multiple substitutions in the way they estimate distances, MP does not even model branch lengths, which leads to problems such as topological inconsistency in extreme cases (Felsenstein 1978). Another lesson from these empirical studies is the importance of improving an initial reconstructed tree through thorough searches in tree space using topological rearrangements such as NNI and SPR (for example (Vinh and von Haeseler 2005)). This is the reason why most modern tree inference methods include this step, as we have seen for FastME (Desper and Gascuel 2002) and FastTree (Price et al. 2009).

In addition to the good trade-off between efficiency and accuracy, another important advantage of distance-based methods is their versatility: they can be used not only with aligned sequence data, but in every context where the available data allow us to estimate a measure of evolutionary distance between taxa. See, for example, section 6.3.2 for alternative approaches to distance estimation.

One area where distance methods prove their versatility is in phylogenomics, where they are often used (on their own or combined with other methods) for inferring species trees based on large collections of alignments from different genomic loci. One approach consists of summarizing the information originating from each locus into a separate distance matrix. This usually produces thousands of distance matrices whose combined analysis is much more efficient that concatenated alignment analysis (Lapointe and Cucumel 1997; Criscuolo et al. 2006; Binet et al. 2016).

Another line of work for species tree inference seeks to account for incomplete lineage sorting (ILS), which can cause phylogenetic trees for different loci to not only have different branch lengths but also different topologies. Several ILS-aware approaches use distance-based inference as a key element. Distances can be derived from gene tree topologies separately inferred for each locus: this is done, for example, in STAR (Liu et al. 2009), NJst (Liu and Yu 2011) and ASTRID (Vachaspati and Warnow 2015). Alternatively distances can be estimated from interspecific gene coalescence times, as in STEAC (Liu et al. 2009), MT (Liu et al. 2010), GLASS (Mossel and Roch 2008) or iGLASS (Jewett and Rosenberg 2012).

All of this demonstrates the usefulness of distance-based methods, which will undoubtedly continue to find new applications whenever computational efficiency and solid theoretical guarantees are paramount.

6.6. References

Atteson, K. (1999). The performance of neighbor-joining methods of phylogenetic reconstruction. *Algorithmica*, 25(2–3), 251–278.

Barthélémy, J.-P. and Guénoche, A. (1991). *Trees and Proximity Representations*. John Wiley & Sons, New York.

Binet, M., Gascuel, O., Scornavacca, C., Douzery, E.J., Pardi, F. (2016). Fast and accurate branch lengths estimation for phylogenomic trees. *BMC Bioinformatics*, 17(1), 23.

Bruno, W.J., Socci, N.D., Halpern, A.L. (2000). Weighted neighbor joining: A likelihood-based approach to distance-based phylogeny reconstruction. *Molecular Biology and Evolution*, 17(1), 189–197.

Bryant, D.J. (2005). On the uniqueness of the selection criterion in neighbor-joining. *Journal of Classification*, 22(1), 3–15.

Bryant, D.J. and Waddell, P.J. (1998). Rapid evaluation of least-squares and minimum-evolution criteria on phylogenetic trees. *Molecular Biology and Evolution*, 15(10), 1346–1359.

Bulmer, M. (1991). Use of the method of generalized least squares in reconstructing phylogenies from sequence data. *Molecular Biology and Evolution*, 8(6), 868–883.

Buneman, P. (1971). The recovery of trees from measures of dissimilarity. In *Mathematics in the Archeological and Historical Sciences*, Hodson, F.R., Kendall, D.G., Tautu, P. (eds). Edinburgh University Press.

Catanzaro, D., Labbé, M., Pesenti, R., Salazar-González, J.-J. (2012). The balanced minimum evolution problem. *INFORMS Journal on Computing*, 24(2), 276–294.

Cavalli-Sforza, L.L. and Edwards, A.W. (1967). Phylogenetic analysis. Models and estimation procedures. *American Journal of Human Genetics*, 19(3 Part 1), 233.

Chakraborty, R. (1977). Estimation of time of divergence from phylogenetic studies. *Canadian Journal of Genetics and Cytology*, 19(2), 217–223.

Criscuolo, A., Berry, V., Douzery, E.J., Gascuel, O. (2006). Sdm: A fast distance-based approach for (super) tree building in phylogenomics. *Systematic Biology*, 55(5), 740–755.

Day, W.H. (1987). Computational complexity of inferring phylogenies from dissimilarity matrices. *Bulletin of Mathematical Biology*, 49(4), 461–467.

Desper, R. and Gascuel, O. (2002). Fast and accurate phylogeny reconstruction algorithms based on the minimum-evolution principle. *Journal of Computational Biology*, 9(5), 687–705.

Desper, R. and Gascuel, O. (2004). Theoretical foundation of the balanced minimum evolution method of phylogenetic inference and its relationship to weighted least-squares tree fitting. *Molecular Biology and Evolution*, 21(3), 587–598.

Desper, R. and Gascuel, O. (2005). The minimum evolution distance-based approach to phylogenetic inference. In *Mathematics of Evolution and Phylogeny*, Gascuel, O. (ed.). Oxford University Press.

Dress, A.W., Huber, K.T., Steel, M. (2012). "Lassoing" a phylogenetic tree I: Basic properties, shellings, and covers. *Journal of Mathematical Biology*, 65(1), 77–105.

Elias, I. and Lagergren, J. (2009). Fast neighbor joining. *Theoretical Computer Science*, 410(21), 1993–2000 [Online]. Available at: http://www.sciencedirect.com/science/article/pii/S0304397508009079.

Evans, J., Sheneman, L., Foster, J. (2006). Relaxed neighbor joining: A fast distance-based phylogenetic tree construction method. *Journal of Molecular Evolution*, 62(6), 785–792.

Felsenstein, J. (1978). Cases in which parsimony or compatibility methods will be positively misleading. *Systematic Zoology*, 27(4), 401–410.

Felsenstein, J. (1997). An alternating least squares approach to inferring phylogenies from pairwise distances. *Systematic Biology*, 46(1), 101–111.

Felsenstein, J. (2004). *Inferring Phylogenies*. Sinauer Associates Sunderland, MA.

Fitch, W.M. and Margoliash, E. (1967). Construction of phylogenetic trees. *Science*, 155(3760), 279–284.

Gascuel, O. (1994). A note on Sattath and Tversky's, Saitou and Nei's, and Studier and Keppler's algorithms for inferring phylogenies from evolutionary distances. *Molecular Biology and Evolution*, 11(6), 961–963.

Gascuel, O. (1997a). BIONJ: An improved version of the NJ algorithm based on a simple model of sequence data. *Molecular Biology and Evolution*, 14(7), 685–695.

Gascuel, O. (1997b). Concerning the NJ algorithm and its unweighted version, UNJ. *Mathematical Hierarchies and Biology*, 37, 149–171.

Gascuel, O. (2000a). Data model and classification by trees: The minimum variance reduction (MVR) method. *Journal of Classification*, 17(1), 67–99.

Gascuel, O. (2000b). On the optimization principle in phylogenetic analysis and the minimum-evolution criterion. *Molecular Biology and Evolution*, 17(3), 401–405.

Gascuel, O. and Steel, M. (2006). Neighbor-joining revealed. *Molecular Biology and Evolution*, 23(11), 1997–2000.

Gascuel, O. and Steel, M. (2016). A "stochastic safety radius" for distance-based tree reconstruction. *Algorithmica*, 74(4), 1386–1403.

Gascuel, O., Bryant, D., Denis, F. (2001). Strengths and limitations of the minimum evolution principle. *Systematic Biology*, 50(5), 621–627.

Gronau, I. and Moran, S. (2007). Optimal implementations of UPGMA and other common clustering algorithms. *Information Processing Letters*, 104(6), 205–210.

Guindon, S. and Gascuel, O. (2003). A simple, fast, and accurate algorithm to estimate large phylogenies by maximum likelihood. *Systematic Biology*, 52(5), 696–704.

Haubold, B. (2014). Alignment-free phylogenetics and population genetics. *Briefings in Bioinformatics*, 15(3), 407–418.

Haubold, B., Klötzl, F., Pfaffelhuber, P. (2015). Andi: Fast and accurate estimation of evolutionary distances between closely related genomes. *Bioinformatics*, 31(8), 1169–1175.

Hein, J.J. (1989). An optimal algorithm to reconstruct trees from additive distance data. *Bulletin of Mathematical Biology*, 51(5), 597–603.

Howe, K., Bateman, A., Durbin, R. (2002). Quicktree: Building huge neighbour-joining trees of protein sequences. *Bioinformatics*, 18(11), 1546–1547.

Huber, K.T. and Kettleborough, G. (2015). Distinguished minimal topological lassos. *SIAM Journal on Discrete Mathematics*, 29(2), 940–961.

Jewett, E.M. and Rosenberg, N.A. (2012). Iglass: An improvement to the glass method for estimating species trees from gene trees. *Journal of Computational Biology*, 19(3), 293–315.

Jukes, T. and Cantor, C. (1969). Evolution of protein molecules. In *Mammalian Protein Metabolism*, Munro, H. (ed.). Academic Press, Cambridge.

Kettleborough, G., Dicks, J., Roberts, I.N., Huber, K.T. (2015). Reconstructing (super)trees from data sets with missing distances: Not all is lost. *Molecular Biology and Evolution*, 32(6), 1628–1642.

Kidd, K.K. and Sgaramella-Zonta, L.A. (1971). Phylogenetic analysis: Concepts and methods. *American Journal of Human Genetics*, 23(3), 235.

Kuhner, M.K. and Felsenstein, J. (1994). A simulation comparison of phylogeny algorithms under equal and unequal evolutionary rates. *Molecular Biology and Evolution*, 11(3), 459–468.

Kumar, S. (1996). A stepwise algorithm for finding minimum evolution trees. *Molecular Biology and Evolution*, 13(4), 584–593.

Lapointe, F.-J. and Cucumel, G. (1997). The average consensus procedure: Combination of weighted trees containing identical or overlapping sets of taxa. *Systematic Biology*, 46(2), 306–312.

Larkin, M.A., Blackshields, G., Brown, N.P., Chenna, R., McGettigan, P.A., McWilliam, H., Valentin, F., Wallace, I.M., Wilm, A., Lopez, R. et al. (2007). Clustal w and clustal x version 2.0. *Bioinformatics*, 23(21), 2947–2948.

Lawson, C.L. and Hanson, R.J. (1974). *Solving Least Squares Problems*. Prentice Hall, Hoboken, NJ.

Lefort, V., Desper, R., Gascuel, O. (2015). Fastme 2.0: A comprehensive, accurate, and fast distance-based phylogeny inference program. *Molecular Biology and Evolution*, 32(10), 2798–2800.

Leimeister, C.-A. (2018). Filtered spaced-word matches: A novel approach to fast and accurate sequence comparison. PhD Thesis, Georg August University of Göttingen.

Leimeister, C.-A. and Morgenstern, B. (2014). Kmacs: The k-mismatch average common substring approach to alignment-free sequence comparison. *Bioinformatics*, 30(14), 2000–2008.

Leimeister, C.-A., Sohrabi-Jahromi, S., Morgenstern, B. (2017). Fast and accurate phylogeny reconstruction using filtered spaced-word matches. *Bioinformatics*, 33(7), 971–979.

Liu, L. and Yu, L. (2011). Estimating species trees from unrooted gene trees. *Systematic Biology*, 60(5), 661–667.

Liu, L., Yu, L., Pearl, D.K., Edwards, S.V. (2009). Estimating species phylogenies using coalescence times among sequences. *Systematic Biology*, 58(5), 468–477.

Liu, L., Yu, L., Pearl, D.K. (2010). Maximum tree: A consistent estimator of the species tree. *Journal of Mathematical Biology*, 60(1), 95–106.

Mailund, T. and Pedersen, C.N. (2004). Quickjoin–fast neighbour-joining tree reconstruction. *Bioinformatics*, 20(17), 3261–3262.

Mihaescu, R. and Pachter, L. (2008). Combinatorics of least-squares trees. *Proceedings of the National Academy of Sciences*, 105(36), 13206–13211.

Mihaescu, R., Levy, D., Pachter, L. (2009). Why neighbor-joining works. *Algorithmica*, 54(1), 1–24.

Mossel, E. and Roch, S. (2008). Incomplete lineage sorting: Consistent phylogeny estimation from multiple loci. *IEEE/ACM Transactions on Computational Biology and Bioinformatics*, 7(1), 166–171.

Murtagh, F. (1984). Complexities of hierarchic clustering algorithms: State of the art. *Computational Statistics Quarterly*, 1(2), 101–113.

Nakhleh, L., Moret, B., Roshan, U., St John, K., Sun, J., Warnow, T. (2002). The accuracy of fast phylogenetic methods for large datasets. *Pacific Symposium on Biocomputing*, 7, 211–222.

Pardi, F. and Gascuel, O. (2012). Combinatorics of distance-based tree inference. *Proceedings of the National Academy of Sciences*, 109(41), 16443–16448.

Pauplin, Y. (2000). Direct calculation of a tree length using a distance matrix. *Journal of Molecular Evolution*, 51(1), 41–47.

Pereira, J.S. (1969). A note on the tree realizability of a distance matrix. *Journal of Combinatorial Theory*, 6(3), 303–310.

Price, M.N., Dehal, P.S., Arkin, A.P. (2009). Fasttree: Computing large minimum evolution trees with profiles instead of a distance matrix. *Molecular Biology and Evolution*, 26(7), 1641–1650.

Rzhetsky, A. and Nei, M. (1993). Theoretical foundation of the minimum-evolution method of phylogenetic inference. *Molecular Biology and Evolution*, 10(5), 1073–1095.

Saitou, N. and Imanishi, T. (1989). Relative efficiencies of the fitch-margoliash, maximum-parsimony, maximum-likehood, minimum-evolution, and neighbor-joining methods of phylogenetic tree construction in obtaining the correct tree. *Molecular Biology and Evolution*, 6(5), 514–525.

Saitou, N. and Nei, M. (1987). The neighbor-joining method: A new method for reconstructing phylogenetic trees. *Molecular Biology and Evolution*, 4(4), 406–425.

Sarich, V.M. and Wilson, A.C. (1967). Immunological time scale for hominid evolution. *Science*, 158(3805), 1200–1203.

Sattath, S. and Tversky, A. (1977). Additive similarity trees. *Psychometrika*, 42(3), 319–345.

Semple, C. and Steel, M. (2003). *Phylogenetics*. Oxford University Press.

Semple, C. and Steel, M. (2004). Cyclic permutations and evolutionary trees. *Advances in Applied Mathematics*, 32(4), 669–680.

Sheneman, L., Evans, J., Foster, J.A. (2006). Clearcut: A fast implementation of relaxed neighbor joining. *Bioinformatics*, 22(22), 2823–2824.

Sibley, C.G. and Ahlquist, J.E. (1984). The phylogeny of the hominoid primates, as indicated by DNA-DNA hybridization. *Journal of Molecular Evolution*, 20(1), 2–15.

Sibson, R. (1973). SLINK: An optimally efficient algorithm for the single-link cluster method. *The Computer Journal*, 16(1), 30–34.

Simonsen, M., Mailund, T., Pedersen, C.N. (2010). Inference of large phylogenies using neighbour-joining. In *International Joint Conference on Biomedical Engineering Systems and Technologies*, Fred, A., Filipe, J., Gamboa, H. (eds). Springer, Berlin, Heidelberg.

Sims, G.E. and Kim, S.-H. (2011). Whole-genome phylogeny of *Escherichia coli/Shigella* group by feature frequency profiles (FFPs). *Proceedings of the National Academy of Sciences*, 108(20), 8329–8334.

Sims, G.E., Jun, S.-R., Wu, G.A., Kim, S.-H. (2009). Alignment-free genome comparison with feature frequency profiles (FFP) and optimal resolutions. *Proceedings of the National Academy of Sciences*, 106(8), 2677–2682.

Sneath, P.H. (1957). The application of computers to taxonomy. *Microbiology*, 17(1), 201–226.

Sokal, R.R. and Michener, C.D. (1958). A statistical method for evaluating systematic relationships. *University of Kansas Science Bulletin*, 38, 1409–1438.

Sokal, R.R. and Sneath, P.H. (1963). *Principles of Numerical Taxonomy*. W. H. Freeman and Company, New York.

Studier, J.A. and Keppler, K.J. (1988). A note on the neighbor-joining algorithm of Saitou and Nei. *Molecular Biology and Evolution*, 5(6), 729–731.

Swofford, D.L. (2003). *PAUP*: Phylogenetic Analysis Using Parsimony (* and Other Methods), Version 4*. Sinauer Associates, Sunderland, MA.

Swofford, D.L., Olsen, G.J., Waddell, P.J., Hillis, D.M. (1990). Phylogeny reconstruction. In *Molecular Systematics*, Hillis, D., Moritz, C., Mable, B. (eds). Sinauer Associates, Sunderland, MA.

Ulitsky, I., Burstein, D., Tuller, T., Chor, B. (2006). The average common substring approach to phylogenomic reconstruction. *Journal of Computational Biology*, 13(2), 336–350.

Vach, W. (1989). Least squares approximation of additive trees. In *Conceptual and Numerical Analysis of Data*, Opitz, O. (ed.). Springer-Verlag, Berlin.

Vachaspati, P. and Warnow, T. (2015). Astrid: Accurate species trees from internode distances. *BMC Genomics*, 16(S10), S3.

Vinga, S. (2014). Information theory applications for biological sequence analysis. *Briefings in Bioinformatics*, 15(3), 376–389.

Vinh, L.S. and von Haeseler, A. (2005). Shortest triplet clustering: Reconstructing large phylogenies using representative sets. *BMC Bioinformatics*, 6(1), 92.

Wheeler, T.J. (2009). Large-scale neighbor-joining with NINJA. In *International Workshop on Algorithms in Bioinformatics*, Salzberg, S.L. and Warnow, T. (eds). Springer-Verlag, Berlin, Heidelberg.

Yang, Z. (2006). *Computational Molecular Evolution*. Oxford University Press.

Zaretskii, K. (1965). Constructing a tree on the basis of a set of distances between the hanging vertices. *Uspekhi Matematicheskikh Nauk*, 20(6), 90–92.

Zaslavsky, L. and Tatusova, T.A. (2008). Accelerating the neighbor-joining algorithm using the adaptive bucket data structure. In *International Symposium on Bioinformatics Research and Applications*, Măndoiu, I., Sunderraman, R., Zelikovsky, A. (eds). Springer-Verlag, Berlin, Heidelberg.

Zielezinski, A., Vinga, S., Almeida, J., Karlowski, W.M. (2017). Alignment-free sequence comparison: Benefits, applications, and tools. *Genome Biology*, 18(1), 186.

Zielezinski, A., Girgis, H.Z., Bernard, G., Leimeister, C.-A., Tang, K., Dencker, T., Lau, A.K., Röhling, S., Choi, J.J., Waterman, M.S. et al. (2019). Benchmarking of alignment-free sequence comparison methods. *Genome Biology*, 20(1), 144.

7

Computing Inference in Phylogenetic Trees

Laurent GUÉGUEN[1,2]
[1] *LBBE, Université Claude Bernard Lyon 1, Villeurbanne, France*
[2] *Swedish Collegium for Advanced Study, Uppsala, Sweden*

7.1. Inferences and modeling

7.1.1. *Inferences*

The purpose of phylogenetic inference is to explain, from current homologous sequences, the evolutionary process that generated them. Evolutionary process designates the mechanisms (mutations, inversions, speciations, transfers, duplications, etc.) that cause events in sequence evolution, as well as the original sequence.

In this chapter, we will see how this inference can be performed in molecular phylogeny. This is based on a model that will provide an objective criterion to select among all of the processes via a score function. Usually, two types of modeling can be employed: maximum parsimony and likelihood. Seeking maximum parsimony means assuming that data are originating from a scenario that is as short as possible in terms of numbers of events or, more generally if there are costs associated with the events, the least expensive possible. Computing the likelihood of these data falls within the framework of a probabilistic model (see Chapter 2) and the goal is then to calculate the data probability according to a given evolutionary model. It may be necessary

Models and Methods for Biological Evolution,
coordinated by Gilles DIDIER and Stéphane GUINDON.

to maximize this likelihood or to use it as part of a Bayesian-based approach (see Chapter 8).

Apart from calculating a score, this approach should be utilized to reconstruct the history that led to the current sequences: what were the ancestral sequences and what steps have they used to evolve into the current sequences? This is known as ancestral inference, we will see which calculations can be used to perform phylogenetic inference and then ancestral inference.

The set of mechanisms known in molecular evolution is too complex to be wholly considered in phylogenetic inference only a few of these mechanisms are therefore taken into consideration. A part of the research in molecular phylogeny aims at taking into account increasingly more mechanisms to infer ever more realistic processes. Here, we only consider two types of mechanisms: speciations and substitutions.

As a matter of fact, it is assumed that the process has taken place according to a dynamic tree (see Chapter 1). It is possible that the topology of this tree is unknown and must be inferred. This can be done by means of methods based on the distance between sequences (see Chapter 6) or during the estimation of the evolutionary process. In this chapter, in order to define the computational methods, we will consider that this topology is given.

7.1.2. *Parsimony and likelihood*

Before treating from a general perspective the algorithmic aspects of the score computation for a process, we will explain this computation in the two most common cases: parsimony and likelihood.

We take a simple example: the data are three sequences $\mathcal{D} = \{\mathtt{S}, \mathtt{T}, \mathtt{U}\}$ originating from an unknown evolutionary process along a tree $\mathcal{A}$ (see Figure 7.1). We denote by $\mathbb{P}$ the set of the possible processes that could have created $\mathcal{D}$ according to $\mathcal{A}$.

For the sake of being general, we will say that a process enables shifting from one state to another, and that a change of state is a substitution. From the most general point of view, a state can be an entire sequence, a site or a set of sites. Let $\mathbb{E}$ be the set of possible states.

7.1.3. *Maximum parsimony*

Given a cost associated with the events, the maximum parsimony, $M(\mathcal{D})$ is the minimum cost over the set of processes:

$$M(\mathcal{D}) = min_{p \in \mathbb{P}} \text{cost}(\mathcal{D}|p, \mathcal{A})$$

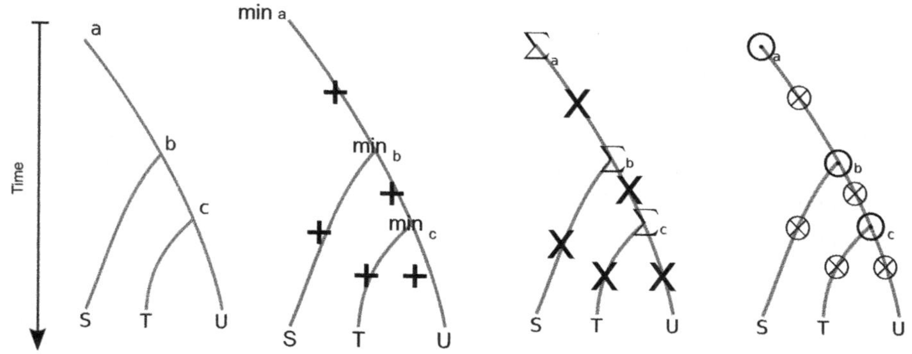

Figure 7.1. *Schemes for process score computations given the current sequences S, T and U. On each tree are placed the operators according to which the computation is performed. From left to right: via given ancestry data, maximum parsimony, likelihood and formal computation*

For simplicity, we denote by $\mathrm{mcost}(\mathcal{D}|\mathcal{A})$ this minimum cost.

To calculate this minimum, we consider that evolution is memoryless. The cost of shifting from a state a to a state b depends on a and b, but a being given, this cost does not depend on states from which a is derived. Therefore, to calculate the cost along a process, one simply has to calculate the costs of the independent shifts between all successive states of this process, and calculating a minimum on $\mathbb{P}$ amounts to calculating a minimum over the set of unknown states.

If we denote by $a, b, c \in \mathbb{E}$, the unknown states at the root and at the tree nodes, and by $a \to b$ the transition from a to b, $\mathrm{mcost}(\mathcal{D}|a, b, c, \mathcal{A})$ is the score given these ancestral states (see Figure 7.1a) in order to calculate the minimum:

$$
\begin{aligned}
M(\mathcal{D}) &= \min_{a,b,c}[\mathrm{mcost}(\mathcal{D}|a,b,c,\mathcal{A})] \\
&= \min_{a,b,c}\Big[\mathrm{mcost}(a \to b) + \Big(\mathrm{mcost}(b \to \mathtt{S}) + \mathrm{mcost}(b \to c) \\
&\quad + \big(\mathrm{mcost}(c \to \mathtt{T}) + \mathrm{mcost}(c \to \mathtt{U})\big)\Big)\Big] \\
&= \min_{a}\Big[\min_{b} \mathrm{mcost}(a \to b) + \Big(\mathrm{mcost}(b \to \mathtt{S}) + \min_{c} \mathrm{mcost}(b \to c) \\
&\quad + \big(\mathrm{mcost}(c \to \mathtt{T}) + \mathrm{mcost}(c \to \mathtt{U})\big)\Big)\Big]
\end{aligned}
$$

It is clear that it is possible to separate the different minima along the tree: the minimum on a does not depend on c, because node b is situated between them. This is due to distributivity $\min(\alpha + \beta, \alpha + \gamma) = \alpha + \min(\beta, \gamma)$.

On each branch, we must compute the minimum cost, over the set of processes, of the transition from the beginning state to the end state of the branch. In absolute terms, this requires explicitly considering all these processes, that is, all possible states that have appeared along the branch: it becomes prohibitive, especially when the branch is long.

Instead, it will be considered that the minimum cost can be calculated directly from the two states located at the ends of this branch, without considering intermediate states. Now the unknown is reduced to the set of states at the nodes. This allows the simplification of the cost function:

$$\forall a, b \in \mathbb{E}, \text{mcost}(a \to b) = \text{cost}(a, b)$$

In section 7.3, we will see which condition the cost function must satisfy in order to always be associated with a minimum over the set of processes.

7.1.3.1. *Likelihood*

We now consider the calculation of the likelihood. We denote by P the probability. The likelihood of the data $P(\mathcal{D})$ is the sum of the likelihoods of D according to all possible processes. By taking the example of Figure 7.1:

$$\begin{aligned} P(\mathcal{D}) &= \sum_{p \in \mathbb{P}} P(\mathcal{D}|p)P(p) \\ &= \sum_{p \in \mathbb{P}} \sum_{a,b,c \in \mathbb{E}} P(\mathcal{D}|a,b,c,p)P(a,b,c|p)P(p) \end{aligned}$$

Assuming also that evolution has no memory, and that each branch has an independent history, a process on a tree can be broken down into independent processes over the different branches[1]:

$$\begin{aligned} P(\mathcal{D}) = &\sum_{a,b,c \in \mathbb{E}} (\sum_p P(a|p)P(p))(\sum_p P(a \to b|p)P(p)) \\ &\times(\sum_p P(b \to \text{S}|p)P(p)) \\ &\times(\sum_p P(b \to c|p)P(p))(\sum_p P(c \to \text{T}|p)P(p))(\sum_p P(c \to U|p)P(p)) \end{aligned}$$

1 Each sum is carried out over the set of processes in the branch under consideration.

By definition, the transition probability between two states a and b is the sum of the probabilities over the set of processes that lead from a to b, so that, for example $\sum_{p \text{process}} P(a \to b|p)P(p) = P(b|a)$:

$$P(\mathcal{D}) = \sum_{a,b,c} \left[P(a) \times P(b|a) \times \Big(P(\mathtt{S}|b) \times P(c|b) \times \big(P(\mathtt{T}|c) \times P(\mathtt{U}|c) \big) \Big) \right]$$
$$= \sum_a P(a) \left[\sum_b P(b|a) \times \Big(P(\mathtt{S}|b) \times \sum_c P(c|b) \times \big(P(\mathtt{T}|c) \times P(\mathtt{U}|c) \big) \Big) \right]$$

Again, it is possible to separate the different sums along the tree. The sum over a does not depend on c, because the node carrying b is between them. This is due to distributivity: $\alpha \times \beta + \alpha \times \gamma = \alpha \times (\beta + \gamma)$.

7.2. Dynamic programming

As the method for calculating the score is the same for both approaches, we express it in a more formal manner in order to focus on the algorithm without having to worry about the criterion itself. We identify the computational operators over the tree, as well as the operator over the set of histories. Regardless of the histories, each transition between two states is associated with a cost σ, and possibly a cost over the states at the root π. The operators are described in Table 7.1.

Location	Computation type		
	Generic	Parcimony	Likelihood
Root	π	1	P
Node	$\oplus$	min	$\sum$
Node	$\odot$	$\sum$	$\times$
Branch	σ	cost	P
Branch	$M\otimes$	min cost $\sum$	$\sum P \times$

Table 7.1. *Synthesis of the operators involved in dynamic programming computations on trees.*

Therefore, reusing the example of Figure 7.1, the calculations consist of evaluating a score S according to the scores σ over the branches:

$$S(\mathcal{D}) = \bigoplus_{a \in \mathbb{E}} \pi(a) \odot \left[\bigoplus_{b \in \mathbb{E}} \sigma(a,b) \odot \Big(\sigma(b,S) \odot \bigoplus_{c \in \mathbb{E}} \sigma(b,c) \odot \big(\sigma(c,\ T) \odot \sigma(c,\mathtt{U}) \big) \Big) \right]$$

Since the calculation of $\bigoplus_x$ in each node depends on the node from which it is derived (in our example the calculation over c depends on b), it is not possible to directly solve the operator $\bigoplus$ at the node where it is applied. It must be calculated at the beginning of its upper branch after calculating $\sigma\odot$ on this branch for all possible states at the end of the branch. For this reason, the computation will be performed from the leaves to the root, going up through the tree, and during the ascension, it will be necessary to store the values calculated for all the states. These values will be stored in vectors v, and the computation will consist of a sequence of actions on these vectors.

A distinction can be made between the computation at the root from the computation at the vectors:

$$\begin{cases} S(\mathcal{D}) &= \bigoplus_{a\in\mathbb{E}} \pi(a) \odot v_{\text{root}}(a) \\ v_{\text{root}}(a) &= \bigoplus_b \sigma(a,b) \odot \Big(\sigma(b,S) \odot \bigoplus_c \sigma(b,c) \odot \big(\sigma(c,\mathtt{T}) \odot \sigma(c,\mathtt{U})\big)\Big) \end{cases}$$

Therefore, let us build the operators that will be applied to the vectors.

7.2.1. *Over the branches*

Over each internal branch, we always have the sequence $\bigoplus\sigma\odot$. These operators will therefore be combined together, and we denote by the product $M\otimes$ their action:

$$M \otimes v = (\bigoplus_{y\in\mathbb{E}} \sigma(x,y) \odot v(y))_{x\in\mathbb{E}}$$

The result of this product is a new vector over the set of possible states, and this allows us to climb up the branches of the tree.

With regard to the terminal branches, it is necessary to define at each leaf a vector v for which $M \otimes v$ equals the vector of the transition costs to the observed value. In our example, we want the following equality to hold:

$$M \otimes v_{\mathtt{U}} = \begin{pmatrix} \sigma(c_1,\mathtt{U}) \\ \vdots \\ \sigma(c_z,\mathtt{U}) \end{pmatrix}$$

To this end, the vector at each leaf comprises the neutral element of $\oplus$ (denoted by 0), except for the value corresponding to the observed state which will be the neutral element of $\odot$ (denoted 1). We recall that the necessary property for our calculation to work in this way is distributivity: $(\alpha\odot\beta)\oplus(\alpha\odot\gamma) = \alpha\odot(\beta\oplus\gamma)$. Consequently, we have that $\forall x \in \mathbb{E}, x = x\odot(0\oplus 1) = (x\odot 0)\oplus(x\odot 1) = (x\odot 0)\oplus x$ and thus $x\odot 0 = 0$: 0 is the absorbing element of $\odot$.

If, for example, $\mathbb{E} = \{\mathtt{A},\ \mathtt{C},\ \mathtt{G},\ \mathtt{T}\}$ and at a leaf the state A is observed, the computation on the terminal branch which leads to this leaf is:

$$M \otimes \begin{pmatrix} 1 \\ 0 \\ 0 \\ 0 \end{pmatrix} = \begin{pmatrix} (\sigma(\mathtt{A},\mathtt{A}) \odot 1) \oplus (\sigma(\mathtt{A},\mathtt{C}) \odot 0) \oplus (\sigma(\mathtt{A},\mathtt{G}) \odot 0) \oplus (\sigma(\mathtt{A},\mathtt{T}) \odot 0) \\ (\sigma(\mathtt{C},\mathtt{A}) \odot 1) \oplus (\sigma(\mathtt{C},\mathtt{C}) \odot 0) \oplus (\sigma(\mathtt{C},\mathtt{G}) \odot 0) \oplus (\sigma(\mathtt{C},\mathtt{T}) \odot 0) \\ (\sigma(\mathtt{G},\mathtt{A}) \odot 1) \oplus (\sigma(\mathtt{G},\mathtt{C}) \odot 0) \oplus (\sigma(\mathtt{G},\mathtt{G}) \odot 0) \oplus (\sigma(\mathtt{G},\mathtt{T}) \odot 0) \\ (\sigma(\mathtt{T},\mathtt{A}) \odot 1) \oplus (\sigma(\mathtt{T},\mathtt{C}) \odot 0) \oplus (\sigma(\mathtt{T},\mathtt{G}) \odot 0) \oplus (\sigma(\mathtt{T},\mathtt{T}) \odot 0) \end{pmatrix}$$

$$= \begin{pmatrix} \sigma(\mathtt{A},\mathtt{A}) \\ \sigma(\mathtt{C},\mathtt{A}) \\ \sigma(\mathtt{G},\mathtt{A}) \\ \sigma(\mathtt{T},\mathtt{A}) \end{pmatrix}$$

This corresponds to the desired values at the top of the branch which leads to A.

7.2.2. *Over the nodes*

After the calculation of the vectors along the branches, let us define the one at the nodes. At each node, the operator $\odot$ is applied between values that share a same state (e.g. $\sigma(c, T) \odot \sigma(c, \mathtt{U})$ for each state c), so that $\odot$ has merely to be extended through a vector term-wise product to produce a new vector:

$$v \odot w = (v(i) \odot w(i))_{i \in \mathbb{E}}$$

7.2.3. *Over the tree*

Moving up along the branches and nodes of the tree, the score can be computed from the leaves up to the root. Let v_n be the vector of values at node n, and $v_{n \to m}$ the vector of values at the top of branch $n \to m$. In our example (see Figure 7.2), the vector at the root is:

$$v_6 = M_{6\to5} \otimes \Big((M_{5\to4} \otimes v_4) \odot \big(M_{5\to3} \otimes ((M_{3\to2} \otimes v_2) \odot (M_{3\to1} \otimes v_1))\big)\Big)$$

And the different steps of the computation can be:

$$\begin{aligned} v_6 &= M_{6\to5} \otimes \Big((M_{5\to4} \otimes v_4) \odot \big(M_{5\to3} \otimes ((M_{3\to2} \otimes v_2) \odot (M_{3\to1} \otimes v_1))\big)\Big) \\ &= M_{6\to5} \otimes \Big(v_{5\to4} \odot \big(M_{5\to3} \otimes (v_{3\to2} \odot v_{3\to1})\big)\Big) \\ &= M_{6\to5} \otimes \big(v_{5\to4} \odot (M_{5\to3} \otimes v_3)\big) \\ &= M_{6\to5} \otimes (v_{5\to4} \odot v_{5\to3}) \\ &= M_{6\to5} \otimes v_5 \end{aligned}$$

7.2.4. *At the root*

Finally, the values of the root vector must be accumulated to obtain a score. Since this accumulation is made over the set of processes, it has the same operator as for the nodes, namely $\odot$. To allow for different weights for the states at the root, a cost π is integrated, and the operator at the root defined:

$$\pi \otimes v = \bigoplus_{x \in \mathbb{E}} \pi(x) \odot v(x)$$

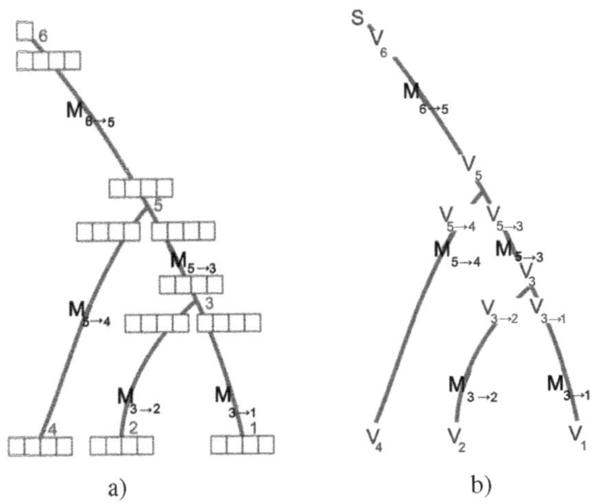

Figure 7.2. *Score vector computation. (a) Representation of the vectors involved in the computation. (b) Representation with the names of the vectors at the nodes and at the top of the branches. For a color version of this figure, www.iste.co.uk/didier/models.zip*

7.2.5. *Recursion relations*

As it is performed in a tree, this computation can be very easily implemented recursively following the tree topology, as shown in Figure 7.3. The node D has three children, A, B and C, which can be internal nodes or leaves. The operators $M\otimes$ enable moving up the branches, and at node D the operator $\odot$ enables branches to be joined. With these operators, we calculate v_D from v_A, v_B and v_C: $v_D = (M_{D\to A} \otimes v_A) \odot (M_{D\to B} \otimes v_B) \odot (M_{D\to C} \otimes v_C)$. This vector v_D will then be used to calculate the scores in the nodes above D.

If $e(n)$ denotes the set of nodes which are directly under node n, the vectors are successively computed from the vectors at the leaves (constructed as explained in section 7.2.1):

$$\begin{cases} v_n &= \bigodot_{m\in e(n)} v_{n\to m} \\ v_{n\to m} &= M_{n\to m} \otimes v_m \\ \\ S(\mathcal{D}) &= \pi \otimes v_{\text{root}} \end{cases}$$

In order for the method to be less memory intensive, the vectors at the top of the branches can be removed ($v_{n\to m}$) and the computations at the nodes grouped together:

$$v_n = \bigodot_{m\in e(n)} M_{n\to m} \otimes v_m$$

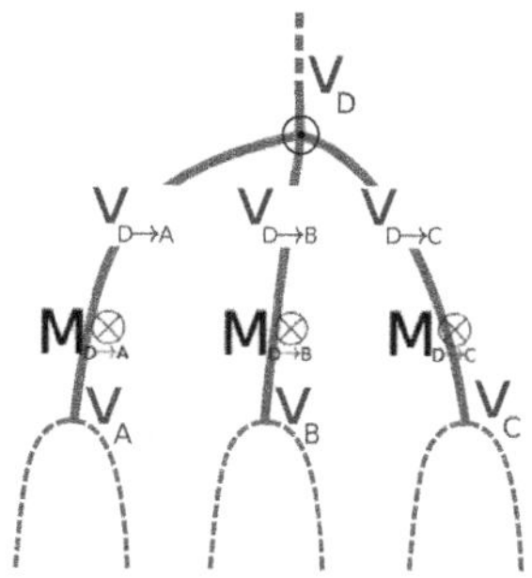

Figure 7.3. *Recursive computation of score vectors from nodes A, B and C toward node D. For a color version of this figure, www.iste.co.uk/didier/models.zip*

In our example, the steps are then:

$$\begin{aligned} v_6 &= M_{6\to 5} \otimes \Big((M_{5\to 4} \otimes v_4) \odot \big(M_{5\to 3} \otimes \; ((M_{3\to 2} \otimes v_2) \odot (M_{3\to 1} \otimes v_1)) \big)\Big) \\ &= M_{6\to 5} \otimes \Big((M_{5\to 4} \otimes v_4) \odot \big(M_{5\to 3} \otimes \qquad v_3 \qquad \big)\Big) \\ &= M_{6\to 5} \otimes \qquad\qquad v_5 \end{aligned}$$

However, the first implementation can be useful when manipulating the scores at the top of the branches (e.g. to test the position of a clade at different locations in the tree, this vector can be used to "walk" at these positions, without having to recalculate the score in this clade). Also, in the likelihood, the vectors $v_{n\to m}$ are useful for the calculation at nodes (see section 7.4.2).

The method presented here can be applied to any type of operators, as long as the distributivity rule between $\odot$ and $\oplus$ is verified. In addition, for the decomposition of the tree into branches to be consistent, it is necessary that the score σ and the operator $\oplus$ are of the same type, namely both a minimum, a sum, etc. over the set of processes. This is not always possible. For example, when the evolutionary mechanisms under consideration are too complex to effectively calculate a cost on any distance (gene transfers, genome rearrangements, etc.) or when the cost function does not simply cumulate along the branch. In this case, each branch should be divided into small pieces from which it will be assumed that zero or one event has taken place, and the costs of all these small events will be accumulated: this is very costly in time and memory.

7.2.6. *Complexity reduction*

As $M\otimes$ is applied to all pairs of states in each branch, the time complexity is linear with the number of branches and quadratic with the number of states[2]. Which states should be considered? If assuming that a state is an entire sequence, the number of possible states grows exponentially with the length of the sequence, and manipulating all these states becomes prohibitive.

The common approach to reduce the number of states is to segment the evolutionary process on sequences into a set of independent evolving processes over portions of these sequences. These portions are called sites. In the finest segmentation, a site is a position of each sequence, the number of states then being the size of the sequence alphabet (4 nucleotides or 20 amino acids). For coding sequences, it makes more sense to take into account the biological reality of the genetic code, and therefore to consider codons as autonomous evolutionary sites (usually 61 non-stop codons). Based on the assumption that all sites are independent, the calculations in the tree can be independently performed at each site, and then "simply" cumulated along the sequences, so that the complexity is now proportional to the number of sites multiplied by squared the size of the alphabet. From a modeling point of view, the considered substitutions are generally substitutions in the biological sense (see Chapter 2), but nothing prevents us from considering different evolutionary mechanisms as long as independence between sites is preserved.

Several independent processes on the data set are then available, whose scores will be combined to form the global score. Since, in the tree calculation, the independent calculations under the nodes are cumulated by the operator $\odot$, this same operator will

2 $\odot$ is applied to every state, thus of linear complexity with the number of states.

be used to accumulate the independent scores of the different sites. As such, if the data $\mathcal{D}$ are divided into sites $\mathcal{D}_1, ..., \mathcal{D}_l$ which are considered independent:

$$S(\mathcal{D}) = \bigodot_{i \in <1,l>} S(\mathcal{D}_i)$$

At the implementation level, at each node the values will be stored as two-dimensional "number of sites × number of states" arrays. It is also possible that the data at several sites be similar. This will be all the more frequent as the alphabet and the tree will be small and the sites numerous. In this case, it is not necessary to maintain the calculations for these different sites but simply one of them, and the calculation of the global value will take into account the number of copies of this site.

The hypothesis of independently evolving sites can be considered as being too strong for the study of certain sequences, or certain mechanisms, or even completely not applicable. For example, in the evolution of proteins this hypothesis prevents from taking into account relationships between amino acids. In the same way, the mutations that involve several neighboring nucleotides (such as CpGs) contradict this hypothesis. In fact, genetic transmission at the individual scale is done through haplotypes, and evolution features can depend on this dynamics (*background selection*, *selective sweeps*). This non-independence between sites is moreover a very important information in population genomics. On the other hand, considering in phylogeny that the sites are independent leads one to assume that at the time scale under consideration, there has been enough recombination events to break the influence of this dependence.

7.2.7. *Root management*

In the example of Figure 7.1, a branch leads from the root to the highest node on the tree. This branch corresponds to the past of the process before the first speciation event. The principle of phylogenetic inference at a point in the tree consists of using information from the comparison between sequences that are on either side of this point. In the absence of information upstream this radical branch, it is not possible to reconstruct the process. Therefore, for phylogenetic inference, this branch is removed from the model and the root is carried by the most ancestral node of the tree.

In the chapters that cover distance-based reconstruction methods and the bayesian approach (see Chapters 6 and 8), the inferred trees are not rooted. This means that we do not have the orientation of the evolutionary process. In this case, under what conditions is it possible to calculate a score or is it always necessary to have a root (i.e. a process orientation)?

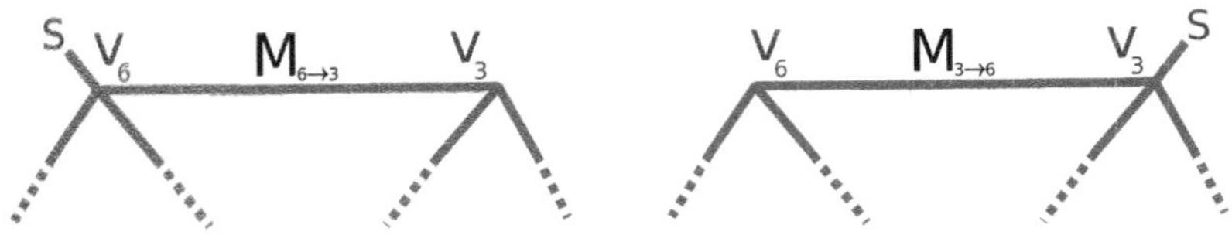

Figure 7.4. *Example of the score computation according to whether the root S is at node 6 or node 3 of the tree. For a color version of this figure, www.iste.co.uk/didier/models.zip*

Let us consider a tree where the process is not oriented. The traveling direction of a branch does not matter in the computational algorithm. In the example of Figure 7.4, this means that $M_{3\to 6} = M_{6\to 3} := M$. In this case, if the root is placed at node 6, the score is $S_6(\mathcal{D}) = \pi \otimes (v_6 \odot (M \otimes v_3))$ and if it is placed at node 3, it is: $S_3(\mathcal{D}) = \pi \otimes (v_3 \odot (M \otimes v_6))$.

In order to calculate a score independently of the position of the root, these two scores must be equal. Since the computation is linear, we can verify for all the components of v_3 and v_6. By denoting δ_i the vector in which all the terms are equal to 0, except for the ith element which is equal to 1, if $v_3 = \delta_i$ and $v_6 = \delta_j$:

$$\begin{aligned} S_6(\mathcal{D}) &= \pi \otimes (\delta_j \odot (M \otimes \delta_i)) \\ &= \pi \otimes \left(\delta_j \odot \begin{pmatrix} M_{1i} \\ \vdots \\ M_{ji} \\ \vdots \\ M_{ni} \end{pmatrix}\right) = \pi \otimes \begin{pmatrix} 0 \\ \vdots \\ M_{ji} \\ 0 \\ \vdots \end{pmatrix} \leftarrow j\text{th line} \\ &= \pi_j \otimes M_{ji} \end{aligned}$$

and $S_3(\mathcal{D}) = \pi_i \otimes M_{ij}$.

The scores are therefore equal if $\forall i, j \in \mathbb{E}, \pi_i \otimes M_{ij} = \pi_j \otimes M_{ji}$. Under this condition, step by step, the root can be moved to all points of the tree, and thereby the score can be computed on an undirected tree. This is called the pulley principle. In practice, to perform the computation, the root is placed at any point of the tree.

If this condition is not satisfied at any point in the tree, the orientation of the process becomes relevant for calculating the score, and it is necessary to identify the position of the root in the tree.

In the following sections, we will apply these calculations to specific maximum parsimony and likelihood-based approaches. We will present specificities for each approach, and how they can be used for ancestral inference.

7.3. Maximum parsimony

Maximum parsimony is used for building the most parsimonious scenario (in other words, with the least amount of changes) that explains the data. The algorithms used for computing this minimum were proposed by Walter Fitch in 1971 and David Sankoff in 1975. We present the Sankoff's approach which is part of the general calculation of dynamic programming and is more generalizable than Fitch's approach. This algorithm can be efficiently extended when the state space is large and the cost function presents certain properties (see Csürös (2013) for an overview).

Here, the algorithm presented in the previous section is performed with the operators $(\min, +)$ for $(\odot, \otimes)$[3]. The matrix of the substitution costs between states must be defined. In the simplest case, each substitution costs 1:

$$M = \begin{pmatrix} 0 & 1 & 1 & 1 \\ 1 & 0 & 1 & 1 \\ 1 & 1 & 0 & 1 \\ 1 & 1 & 1 & 0 \end{pmatrix}$$

At the leaves, the vectors are equal to ∞ for all states, except for observed states where the value is 0. As a matter of fact, ∞ is the neutral element of min ($\min(x, \infty) = x$) and 0 is the neutral of $+$ (see section 7.2.1).

In the example of Figure 7.5, if the states are taken following the order A, C, G, T: $v_1 = (0, \infty, \infty, \infty)^{\mathrm{T}}$, $v_2 = (\infty, 0, \infty, \infty)^{\mathrm{T}}$ and:

$$v_{3\to 1} = \begin{pmatrix} 0 & 1 & 1 & 1 \\ 1 & 0 & 1 & 1 \\ 1 & 1 & 0 & 1 \\ 1 & 1 & 1 & 0 \end{pmatrix} \otimes \begin{pmatrix} 0 \\ \infty \\ \infty \\ \infty \end{pmatrix} = \begin{pmatrix} \min(0, \infty, \infty, \infty) \\ \min(1, \infty, \infty, \infty) \\ \min(1, \infty, \infty, \infty) \\ \min(1, \infty, \infty, \infty) \end{pmatrix} = \begin{pmatrix} 0 \\ 1 \\ 1 \\ 1 \end{pmatrix}$$

$$v_{3\to 2} = \begin{pmatrix} 0 & 1 & 1 & 1 \\ 1 & 0 & 1 & 1 \\ 1 & 1 & 0 & 1 \\ 1 & 1 & 1 & 0 \end{pmatrix} \otimes \begin{pmatrix} \infty \\ 0 \\ \infty \\ \infty \end{pmatrix} = \begin{pmatrix} \min(1, \infty, \infty, \infty) \\ \min(0, \infty, \infty, \infty) \\ \min(1, \infty, \infty, \infty) \\ \min(1, \infty, \infty, \infty) \end{pmatrix} = \begin{pmatrix} 1 \\ 0 \\ 1 \\ 1 \end{pmatrix}$$

3 Compared to the usual computations in $(\mathbb{R}, +, \times)$, min replaces "+", "+" replaces "×".

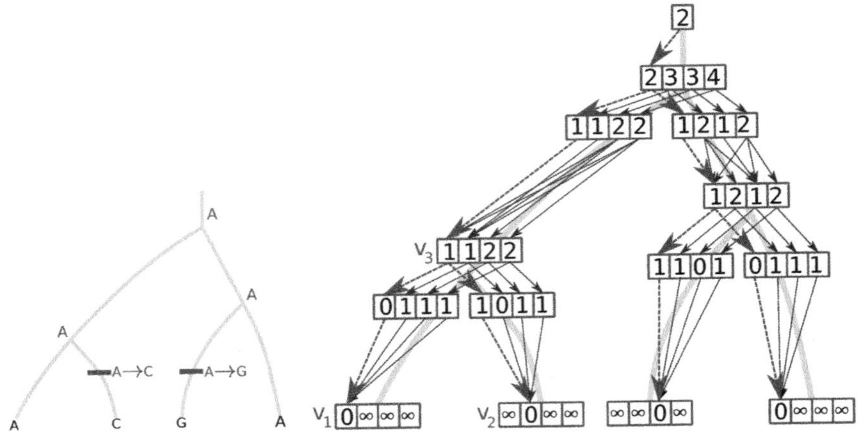

Figure 7.5. *Example of parsimony and reconstruction computations for the most parsimonious scenario. (a) The letters at the leaves and the states inferred by the most parsimonious scenario. Both red bars represent the substitutions. (b) The calculated vectors and in dotted lines the links of the minimum cell arguments. For a color version of this figure, www.iste.co.uk/didier/models.zip*

Hence:

$$v_3 = v_{3\to 1} \odot v_{3\to 2} = \begin{pmatrix} 0 \\ 1 \\ 1 \\ 1 \end{pmatrix} + \begin{pmatrix} 1 \\ 0 \\ 1 \\ 1 \end{pmatrix} = \begin{pmatrix} 1 \\ 1 \\ 2 \\ 2 \end{pmatrix}$$

Since a substitution costs 1, each value of the vector v_3 represents the number of substitutions needed to explain the data from a starting state. For example, if at node 3 the state is G, two substitutions are required (one for each branch) to explain the data under this node. If the state is C, one substitution is enough (on the branch leading to leaf 1).

As shown in Figure 7.5, the cost necessary to explain the data according to the state present in each node is calculated. Finally, if there is a priori no preference on the state at the root, $\pi = 0$ and the minimum of the values of v_{root} is kept.

It is quite possible to put other values in M such that different substitution costs between states are taken into account. Nevertheless, it should be remembered that the cost over a whole branch is the minimum of the costs of all possible processes over this branch. Consequently, in the case where this cost is the same regardless of the branch (e.g. it does not depend on length), if M is the cost matrix, $M \otimes M = M$

because the introduction of the calculation of an intermediate vector within a branch must not change the result. This means that:

$$\forall a, b \in \mathbb{E}, M(a,b) = \min_{c \in \mathbb{E}}(M(a,c) + M(c,b))$$

In particular, the triangular inequality must be respected:

$$\forall a, b, c \in \mathbb{E}, M(a,b) \leqslant M(a,c) + M(c,b)$$

Generally, the cost corresponds to a distance between states and is symmetric ($M(a,b) = M(b,a)$). It makes no difference between the process orientations. There is also generally no preferences on the root states, so that the parsimony computation does not depend on the location of the root in the tree. In this case, this calculation can be done on unrooted trees.

7.3.1. *Ancestral interference*

Since the parsimony computation constructs a minimum, it is directly possible to identify the most parsimonious scenario(s), that is, both the ancestral sequences and the events that minimize the number of substitutions on the tree. For each minimum calculation, the state from which this minimum has been reached should be simply memorized (or several states in case of equality). Then, starting from the root, the reverse path is traveled downwards the tree to obtain the succession of substitutions that have reached the minimum.

For example, in Figure 7.5, the minimum score is equal to 2. During the calculation, in each vector cell, the references of the cell that provided the minimums are stored (descending arrows). Then, we proceed by following the arrows from the minimum stored at the root up to the leaves (red arrows). If an arrow points different vector cells, it means that there has been a substitution in the corresponding branch. For example, in Figure 7.5, the red arrow on branch $3 \rightarrow 2$ corresponds to a substitution from state A to state C.

Therefore, the succession of ancestral sequences and the successive substitutions between them can be quickly reconstructed.

7.4. Likelihood

In this part, we come back to the calculation of the data score within the context of a probabilistic process: this score is the probability that the data were generated by this process, which is called likelihood. This calculation is achieved in the same way as parsimony, but this time in the standard $(+,.)$ algebra.

The transition costs between states are probabilities. Due to lack of memory, the processes are Markovian, and the score matrices are conditional probability matrices. According to the same principle as for parsimony, when calculating these probabilities, it is necessary to respect the fact that the transition probability between two states over an entire branch is the sum of the probabilities of all possible processes that connect these two states on that branch. This means that if a branch β is split into two branches β_1 and β_2, the transition probability between two states a and b on β must be equal to the sum, for all states c at the break, of the transition probabilities from a to c on β_1, then from c to b on β_2. In other words, if M, M_1 and M_2 are the transition matrices on these branches, then $M = M_1.M_2$. For this purpose, the usual approach consists of modeling the evolution by continuous-time Markov processes (see Chapter 2): along a branch of length l, if Q is the generator of the model in this branch, the transition matrix $M(l) = e^{Q.l}$.

As presented in section 7.2.1, the likelihood calculation of an alignment $\mathcal{D}$ is achieved from the leaves to the root. The algorithm was described by Joe Felsenstein in 1973 (Felsenstein 1973, 1981). At a node n, the values in the vector v_n are calculated on the basis of the data $\mathcal{D}_n$ which are under this node. Denoting by X_n the random state at node n, let $x \in \mathbb{E}$, $v_n(x)$ (respectively, $v_{n\to m}(x)$) is the likelihood of $\mathcal{D}_n$ (respectively, $\mathcal{D}_m$) if the state at node n is x:

$$v_n(x) = P(\mathcal{D}_n|X_n = x) \text{ and } v_{n\to m}(x) = P(\mathcal{D}_m|X_n = x)$$

This is called conditional likelihood.

To compute these likelihoods, $\otimes$ is the usual matrix product and $\odot$ is the term-wise product of the vector elements. At the leaves, the vectors are equal 0 for all states, except for the observed states where the value is 1.

In the example in Figure 7.6, the branches $3 \to 1$ and $3 \to 2$, respectively, have lengths 1 and 0.5. The generator of the transition model is set:

$$Q = \begin{matrix} & A & C & G & T \\ A & -0.652 & 0.174 & 0.406 & 0.0725 \\ C & 0.609 & -1.52 & 0.652 & 0.261 \\ G & 0.261 & 0.652 & -1.52 & 0.609 \\ T & 0.0725 & 0.406 & 0.174 & -0.652 \end{matrix}$$

The transition matrices in these branches are, respectively:

$$M_{3\to 1} = e^{0.5Q} = \begin{matrix} & A & C & G & T \\ A & 0.583 & 0.124 & 0.182 & 0.111 \\ C & 0.273 & 0.323 & 0.217 & 0.187 \\ G & 0.187 & 0.217 & 0.323 & 0.273 \\ T & 0.111 & 0.182 & 0.124 & 0.583 \end{matrix}$$

and:

$$M_{3\to 2} = e^Q = \begin{array}{c} \\ A \\ C \\ G \\ T \end{array}\begin{array}{c} \begin{array}{cccc} A & C & G & T \end{array} \\ \begin{pmatrix} 0.742 & 0.074 & 0.132 & 0.052 \\ 0.198 & 0.512 & 0.179 & 0.112 \\ 0.112 & 0.179 & 0.512 & 0.198 \\ 0.052 & 0.132 & 0.074 & 0.742 \end{pmatrix} \end{array}$$

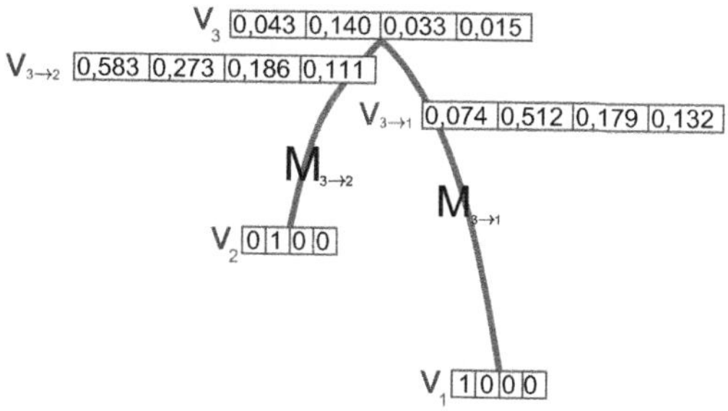

Figure 7.6. *Likelihood computation at node* 3 *from the leaves* 1 *and* 2. *For a color version of this figure, www.iste.co.uk/didier/models.zip*

To calculate v_3, one first defines $v_1 = \begin{pmatrix} 1 \\ 0 \\ 0 \\ 0 \end{pmatrix}$ and $v_2 = \begin{pmatrix} 0 \\ 1 \\ 0 \\ 0 \end{pmatrix}$.

Then:

$$v_{3\to 1} = M_{3\to 1}.v_1 = \begin{pmatrix} 0.583 \\ 0.273 \\ 0.186 \\ 0.111 \end{pmatrix}$$

and:

$$v_{3\to 2} = M_{3\to 2}.v_2 = \begin{pmatrix} 0.074 \\ 0.512 \\ 0.179 \\ 0.132 \end{pmatrix}$$

and finally:

$$v_3 = v_{3\to1} \odot v_{3\to2} = \begin{pmatrix} 0.583 \\ 0.273 \\ 0.186 \\ 0.111 \end{pmatrix} \odot \begin{pmatrix} 0.074 \\ 0.512 \\ 0.179 \\ 0.132 \end{pmatrix} = \begin{pmatrix} 0.043 \\ 0.140 \\ 0.033 \\ 0.015 \end{pmatrix}$$

Going up the tree, the conditional likelihood vector is thus calculated from the data set. Finally, to calculate the overall likelihood, this vector must be multiplied by the *prior* vector π of the state probabilities (that is without the knowledge of the data): $P(\mathcal{D}) = \pi^{\mathrm{T}}.v_{\mathrm{root}}$.

At the scale of an alignment $\mathcal{S}$, given that the sites are considered independent, the likelihood is the product of site-wise likelihoods: $P(\mathcal{S}) = \prod_{i\in<1,l>} P(\mathcal{D}i)$. In practice, if the tree is large or the alignment too long, the likelihood is too close to zero and there may be problems with the accuracy in its computation. It will then be preferable to calculate the log-likelihood: $\log P(\mathcal{S}) = \sum_{i\in<1,l>} \log P(\mathcal{D}i)$.

7.4.1. *Root management*

To apply the principle of the pulley to the likelihood case, it is necessary that for each generator Q used by the process: $\forall i, j, Q_{ij}\pi_i = Q_{ji}\pi_j$. This involves that the process is stationary, namely that π is the equilibrium distribution of the different models along the process. In particular, the sequences, both observed and ancestral, are all at this equilibrium, which is a very strong assumption.

In addition, this equality means that Q is reversible: global equilibrium is maintained by a balance between transitions within all state pairs.

For a process out of equilibrium, or whose equilibrium is not maintained by the generator reversibility, the location of the root will be important for computing the likelihood. More specifically, it will be possible to optimize the location of this root during phylogenetic inference.

7.4.2. *Computation at the nodes*

The likelihood can be computed at any node n of the tree. To this end, it is necessary to differentiate the data that are under this data node ($\mathcal{D}_n$) and those that are on the other side (the rest of the data), denoted by $\mathcal{D}^n = \mathcal{D}\backslash\mathcal{D}_n$. Since $\mathcal{D}^n$ and $\mathcal{D}_n$ are independent, and still denoting by X_n the random state variable at node n:

$$\begin{aligned} P(\mathcal{D}) &= \textstyle\sum_{x\in\mathbb{E}} P(\mathcal{D}|X_n = x)P(X_n = x) \\ &= \textstyle\sum_{x\in\mathbb{E}} P(\mathcal{D}_n|X_n = x)P(\mathcal{D}^n|X_n = x)P(X_n = x) \\ &= \textstyle\sum_{x\in\mathbb{E}} P(\mathcal{D}_n|X_n = x)P(\mathcal{D}^n, X_n = x) \end{aligned}$$

The last thing to calculate is the joined likelihood $P(\mathcal{D}^n, X_n = x)$. This computation will involve the joined probability vectors at the nodes and branches:

$$\begin{cases} v^n &= (P(\mathcal{D}^n, X_n = x))_{x \in \mathbb{E}} \\ v^{p \to n} &= (P(\mathcal{D}^n, X_p = x))_{x \in \mathbb{E}} \end{cases}$$

Calculating v^n means traveling the tree from the leaves of $\mathcal{D}^n$ up to node n, the branches going from the root to n being taken backwards (see Figure 7.7). The joint probabilities can be calculated along this backwards path, that is to say, from the parent to the child, where the conditional probabilities allowed the calculation from the child to the parent. More precisely, denoting by p the parent of a node n and the vectors:

$$\begin{cases} w_p^n = (P(\mathcal{D}^n | X_p = x))_{x \in \mathbb{E}} \\ z_n \;= (P(X_n = x))_{x \in \mathbb{E}} \end{cases}$$

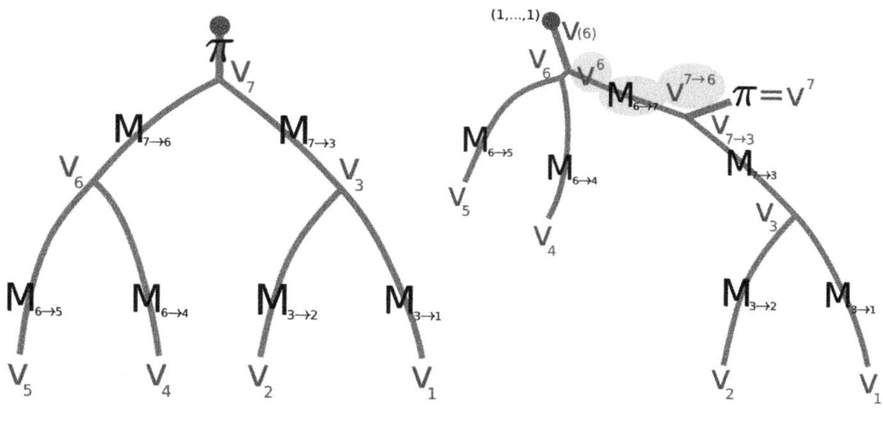

Figure 7.7. *Likelihood computation at the root (a) and at node 6 (b). In yellow, the new necessary elements for calculating the joint probabilities ($M_{6\to7} = {}^T M_{7\to6}$). For a color version of this figure, www.iste.co.uk/didier/models.zip*

The transition matrices always allow the conditional likelihoods to be linked: $w_n^n = M_{n\to p}.w_p^n$.

Since:

$$\begin{cases} v^n &= z_n \odot w_n^n \\ v^{p \to n} &= z_p \odot w_p^n \end{cases}$$

we can calculate[4] υ^n:

$$\begin{aligned}\upsilon^n &= z_n \odot w_n^n = diag(z_n).w_n^n \\ &= diag(z_n).M_{n\to p}.w_p^n\end{aligned}$$

Now, $\forall x, y, P(X_p = y|X_n = x)P(X_n = x) = P(X_n = x|X_p = y)P(X_p = y)$, which means that $diag(z_n).M_{n\to p} = (M_{p\to n}^{\mathrm{T}}).diag(z_p)$. Therefore:

$$\begin{aligned}\upsilon^n &= (M_{p\to n}^{\mathrm{T}}).diag(z_p).w_p^n \\ &= (M_{p\to n}^{\mathrm{T}}).(z_p \odot w_p^n) = (M_{p\to n}^{\mathrm{T}}).\upsilon^{p\to n}\end{aligned}$$

On the branch $p \to n$, the fact that $D^n = D^p \uplus \biguplus_{m\in e(p);m\neq n} D_m$ is used:

$$\begin{aligned}\upsilon^{p\to n} &= z(p) \odot w_p^n \\ &= z(p) \odot w_p^p \odot \bigodot_{m\in e(p);m\neq n} \upsilon_{p\to m} \\ &= \upsilon^p \odot \bigodot_{m\in e(p);m\neq n} \upsilon_{p\to m}\end{aligned}$$

where υ^p is the likelihood vector above p. If p is the root, there are no data above p, and thereby υ^p is equal to the prior state distribution: π.

Therefore, the calculation of the joint probabilities is carried out by moving through the tree up to the node of interest and applying the computations:

$$\begin{cases}\upsilon^n &= M_{p\to n}^{\mathrm{T}}.\upsilon^{p\to n} \\ \upsilon^{p\to n} &= \upsilon^p \odot \bigodot_{m\in e(p);m\neq n} \upsilon_{p\to m}\end{cases}$$

Another way to represent this calculation is to consider the tree such that n is at the top, with a fake leaf corresponding to the root π. The likelihood is calculated as previously, with υ_n and υ^n on either side of n, by multiplying by the matrix M^T when descending a branch and by the M matrix when ascending a branch.

Finally, at node n, the joint likelihoods are

$$\upsilon(n) = (P(D, X_n = x))_x = \upsilon^n \odot \upsilon_n \text{ and } P(\mathcal{D}) = (1, \ldots, 1).\upsilon(n) = \upsilon_n^{\mathrm{T}}.\upsilon^n$$

7.4.3. *Maximization, differentiation*

Generally, the likelihood $P(\mathcal{D})$ is used to find the processes that "correctly" model the data, either using a Bayesian approach, that is, the processes are sampled

4 We denote by $diag(z)$ the diagonal matrix whose diagonal terms are those of the vector z and utilize the fact that for two vectors z and w, $z \odot w = diag(w).z$.

according to their *posterior* probability (see Chapter 8), or according to a maximum likelihood-based approach, that is, by maximizing $P(\mathcal{D}|\Theta)$ over a set Θ of given parameters that define the processes. These parameters are continuous, such as modeling parameters (substitution rates, equilibrium distributions, etc.), branch lengths, or discrete, such as the topology of the tree.

For maximum parsimony, the score calculation itself defines an optimum. On the other hand, the log-likelihood calculation is a sum and does not define an optimum itself. Likelihood maximization is an optimization problem on which many numerical optimization techniques can be applied.

Many of these techniques are gradient-based methods that require knowledge of the likelihood derivatives with respect to the parameters of interest. In general, analytically calculating these derivatives proves too difficult or time consuming (for instance, for modeling parameters), and it is preferable to resort to numerical derivatives. Nevertheless, when the differentiation parameter is the branch length, it is possible to simply calculate the analytical derivative.

The likelihood over a set of sites is the product of the local likelihoods, and the differentiation of a long product proves too expensive. It will be preferable to differentiate (and optimize) the log-likelihood, which is a sum over the sites:

$$\frac{d\log P(\mathcal{S})}{d\theta} = \sum_{i \in <1,l>} \frac{dP(\mathcal{D}i)}{d\theta} . \frac{1}{P(\mathcal{D}i)}$$

and:

$$\frac{d^2\log P(\mathcal{S})}{d\theta^2} = \sum_{i \in <1,l>} \frac{\frac{d^2P(\mathcal{D}i)}{d\theta^2}P(\mathcal{D}i) - \left(\frac{dP(\mathcal{D}i)}{d\theta}\right)^2}{P(\mathcal{D}i)^2}$$

Lastly, $\frac{dP(\mathcal{D}i)}{d\theta}$ and $\frac{d^2P(\mathcal{D}i)}{d\theta^2}$ have to be calculated at each site.

7.4.3.1. *Numerical differentiation*

The principle behind the numerical derivative is to make the difference between likelihoods around the value of interest. For example, around the value θ_0 of the parameter θ by setting a small value $h > 0$:

$$\left.\frac{dP(\mathcal{D})}{d\theta}\right|_{\theta_0} \simeq \frac{P(\mathcal{D}|\theta = \theta_0 + h) - P(\mathcal{D}|\theta = \theta_0 - h)}{2h}$$

$$\left.\frac{d^2P(\mathcal{D})}{d\theta}\right|_{\theta_0} \simeq \frac{P(\mathcal{D}|\theta = \theta_0 + h) - 2P(\mathcal{D}|\theta = \theta_0) + P(\mathcal{D}|\theta = \theta_0 - h)}{h^2}$$

There are, of course, other numerical methods, but this one has the advantage of not calculating the likelihood too often. Actually, since optimization may require a large number of steps for approximating the maximum and given that the computation of the likelihood is potentially time consuming, it is critical to implement efficient auxiliary computations. This question is also valid within the context of Bayesian inference, which generally requires numerous likelihood calculations.

In absolute terms, the calculation should be contained to the places in the tree where the parameter is directly involved. For example, if it is involved in the construction of a model, the parameter will have an action on the branches where the model is applied. If this model is used everywhere, it will be necessary to recalculate the likelihood over the whole of the tree, but if it is used on a single branch, using conditional and joint likelihoods, the computation can be limited to this branch. In the example in Figure 7.7, for calculating likelihoods by modifying the transition probabilities between nodes 6 and 7, since v_6 and $v^{7\to 6}$ do not depend on $M_{6\to 7}$, one simply uses:

$$P(\mathcal{D}) = v_6^{\mathrm{T}}.M_{6\to 7}.v^{7\to 6}$$

with the different matrices $M_{6\to 7}$ that are to be tested.

7.4.3.2. *Analytical differentiation*

Let us consider an arbitrary parameter θ according to which the likelihood is to be derived. The computation at the root is:

$$\begin{cases} \mathbb{P}(\mathcal{D}) = \pi^{\mathrm{T}}.v_{\text{root}} \\ v_n = \bigodot_{m\in e(n)} M_{n\to m}.v_m \end{cases}$$

and the differentiation with respect to θ:

$$\begin{cases} \frac{dP(\mathcal{D})}{d\theta} = \pi^{\mathrm{T}}.\frac{dv_{\text{root}}}{d\theta} + \frac{d\pi^{\mathrm{T}}}{d\theta}.v_{\text{root}} \\ \frac{dv_n}{d\theta} = \sum_{m\in e(n)} \left((M_{n\to m}.\frac{dv_m}{d\theta} + \frac{dM_{n\to m}}{d\theta}.v_m) \odot \left(\bigodot_{\substack{z\in e(n)\\ z\neq m}} M_{n\to z}.v_z \right) \right) \end{cases}$$

It is thus possible to calculate $\frac{dv_n}{d\theta}$ from the vectors v_m and $\frac{dv_m}{d\theta}$ for the children m of n. Progressively up to the leaves, this equation can be implemented in an algorithm similar to the one used for calculating v_n. In general, the expression for the derivative of M with respect to a model parameter is not easy to express (mainly because of the generator normalization factor). On the other hand, for the purpose of differentiating with respect to the length of a branch, length denoted by $t_{n\to m}$, the expression is simple: $\frac{dM_{n\to m}}{dt_{n\to m}} = Q_{n\to m}.M_{n\to m}$ and for the other branches b, $\frac{dM_b}{dt_{n\to m}} = 0$, such that:

$$
\begin{cases}
\dfrac{dP(\mathcal{D})}{dt_{n\to m}} = \pi^{\mathrm{T}}.\dfrac{dv_{\text{root}}}{dt_{n\to m}} \\[2ex]
\text{If } n' \neq n \text{ is not above } n \\
\qquad \dfrac{dv_{n'}}{dt_{n\to m}} = 0 \\[2ex]
\text{If } n' \neq n \text{ is above } n\text{, denoting by } m' \text{the child of } n' \text{above } n \\
\qquad \dfrac{dv_{n'}}{dt_{n\to m}} = (M_{n'\to m'}.\dfrac{dv_{m'}}{dt_{n\to m}}) \odot \left(\bigodot_{\substack{z\in e(n') \\ z\neq m'}} M_{n'\to z} v_z \right) \\[3ex]
\dfrac{dv_n}{dt_{n\to m}} = (Q_{n\to m}.M_{n\to m}.v_m) \odot \left(\bigodot_{\substack{z\in e(n) \\ z\neq m}} M_{n\to z} v_z \right)
\end{cases}
$$

In fact, the calculation of $\frac{dP(\mathcal{D})}{dt_{n\to m}}$ is performed exactly as for $P(\mathcal{D})$, except that at the nodes above the derived branch, it is necessary to manage the computation of the derivative vectors instead of the conditional likelihood vectors (Figure 7.8).

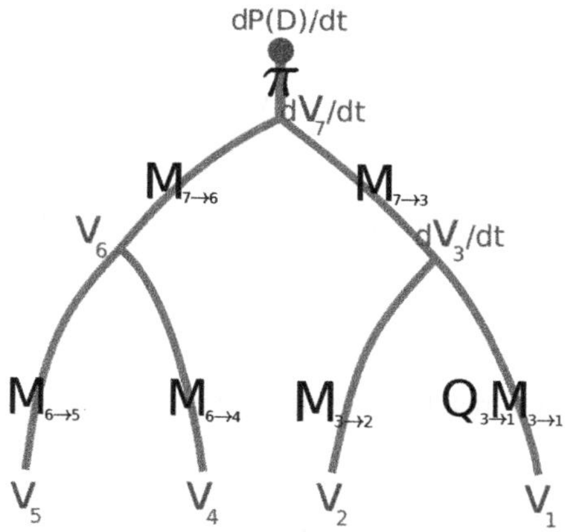

Figure 7.8. *Computation of the likelihood derivative with respect to the branch length $t_{3\to 1}$. For a color version of this figure, www.iste.co.uk/didier/models.zip*

This computation is inexpensive in terms of memory because it does not make use of the vectors $v_{n\to m}$ at the top of the branches. However, the approach is also valid if using these vectors and will be faster. In the example in Figure 7.8, the use of vectors $v_{7\to 6}$ and $v_{3\to 2}$ leads to avoiding the multiplication of the vectors at the bottom of their branch with the transition matrices.

In addition, since the branch length is involved only over one branch, the joint probabilities can be used to save a lot of computations as for the numerical derivatives. In our example:

$$\frac{dP(\mathcal{D})}{dt_{3\to 1}} = (v^{3\to 1})^{\intercal} Q_{3\to 1}.M_{3\to 1}.v_1$$

The approach will be similar to calculate the second derivatives with respect to the branch lengths.

7.4.4. *Ancestral interference*

In addition to computing the overall likelihood, likelihood vectors at the nodes are mainly used for ancestral inference. It is indeed possible to calculate at any node the probabilities of the different ancestral sequences and thus to describe the evolutionary events along the branches.

Let n denote a node where the conditional likelihood vector $v(n) = (P(\mathcal{D}|X_n = x))_{x\in\mathbb{E}}$ is available. By using Bayes' law:

$$P(X_n = x|\mathcal{D}) = \frac{P(\mathcal{D}|X_n = x).P(X_n = x)}{P(\mathcal{D})} \propto P(\mathcal{D}|X_n = x)P(X_n = x)$$

If the *prior* over the states is uniform, we normalize to have probabilities:

$$P(X_n = x|\mathcal{D}) = \frac{v(n)_x}{\sum_{x\in\mathbb{E}} v(n)_x}$$

This simple approach results in calculating the ancestral probabilities on a sequence. By taking the most probable state at every site, the most probable sequence is thus constructed. On the other hand, in order to build the most probable set of sequences over several nodes, the dependency between these sequences must be taken into account. For this purpose, a method is presented in Pupko et al. (2002).

With regard to substitutions along the branches, we can calculate the expectation (over the set of processes) of the number of substitutions of any type between two nodes, as well as the expectation of the time spent in any state between two nodes[5]. This is known as substitution mapping.

We will not explain here in too much detail the calculations that lead to these counts, but we outline the principle.

5 It is also possible to calculate the variances and other moments of these values (see Dhar and Minin 2017).

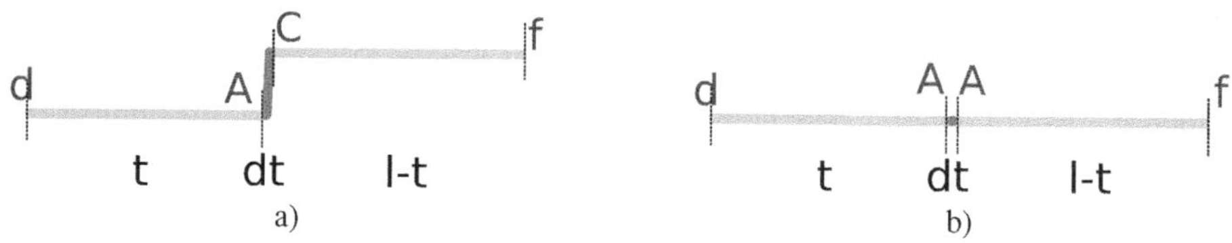

Figure 7.9. *Elementary paths to count the number of transitions from* A *to* C *(a), and the time spent at state* A *(b). For a color version of this figure, www.iste.co.uk/didier/models.zip*

The calculations consist of two steps.

First, we consider a process on a branch $n \to m$ of length l that starts with state d and ends by state f following a generator model Q. We want to measure the expectation of the number of substitutions from A to C on this branch (see Figure 7.9): $E(N_{AC}|X_n = d, X_m = f)$.

The calculation is carried out over the set of processes that end in f, knowing that they start at d, a set whose probability is $(e^{Ql})_{df}$.

Let us consider the set of processes starting from d whose state is A at time t. The probability of this set is $(e^{Qt})_{d\mathtt{A}}$. Then, the number of these processes that move from A to C for a small time dt is $Q_{\mathtt{AC}}dt$. Next, the probability of the set of processes that travel from C to f over the rest of the branch is $(e^{Q(l-t)})_{\mathtt{C}f}$. Therefore, among the processes from d to f, the expectation of the number of transitions during dt from A to C at time t is $(e^{Qt})_{d\mathtt{A}}.Q_{\mathtt{AC}}(e^{Q(l-t)})_{\mathtt{C}f}dt$. We integrate this relation for all possible t between 0 and l, and divide by the probability of the processes from d to f, we obtain:

$$E(N_{AC}|X_n = d, X_m = f) = \frac{Q_{\mathtt{AC}}}{(e^{Ql})_{df}} \int_{t=0}^{l} (e^{Qt})_{d\mathtt{A}}.(e^{Q(l-t)})_{\mathtt{C}f}dt$$

The same procedure is used to measure the average time spent at state A on this branch between states d and f: $E(T_{\mathtt{A}}|X_n = d, X_m = f)$. During a time dt, if the state at time t is A, the time spent at state A is dt. The expectation of the time spent at A during dt is therefore $(e^{Qt})_{d\mathtt{A}}.(e^{Q(l-t)})_{\mathtt{A}f}dt$, and thus:

$$E(T_A|X_n = d, X_m = f) = \frac{1}{(e^{Ql})_{df}} \int_{t=0}^{l} (e^{Qt})_{d\mathtt{A}}.(e^{Q(l-t)})_{\mathtt{A}f}dt$$

In both cases, it is necessary to calculate $I_{df}^{\mathtt{XY}} = \int_{t=0}^{l}(e^{Qt})_{d\mathtt{X}}.(e^{Q(l-t)})_{\mathtt{Y}f}dt$. This can be analytically achieved (see for example (Minin and Suchard 2008; Tataru and Hobolth 2011)).

Second, we integrate this calculation over the probabilities calculated in the tree knowing the data. Let us revisit the example of the substitutions from A to C over a branch $n \to m$ of length l. The point is to calculate the expectation of the number of these substitutions knowing the data $\mathcal{D}$: $E_{n\to m}(N_{AC}|\mathcal{D})$:

$$\begin{aligned} E_{n\to m}(N_{AC}|\mathcal{D}) &= \textstyle\sum_{d,f} E(N_{\mathtt{AC}}|X_n = d, X_m = f).P(X_n = d, X_m = f|\mathcal{D}) \\ &= \textstyle\sum_{d,f} E(N_{\mathtt{AC}}|X_n = d, X_m = f).(e^{Ql})_{df}.P(X_n = d|\mathcal{D}) \\ &= Q_{\mathtt{AC}} \textstyle\sum_{d,f} I_{df}^{\mathtt{AC}} P(X_n = d|\mathcal{D}) \end{aligned}$$

And likewise, the sojourn time at state A is:

$$E_{n\to m}(T_A|\mathcal{D}) = \sum_{d,f} I_{df}^{\mathtt{AA}} P(X_n = d|\mathcal{D})$$

In practice, not all $I_{df}^{\mathtt{XY}}$ are explicitly computed, this would be far too time consuming. But their calculation can be broken down into smaller pieces that will be assembled in an ad hoc manner when calculating the expectations (Tataru and Hobolth 2011).

Based on this approach, the expectation of the different types of events that took place throughout the process can be calculated, in addition to the different ancestral states. This thus gives access to the same type of information as for parsimony, but within a probabilistic context.

7.5. References

Csürös, M. (ed.) (2013). Models and algorithms for genome evolution. In *How to Infer Ancestral Genome Features by Parsimony: Dynamic Programming Over an Evolutionary Tree*. Springer-Verlag, Berlin.

Dhar, A. and Minin, V.N. (2017). Calculating higher-order moments of phylogenetic stochastic mapping summaries in linear time. *Journal of Computational Biology*, 24(5), 377–399.

Felsenstein, J. (1973). Maximum-likelihood estimation of evolutionary trees from continuous characters. *American Journal of Human Genetics*, 25(5), 471–492.

Felsenstein, J. (1981). Evolutionary trees from DNA sequences: A maximum likelihood approach. *Journal of Molecular Evolution*, 17(6), 368–376.

Fitch, W. (1971). Toward defining the course of evolution: Minimum change for a specified tree topology. *Systematic Zoology*, 20(4), 406–416.

Minin, V. and Suchard, M. (2008). Fast, accurate and simulation-free stochastic mapping. *Philosophical Transactions of the Royal Society B*, 363, 3985–3995.

Pupko, T., Pe'er, I., Hasegawa, M., Graur, D., Friedman, N. (2002). A branch-and-bound algorithm for the inference of ancestral amino-acid sequences when the replacement rate varies among sites: Application to the evolution of five gene families. *Bioinformatics*, 18(8), 1116–1123.

Sankoff, D. (1975). Minimal mutation trees of sequences. *SIAM Journal on Applied Mathematics*, 28, 35–42.

Tataru, P. and Hobolth, A. (2011). Comparison of methods for calculating conditional expectations of sufficient statistics for continuous time Markov chains. *BMC Bioinformatics*, 12, 465.

8

The Bayesian Paradigm in Molecular Phylogeny

Nicolas RODRIGUE
Carleton University, Ottawa, Canada

8.1. Introduction

Since the turn of the century, probabilistic phylogenetic inference methods have superseded canonical approaches based on the notion of maximum parsimony (Felsenstein 2004). Historically, the first applications of such probabilistic methods were initially developed within a maximum likelihood (ML) framework (see Chapter 7). Accommodating for multiple substitutions along a branch in a phylogenetic tree is a major advantage of probabilistic methods. Indeed, such a phenomenon introduces a non-phylogenetic signal that leads to confusing inference and negatively impacts parsimony approaches in particular. Probabilistic methods may also likely cause erroneous phylogenetic inference in the presence of non-phylogenetic signals, however these errors can ultimately be traced back to violations of the model employed for the inference itself. The construction of richer models, which consider the variety of the sources of non-phylogenetic signals should provide a more robust inference system. Finally, given the sheer quantity of molecular data, it becomes possible to apply increasingly complex models.

In this chapter, we will see the technical limitations of the ML framework in building rich molecular evolutionary models, and how the computational development environment of Bayesian models overcomes them. We will see the great

Models and Methods for Biological Evolution,
coordinated by Gilles DIDIER and Stéphane GUINDON.

flexibility that they enable, allowing the study of substitution models currently unavailable in an ML framework. We will first introduce the basic principles of Bayesian phylogenetic inference, namely the Monte Carlo–based sampling numerical methods commonly used for approximating the probabilities involved, and possible ways to summarize the model posterior distribution parameters. We will next explain the principle of demarginalization based on two examples, often resulting in faster Monte Carlo sampling, as well as the implementation of substitution models including a non-analytic likelihood function. We will conclude with a discussion of the possible areas for future research in Bayesian molecular phylogeny and the necessary work to access its full potential at the genomic scale.

8.2. General principles of the Bayesian approach in phylogeny

At the heart of probabilistic approaches to phylogeny, whether within an ML or Bayesian framework, we have Markovian substitution models (see Chapter 2). These models specify the substitution rate for each possible type of event. For nucleotide-level models (with four states), these substitution rates are gathered in a 4×4 matrix, denoted by Q, whereas for amino acid level data sets (with 20 states), this matrix becomes 20×20. For most commonly used models, the substitution rate matrix takes the following form:

$$Q_{ab} = \rho_{ab}\pi_b \qquad [8.1]$$

where $\rho = (\rho_{ab})_{1 \leq a,b \leq 4}$ is a (symmetric) set of relative interchangeability parameters between nucleotides (generically represented by the subscripts a and b), and $\pi = (\pi_a)_{1 \leq a \leq 4}$ is a set of nucleotide frequency parameters (or stationary Markov process probabilities). At the same time, it should be noted that a process in the space of amino acids can be similarly specified with the indices on ρ and π up to 20 rather than 4. In addition, since ρ and π contain relative values, their respective sums are equal to 1. Finally, the matrix Q can be normalized, for example, by dividing each of its entries by $2 \times \sum_{1 \leq a \leq b \leq 4} \rho_{ab}\pi_a\pi_b$, so that the units of branch lengths can be interpreted as the expected number of substitutions per site.

Given a substitution matrix specified in this way, and a tree having a topology τ, with a set of branch lengths $\lambda = (\lambda_j)_{1 \leq j \leq 2S-3}$, where S is the number of species in the alignment, the probability of observing our data alignment D, which is expressed as $p(D \mid \rho, \pi, \tau, \lambda)$ can be computed. For simplicity, one can write $p(D \mid \theta)$ with θ representing the set of parameters under the model (the present case being $\theta = \rho, \pi, \tau, \lambda$). The parameter *likelihood* is often referred to, expressed as $L(\theta \mid D)$, and the literature swaps the meaning of the *likelihood function* between the two notations, but we will work with the probabilistic expression $p(D \mid \theta)$ in what follows, or one may write $p(D \mid \theta, M)$, where M represents the way the parameters are gathered in order to be explicit about the fact that data probability is conditional on the model mentioned.

The core of the likelihood computation stands in the expression of the transition probability from one state a to another state b, separated by a branch length λ_j which is read as the probability of reaching state b knowing the evolutionary distance λ_j, the substitution model parameters ρ and π and the initial state a:

$$p(b \mid a, \theta) = [e^{\lambda_j Q}]_{ab} \quad [8.2]$$

$$= \int_{\Xi} p(b, \xi \mid a, \theta) d\xi \quad [8.3]$$

where ξ represents a detailed substitutional history including the nature of all the events along the branch and the waiting times between them. These two equations show that the probability $p(b \mid a, \theta)$ can be calculated with the basic tools of linear algebra for matrix exponentiation [8.2], but it can also be expressed as the integral over all possible substitutional histories (Ξ, in equation [8.3]). Certainly, relatively few of the possible substitutional histories significantly contribute to the integral in [8.3], but the approach nevertheless equips us with a flexible formal framework that recognizes the possibility of substitutional histories with multiple events, with an explicit probability density which we will revisit further on in this chapter.

We assume, for the moment, that we had only one pair of sequences, $S = 2$. The likelihood computation at a given site, i, becomes only the probability of being in state a at stationarity, multiplied by the transition probability from state a to state b:

$$p(D_i \mid \theta) = p(a \mid \theta) p(b \mid a, \theta)$$

This last equation illustrates some important properties of the model. First, the stationary probability distribution of the substitution process depends only on the frequency parameters $p(a \mid \theta) = \pi_a$. Second, the direction according to which the substitution process takes place is irrelevant in the form of the *reversible* Markov model under study. However, the notion of reversibility may be misleading for some: it somewhat cognitively tricks our perception giving a (false) impression of symmetry in the transition probability:

$$p(a \mid b, \theta) \neq p(b \mid a, \theta)$$

Rather, process reversibility is illustrated in a subjective factorization of the likelihood function at a site i:

$$p(D_i \mid \theta) = p(a \mid \theta) p(b \mid a, \theta)$$

$$= p(b \mid \theta) p(a \mid b, \theta)$$

The overall likelihood function, for the set of sites of the sequence pair, is expressed as a product between sites under the assumption of independence between them:

$$p(D \mid \theta) = \prod_i p(D_i \mid \theta)$$

With more than two alignment sequences and several branches on a tree τ, the internal nodes of the tree represent bifurcation points in species evolution, and calculating likelihood becomes more complicated. We assume for the moment that we know the set of ancestral molecular states at the internal nodes of the tree for any site. The likelihood function would then only be the product of the state stationary probability at the root of the tree (which may be an arbitrarily chosen node knowing the reversibility characteristic) by the product of the transition probabilities along the branches of the tree from the root to the leaves. Yet, not knowing the ancestral states, we must reuse the product described above for the set of possible ancestral states and sum them to finally obtain $p(D \mid \theta)$. The enlightening article by Felsenstein (1981) explains how family algorithms based on *dynamic programming* should be used to efficiently compute this sum. Therefore, knowing that the products of transition probabilities represent integrals over all substitutional histories between ancestral states and that the sum of these products is computed over all possible ancestral state configurations, the computation of the global likelihood is an integral over all possible histories along the phylogenetic tree. We will come back to this last property later in this chapter.

In an ML context, the purpose of inference is to obtain a point estimation of the parameter values that maximize the likelihood function. In contrast, Bayesian inference does not imply a point estimate, but a computation of the posterior distribution, given a *prior* distribution, denoted by $p(\theta)$, or still $p(\theta \mid M)$, and the likelihood function, obtained by Bayes' theorem:

$$p(\theta \mid D, M) = \frac{p(D \mid \theta, M)p(\theta \mid M)}{p(D \mid M)} \qquad [8.4]$$

The designations "posterior" and "prior" reflect their interpretations as *confidence states* related to the parameter values; the *prior* being our state of confidence about the parameter values *before* having considered the data, but knowing the general construction of the model $p(\theta \mid M)$, while the posterior is our state of confidence *after* having considered (knowing) the data, as well as the construction of the model $p(\theta \mid D, M)$. The denominator of Bayes' theorem, sometimes called *marginal likelihood*, is a normalization factor such that the posterior probability integrates to 1:

$$p(D \mid M) = \int_\Theta p(D \mid \theta, M)p(\theta \mid M)d\theta \qquad [8.5]$$

One should not forget that the integral of equation [8.5] is multidimensional (and in fact, on τ, it is rather a sum given that the space of topologies is discrete) and non-analytical, which complicates the application of Bayesian methods. In practice, this issue was only solved by the end of the 20th century with the combination of increasing computational capacity and awareness of the generality of Monte Carlo algorithms based on the principles of Markov chains.

8.2.1. *Markov chain Monte Carlo sampling*

In order to avoid explicitly having to compute the posterior distribution, the Monte Carlo strategy consists of drawing a very large number (K) of samples of parameter values, denoted as $(\theta^h)_{1\leq h\leq K}$, from the posterior distribution. Then, the expectation of any statistic of interest as a function of θ, denoted as $T(\theta)$, can be approximated from the following Monte Carlo relation:

$$\begin{aligned}\langle T\rangle &= \int_{\Theta} T(\theta)p(\theta \mid D, M)d\theta \\ &\simeq \frac{1}{K}\sum_{h=1}^{K} T(\theta^{(h)})\end{aligned}$$

where $\langle.\rangle$ represents an expectation with respect to [8.4].

In order to draw the sample $(\theta^h)_{1\leq h\leq K}$ from the posterior, a simulation is created starting from any set of parameter values, θ, drawn randomly from the *prior*. Next, the idea consists of simulating a random walk in the space of possible parameter values. The walk is biased according to the Metropolis–Hastings rule, (Metropolis et al. 1953; Hastings 1970) which proposed a new set of values (θ') given the current ones (θ) and accept it with a probability ϑ:

$$\vartheta = \min\left\{1, \frac{p(\theta' \mid D, M)}{p(\theta \mid D, M)}\frac{q(\theta', \theta)}{q(\theta, \theta')}\right\} \quad [8.6]$$

The key idea underlying the algorithm is that the denominator of Bayes' theorem is simplified and vanishes in the ratio $p(\theta' \mid D, M)/\ p(\theta \mid D, M)$. This ratio is the main kernel of the acceptance probability. The ratio $q(\theta', \theta)/q(\theta, \theta')$, which is called Hastings ratio (Hastings (1970)), is utilized as a correction for biases in the mechanisms for proposing new parameter values (θ') knowing the current values (θ). If we ignore the latter ratio for the moment, the acceptance probability can be viewed as two parts: either the new set of parameter values increases the posterior probability, in which case it is accepted, or it decreases the posterior probability, in which case it is sometimes accepted, sometimes rejected; it will be accepted with a probability equal to the Metropolis–Hastings ratio. Under certain conditions (for more details, see Robert and Casella (2004)) repeating these steps in which new parameter values are proposed, followed by acceptance or rejection according to the Metropolis–Hastings ratio, will produce a Markov chain having the posterior distribution of interest as a stationary distribution, and by recording the visited values we will create our sample of inferred parameter values. The large number of iterations used for the approximation, as well as its random walk nature, explains its designation as *Markov chain Monte Carlo approximation* (MCMC).

In practice, the algorithm requires several initial cycles before the chain reaches stationarity. These initial cycles are usually discarded (the so-called chain *burn-in*), after which the parameter approximation relies on the following idea: the frequency with which any parameter value (or range of values) is visited during the iterations of the stationarity algorithm is proportional to the posterior probability of that value (or range). For a continuous parameter such as π_A, for example, if the sampler records a value in the interval [0.20: 0.25] in 8,880 out of 10,000 iterations, the posterior probability for π_A between 0.2 and 0.25 will be estimated as 88.8%; this estimate will be accurate in the limit where the sample size tends to infinity. Consequently, an approximation can be constructed for the entire posterior distribution of the continuous parameters. For the tree itself, which defines a discrete space, the posterior probability of any topology τ is simply approximated by the relative frequency with which that topology occurs in the sample.

Besides these theoretical tools, the implementation of MCMC approaches for data analysis also relies on a certain level of creativity in the way how changes in parameter values are proposed. In practice, proposal mechanisms operate only on a single subset of parameters, or even a single parameter, independently applying the Metropolis–Hastings rule [8.6] in turn in a series of operator blocks. We briefly describe below some of the operators allowing for implementing most of the phylogenetic models applied today, focusing on those applicable to continuous parameters:

– Additive: Let us assume here that θ is univariate and real. This operator proposes a new value $\theta' = \theta + \delta(U - 1/2)$ where U is a random draw on a uniform continuous distribution in the interval [0, 1] and where δ is a tuning parameter; a larger δ value will produce much adventurous propositions. Since U is between 0 and 1, $U - 1/2$ will be between -0.5 and 0.5. For a modest setting of $\delta = 0.001$, the operator thus consists of adding a small value (positive or negative) to the current value of θ. This type of operator is typically applied for continuous parameters that can take any real value (positive or negative), but if there are bounds defined by the *prior*, retroreflection can be applied in order to remain within the range of allowed values. The proposition does not imply any bias, in the sense that proposing θ' knowing θ would be just as probable as proposing θ knowing θ'. This symmetry is represented by a Hastings ratio of 1.

– Multiplicative: when θ must be strictly positive, a new value can be proposed as follows: $\theta' = \theta e^{\delta(U-1/2)}$. The Hastings ratio of such an operator is θ'/θ. Here, the idea is thus to multiply the current value by a factor close to 1, either slightly above or below.

– SimpleProfil: for profile type parameters, with values that sum to 1 such as ρ and π, this operator is an alternative to the one described by Larget and Simon (1999). The operator first randomly selects a pair of inputs in the profile, and their sum is recorded; then, an additive operator is applied to one of the two inputs, with retroreflection if the value exceeds the recorded sum or if it becomes negative; the other entry of the pair is then set under the constraint that the sum of the pair remains invariant. It should be noted that the operator can be applied to several pairs of inputs at once in the original profile, and since the core of the operator is a call to the additive operator, the Hastings ratio is always 1. The number of pairs on which the operator is applied then becomes another tuning element.

Tuning the operators, either δ or the number of pairs, in the case of SimpleProfil is empirical work. If the tunings propose changes in parameter values that are too significant, the probability of acceptance will often be very low, and a lot of computational time will be wasted without accumulating a good sample of the posterior. Conversely, if the tunings lead to the operators proposing minimal changes, almost all propositions will be accepted; on the other hand, the different values visited from one iteration to the next will be highly correlated, and thereby several rounds of operators will be necessary to reach a large sample of points that are effectively independent of each other. The right balance will thus have to be found, typically with acceptance rates ranging from 30% to 70%. In addition to the tuning of the operators, the relative frequencies of the calls to the operators must also be set for each parameter. Most Bayesian phylogenetic inference software programs have been tuned by developers, based on numerous tests. Others, on the other hand, automate this tuning with a pre-burn-in sampling period, during which the operators are called with different settings in order to find those that would give the targeted acceptance rates, and their relative frequencies are adjusted to decorrelate the successive points recorded during the algorithm (see, for example, Baele et al. 2017).

With the operators previously presented, it is possible to sample within the posterior, substitution model parameters and branch lengths, given an arbitrary topology τ. There are several operators that make it possible to sample within the space of tree topologies, and the reader should refer to the works by Larget and Simon (1999), Lartillot and Philippe (2004), Lakner et al. (2008) and Lartillot et al. (2013) for explanations on these principles with several examples.

8.2.2. *Summary of posterior distribution and sampling*

One of the central motivations of the Bayesian paradigm is to take into account the inherent uncertainty associated with the inference of model parameters. When researchers want to focus on a parameter in particular, or even when they want to find the probability of a particular value (e.g. any tree τ) and in case the other parameters are not of interest (which are then called *nuisance parameters*), the latter will be marginalized, thus taking the associated uncertainty into account.

A summary of the posterior distribution is often presented as simple histograms on continuous parameters, and a consensus tree of the topology sample drawn by MCMC. There are existing tools (for instance, Rambaut et al. 2018) for processing MCMC outputs in a user-friendly way, producing this summary while presenting statistics on the sample quality. By repeating several independent MCMC instances and observing that the summary for each one is similar to the others, with good sampling statistics one becomes increasingly more confident that the algorithm has converged to the desired stationary distribution. Differences between these independent computations are an indication that it is necessary to let the algorithm extract a larger sample, or that it is necessary to modify the settings, or even to invent new ones.

8.3. Demarginalization of the likelihood function

The substitution model under consideration in this chapter is the most general one which have the reversibility (or invariance at the tree root position) property. In the literature, this model is called *general time reversible* (GTR, see Chapter 2) (Lanave et al. 1984), and the likelihood function does not involve any non-analytic integral. On the other hand, other models, richer in parameters than the GTR, involve a non-analytical integral and therefore require different computational approaches. In this section, we will cover a method that is naturally derived from the MCMC principle to manage these cases, and which can bring advantages even when an analytical solution of the likelihood function is available, due to effects on the speed of MCMC-based sampling.

8.3.1. *Parameter expansion*

Although the GTR model is widely used, another aspect of the modeling which is generally coupled therewith is to consider the overall rates (regardless of initial or final states) between sites as random variables obtained from a gamma distribution of mean 1 and variance α, where α is an adjustable parameter (Yang 1993, 1994) either by ML, or equipped with a prior distribution and sampled by MCMC with the rest of a Bayesian inference. This model introduces these random variables as branch length multipliers in the exponentiation of the substitution matrix: at site i, the variable is denoted by r_i, and the exponentiation becomes $e^{r_i \lambda_j Q}$. This means that if the rate at site i is greater than 1, the transition probability along the j branch will be calculated not with a branch length of value λ_j, but rather a higher value, since λ_j will be multiplied with a value greater than 1. Conversely, a r_i rate smaller than 1 will correspond to a shorter branch j.

However, until a dataset is considered, it is not known which site should have which rate. The principle of random variables says that the likelihood at a site i in such a case becomes an integral over the set of all possible rates under the gamma law:

$$p(D_i \mid \theta, M) = \int_{r_i} p(D_i \mid \theta, r_i, M) p(r_i \mid \alpha) \mathrm{d}r_i \qquad [8.7]$$

This is therefore an expectation of the likelihood with respect to the gamma distribution of mean 1 and variance $1/\alpha$. The integral given in [8.7] is not analytic, and the adopted solution in most implementations of this model, which is called GTR+Γ, consists of discretizing the gamma distribution into a few categories, typically four (see Chapter 2). The integral then becomes a simple sum requiring four likelihood calculations under four multipliers of different branch lengths, which will be weighted according to the mass of each category; in practice, the discretization is often carried out so that each category has equal weight, and thus the likelihood is a simple average over the four categories.

On the other hand, the Bayesian MCMC paradigm provides an approach where the continuous gamma distribution is applied by means of new auxiliary variables, $r = (r_i)_{1 \leq i \leq N}$ with N being the number of sites in the alignment. The objective is then to apply these Metropolis–Hastings multiplicative operators with the gamma distribution as *prior*. To better understand this, it should be noted that marginal and joint probabilities on θ and r have the following relation:

$$\begin{aligned} p(\theta \mid D, M) &= \int_{\mathbf{r}} p(\theta, r \mid D, M) \mathrm{d}r \\ &\propto \int_{\mathbf{r}} p(D \mid \theta, r, M) p_\alpha(r) \mathrm{d}r \end{aligned}$$

The key idea behind the concept of *parameter expansion* is that if a sample $(\theta^{(h)}, r^{(h)})_{1 \leq h \leq K}$ is drawn from the joint distribution $p(\theta, r \mid D, M)$, then the component θ of this sample, $(\theta^{(h)})_{1 \leq h \leq K}$, is distributed according to $p(\theta \mid D, M)$. Therefore, to obtain a sample of $p(\theta \mid D, M)$, one merely has to draw a sample of $p(\theta, r \mid D, M)$, and then, if θ is the only variable of interest, ignore the r component of the sample. It is however also possible to study the posterior distribution of r

within this framework from the sample accumulated by MCMC (for this type of applications with random variables, see, for example, (Huelsenbeck et al. 2006; Rodrigue et al. 2007)).

While the ML framework requires that we solve [8.7], or that we approximate it by discretization, the Bayesian MCMC framework addresses the integral by means of a simple sampling of a new set of random variables. Such a system certainly involves even more sampling than the option based on discretization. However, this model illustrates the way in which new, richer and more complex models can be developed, without being overly concerned about the underlying analytical aspects. A similar parameter expansion is applied for the implementation of mixture models capturing the compositional heterogeneity between sites (Lartillot and Philippe 2004) where the new random variables introduced are allocations of each site to a nucleotide or amino acid frequency profile among several; on the one hand, this avoids calculating the likelihood as a weighted sum of the likelihoods under each mixture profile for any call to the Metropolis–Hastings operators; on the other hand, it is now necessary to perform updates on the allocations themselves. This additional cost can be seen as largely worthwhile in practice, and the CAT+Γ, CAT-GTR+Γ-type models would not be usable without making use of such a parameter expansion-based strategy.

8.3.2. *Data augmentation*

Along a similar line as the Monte Carlo strategy based on the use of a demarginalized version of the likelihood function, the principle of *data augmentation* in the context of probabilistic phylogenetics consists of working with a likelihood augmented by a detailed substitutional history, namely the integrand of equation [8.3], for a GTR-based model:

$$p(b, \xi_{ij} \mid a, \theta) = \left(\prod_{k=1}^{z} Q_{s_{k-1}s_k} e^{-(t_k - t_{k-1}) q_{s_{k-1}}} \right) \times e^{-(\lambda_j - t_z) q_{s_z}} \qquad [8.8]$$

The notational elements of equation [8.8] are as follows:

– $q_a = -Q_{aa}$ is the total rate of substitution, starting from state a;

– ξ_{ij} represents the substitutional history at site i along the branch j, consisting of the following three aspects;

– z is the total number of state change(s) along the branch at this site;

– t_k is the time associated with the substitution event k ($t_0 = 0, 0 \leq t_1 \leq ... \leq t_z \leq \lambda_j$);

– s_{k-1} and s_k represent the states before and after the event k ($s_k \neq s_{k-1}$, $s_0 = a$, and $s_z = b$).

A detailed sampling of substitutional histories over the whole tree is explained in Rodrigue et al. (2008b); however, it is easy to see that, by having an explicit solution of the integral in [8.3] by the exponentiation of matrices [8.2], it is relatively easy to derive instances of substitutional histories thereof, namely for all sites, for the whole tree. The overall increased likelihood is denoted $p(D, \xi \mid \theta, M)$, and the principle of data augmentation is to use the data in the Metropolis–Hastings rule rather than the integrated likelihood on the substitutional histories $p(D \mid \theta, M)$: having drawn a substitutional story, several cycles of calls to MCMC operators on θ are carried out, and following this series of calls, a new substitution history is drawn and so on.

The result is a significantly more efficient MCMC sampler since the increased likelihood $p(D, \xi \mid \theta, M)$ is extremely fast to calculate compared to $p(D \mid \theta, M)$. Once again, models such as CAT-GTR-Γ would be inaccessible in practice without this kind of data augmentation technique (Lartillot 2006). Even for models where MCMC sampling based on $p(D \mid \theta, M)$ is possible, the efficiency of a system by data augmentation can decrease the calculation time by several orders of magnitude. In other cases, data augmentation enables the study of models for which the exponentiation of substitution matrices is impossible (Robinson et al. 2003; Rodrigue et al. 2005, 2006, 2009).

For all its computational advantages, data augmentation within this phylogenetic context is only applicable with a fixed topology. In other words, it is currently not possible to apply operators on the topology based on augmented likelihood: in order to do so, some method would be necessary to propose both a new topology and a new substitutional history compatible with the latter. For now, implementations employing data augmentation revert to using traditional likelihood for the topological operators. After updating the topology, a new substitutional history is sampled and the operators using the augmented likelihood are re-triggered to sample over the set of parameters and so on. Although with a traditional MCMC system the computationally limiting aspect in the CAT-GTR-Γ model involves the operators on the parameters of the shuffle model π, with data increasing, this sampling component becomes greatly accelerated, and the limiting aspect with modern datasets is generally sampling on topology.

8.4. Bayesian selection of substitution models

The Bayesian framework provides as many approaches to selecting models as the ML framework, often with more natural and direct interpretations, while being more general. There are two main directions of model evaluation. The first, discussed in section 8.4.1, seeks a *relative comparison* of the different models available: which model enables the greater predictive power of real data? The second direction seeks an *absolute evaluation* of model performance in its ability to capture specific aspects of the actual data.

8.4.1. *Relative model comparison* via *the Bayes factor*

The comparison of two models (M_0 and M_1) within the Bayesian framework typically aims to compute the *Bayes factor* (B_{01}) defined as the ratio of their respective marginal likelihoods (Jeffreys 1935; Kass and Raftery 1995):

$$B_{01} = \frac{p(D \mid M_1)}{p(D \mid M_0)} \quad [8.9]$$

A Bayes factor greater (less) than 1 is considered in support of M_1 (M_0). The Bayes factor does not require the models to be nested and thus penalizes for a larger number of free parameters. The fact that the marginal likelihood is a weighted average of the likelihood function (weighted by the *prior*) leads to several configurations of parameter values inducing very low likelihoods, and since models with more parameters have more *prior* configurations with low likelihood, one comes across a natural Ockham effect.

Unfortunately, since the MCMC sampling methods described above explicitly avoid the calculation of marginal likelihood, even more elaborate methods are needed for its computation. One of the first general and reliable methods for calculating the Bayes factor has been proposed by Lartillot and Philippe (2006), derived from classical methods used for the calculation of free energy differences, named *thermodynamic integration*. In particular, the *model-switch* method seeks to define a path connecting the two models being compared. Let us assume that θ now represents the union of the parameters of both models, some of which are relevant to both models (e.g. topology and branch lengths). The path connecting the two models is constructed by defining the following relations:

$$p(D \mid \theta, M_\beta) = e^{(1-\beta)\ln p(D|\theta,M_0)+\beta \ln p(D|\theta,M_1)}$$

$$p(\theta \mid M_\beta) = e^{(1-\beta)\ln p(\theta|M_0)+\beta \ln p(\theta|M_1)} \quad [8.10]$$

$$p(\theta \mid D, M_\beta) = \frac{p(D \mid \theta, M_\beta)p(\theta \mid M_\beta)}{p(D \mid M_\beta)} \quad [8.11]$$

The Metropolis–Hastings kernel then becomes:

$$\vartheta = \min\left\{1, \frac{p(\theta' \mid D, M_\beta)}{p(\theta \mid D, M_\beta)} \frac{q(\theta', \theta)}{q(\theta, \theta')}\right\} \quad [8.12]$$

Here, the new parameter β is used to control the route along the path connecting the two models. For any value $0 < \beta < 1$, the Metropolis–Hastings decision kernel in equation [8.12] enables sampling a posterior distribution being in fact a semi-morphosis between models M_0 and M_1, without knowing the corresponding normalization factor $p(D \mid M_\beta)$. The quasistatic method described by Lartillot and Philippe (2006) starts from a sampler with $\beta = 0$, so that the MCMC algorithm has the posterior of the parameters of M_0 as stationary distribution. The value of β is then incremented by a small value $\delta\beta$ (typically $\delta\beta = 0.001$ or 0.0001) after having achieved a number of MCMC cycles, until $\beta = 1$; the sampler then proceeds by drawing from the posterior under M_1 at stationarity.

It should be noted that previous studies have not explored scenarios of comparisons between models with different *priors* with the same parameters, and thus have been able to avoid the semi-morphosis of *priors* (see equation [8.10]), simply substituting $p(\theta \mid M_0, M_1)$ therein. The *prior* can then be further simplified during calls to the operators on the elements of θ that are relevant only to the model M_0 at $p(\theta \mid M_0, M_1) = p(\theta \mid M_0)$, and similarly on operators that are relevant only to the model M_1, in that case to $p(\theta \mid M_0, M_1) = p(\theta \mid M_1)$. Based on a sample collected along the whole path between the two models, $(\theta^{(h)})_{0 \le h \le K}$, with the hth draw being associated with β_h ($\beta_0 = 0$, $\beta_K = 1$ and $\forall h$, $0 \le h < K$, $\beta_{h+1} - \beta_h = \delta\beta$), the natural logarithm of the Bayes factor between M_0 and M_1 can be estimated using the following Monte Carlo relation:

$$\ln B_{01} = \ln p(D \mid M_1) - \ln p(D \mid M_0)$$

$$= \int_0^1 \langle \ln p(D \mid \theta, M_1) - \ln p(D \mid \theta, M_0) \rangle_\beta d\beta \quad [8.13]$$

$$\simeq \frac{1}{K}\left[\frac{1}{2}\Big(\ln p(D \mid \theta^{(0)}, M_1) - \ln p(D \mid \theta^{(0)}, M_0)\Big)\right.$$
$$+\Big(\sum_{h=1}^{K-1} \ln p(D \mid \theta^{(h)}, M_1) - \ln p(D \mid \theta^{(h)}, M_0)\Big)$$
$$\left.+\frac{1}{2}\Big(\ln p(D \mid \theta^{(K)}, M_1) - \ln p(D \mid \theta^{(K)}, M_0)\Big)\right] \quad [8.14]$$

where $\langle . \rangle_\beta$ represents an expectation with respect to [8.11].

The overall accuracy of the method depends on several factors such as the step size of the quasi-static traversal ($\delta\beta$), the number of cycles between steps on β to allow the sampler to readjust itself to [8.11], and especially the distance between the two compared models. These issues need to be empirically explored, by applying the quasi-static approach in both directions (from $\beta = 0$ to $\beta = 1$, as well as from $\beta = 1$ to $\beta = 0$) for different values of $\delta\beta$ with different blocks of operators between increments on β, and Lartillot and Philippe (2006) explains how these different sources of error should be evaluated.

In practice, these methods have found that the richest models are not always those that have the highest predictive power on real data (e.g. Lartillot and Philippe 2006; Rodrigue et al. 2006, 2008). However, the trend with modern, often very large-sized data sets, is that very rich models such as CAT-GTR-Γ can be applied in a reliable manner and outperform single substitution matrix-based models.

The ideas of thermodynamic integration have been further developed since their introduction, leading to computationally faster strategies (Fan et al. 2011; Xie et al. 2011) but it should be noted that all of these approaches are based on MCMC sampling with conventional likelihood. It is not possible to "heat up" or sample in semi-morphosis as described in this section with data augmentation. There are existing notions that approach semi-morphosis with the same substitution model (Rodrigue et al. 2006; Rodrigue and Aris-Brosou 2011), but these methods are specific to the models under study. The cross-validation approach is a another very general model comparison method that is increasingly used within the Bayesian framework since the studies of Lartillot et al. (2007); the reader should refer to the latter article for a careful educational introduction.

8.4.2. *Absolute evaluation of models* via *predictive posterior simulation*

Although the Bayes factor makes it possible to establish a relative ranking of the predictive power of a collection of models, it tells us nothing about the absolute performance of any model or what are the aspects of the real data that are not captured by the model. This type of question is addressed within the Bayesian framework with the posterior predictive simulation approach (Rubin 1984; Meng 1994; Gelman et al. 1996). Bayesian version of parametric bootstrap, the central idea behind the predictive posterior simulation is to choose any statistic that can be computed from the real dataset. A number of simulations are then executed utilizing the parameter values drawn from the posterior distribution: for each MCMC run, a simulation of a new artificial dataset is achieved. The statistic of interest is then computed from these artificial sets, which gives a distribution of this statistic and which will act as null distribution. The question then becomes: Is the value of the statistic of interest calculated from the actual data would be expected in the model? A posterior p-value can be calculated simply as the proportion of simulations giving a value to the statistics of interest larger than what is obtained on the real alignment. One should note that this p-value does not necessarily follow a uniform distribution under the null hypothesis (Gelman 2013), but is simply used as a preliminary quantification of the potential mismatch between model and data.

The choice of statistics to be compared from real and simulated data is based on a certain creativity, but it should capture a meaningful aspect of the real data. An example for applying the posterior predictive approach includes the work by

Huelsenbeck et al. (2001) and Foster (2004) which use a χ^2 nucleotide composition homogeneity statistic between alignment sequences. On data simulated with a model based on GTR-Γ, or CAT-GTR-Γ, which does not involve any compositional heterogeneity along the tree branches, which can induce differences in nucleotide frequencies between alignment sequences, the χ^2 statistic will tend to be very small. If the real data indeed have a compositional homogeneity, it should be expected that the χ^2 statistic that is calculated is found somewhere in the central region of the distribution of χ^2 values obtained from the simulated data. As demonstrated by Huelsenbeck et al. (2001) and Foster (2004), there are cases where the hypothesis of compositional homogeneity is clearly violated, which motivates the development of models with time variations of frequencies (Foster 2004; Blanquart and Lartillot 2006). Another example of a statistic that has been studied includes the number of distinct states observed at each alignment site, and Lartillot et al. (2007) demonstrate that the classical GTR-Γ-based models largely overestimate this number, while the approach of the CAT model (Lartillot and Philippe 2004) behaves much better in this case.

An extension of the posterior predictive approach uses substitutional histories to compute any statistic (Nielsen 2002). For example, the variance of the number of substitutions between sites can easily be calculated. This variance is typically significant for substitutional histories constrained to alignment states given the heterogeneity of rates between sites: some sites must have large numbers of substitutions to be compatible with the alignment, whereas others may have none, or only a few. Nonetheless, simulated substitutional histories without being conditional on data will generate a low variance in the number of substitutions between sites if the model does not accommodate any rate heterogeneity between sites (for instance, the GTR model). The result will then be two very distinct distributions: the first distribution, the predictive posterior variances between sites will be very small, and a second, the "observed" posterior distribution of variances between sites far to the right of the former. By itself, the approach reveals the importance of modeling rate heterogeneity between sites. The studies of Rodrigue et al. (2009) have shown that explicit modeling of rates between sites makes both distributions very similar. Other examples of the predictive posterior approach with substitutional histories are discussed in the works of Bollback (2005), as well as Brown and Thomson (2018).

8.5. Impacts and future directions

The Bayesian approach has rapidly progressed since the beginning of the 20th century, and its impact is very significant in molecular phylogeny. In particular, CAT-Γ and CAT-GTR-Γ models have played an important role in inferring deep phylogenies such as the metazoa (Simion et al. 2017), the eukaryote (Brown et al.

2018), the archaea (Adam et al. 2017) and the eubacteria (Antunes et al. 2016) trees. But while the primary motivations for adopting the Bayesian framework were based on the more intuitive interpretation of the support for any type of tree (see also Chapter 9), in contrast to the difficult interpretation and the computational cost of non-parametric bootstrap, its success is rather due to the great modeling flexibility that results thereof. Without the concerns of an analytical likelihood function, and due to the underlying Monte Carlo methods, the Bayesian framework has updated models whose approaches were otherwise very demanding. Developments of increasingly more sophisticated models continue to emerge, for example, for the detection of adaptive regimes (Rodrigue and Lartillot 2016) or for the integration of several data sources of distinct types (Lartillot and Poujol 2010).

Much effort is now needed to adapt the Bayesian framework to the genomic era where the size of data sets makes MCMC-based sampling very cumbersome, especially for rich models such as CAT-GTR-Γ. Further work is also needed to better manage conflicts between gene trees and species trees often neglected in current applications. In terms of integrative modeling, much work remains to be done in order to be able to combine the variety of information sources accumulating at the molecular sequence level, but also relatively to morphological data of gene expression levels, species life histories, etc. After all, despite the flexibility provided by the Bayesian framework, important technical research remains to be conducted to fully exploit its potential.

8.6. References

Adam, P.S., Borrel, G., Brochier-Armanet, C., Gribaldo, S. (2017). The growing tree of Archaea: New perspectives on their diversity, evolution and ecology. *The ISME Journal*, 11, 2407–2425.

Antunes, L.S., Poppleton, D., Klingl, A., Criscuolo, A., Dupuy, B., Brochier-Armanet, C., Beloin, C., Gribaldo, S. (2016). Phylogenomic analysis supports the ancestral presence of LPS-outer membranes in the firmicutes. *Elife*, 5, e14589.

Baele, G., Lemey, P., Rambaut, A., Suchard, M.A. (2017). Adaptive mcmc in Bayesian phylogenetics: An application to analyzing partitioned data in BEAST. *Bioinformatics*, 33, 1798–1805.

Blanquart, S. and Lartillot, N. (2006). A Bayesian compound stochastic process for modeling nonstationary and nonhomogeneous sequence evolution. *Molecular Biology and Evolution*, 23, 2058–2071.

Bollback, J.P. (2005). Posterior mapping and posterior predictive distributions. In *Statistical Methods in Molecular Evolution*, Nielsen, R. (ed.). Springer, New York.

Brown, J.M. and Thomson, R.C. (2018). Evaluating model performance in evolutionary biology. *Annual Review of Ecology, Evolution and Systematics*, 49, 95–114.

Brown, M.W., Heiss, A.A., Kamikawa, R., Inagaki, Y., Yabuki, A., Tice, A.K., Shiratori, T., Ishida, K.-I., Hashimoto, T., Simpson, A.G. et al. (2018). Phylogenomics places orphan protistan lineages in a novel eukaryotic super-group. *Genome Biology and Evolution*, 10, 427–433.

Fan, Y., Wu, R., Chen, M.-H., Kuo, L., Lewis, P.O. (2011). Choosing among partition models in Bayesian phylogenetics. *Molecular Biology and Evolution*, 28, 523–532.

Felsenstein, J. (1981). Evolutionary trees from DNA sequences: A maximum likelihood approach. *Journal of Molecular Evolution*, 17(6), 368–376.

Felsenstein, J. (2004). *Inferring Phylogenies*. Sinauer Associates, Sunderland, MA.

Foster, P.G. (2004). Modeling compositional heterogeneity. *Systematic Biology*, 53(3), 485–495.

Gelman, A. (2013). Two simple examples for understanding posterior p-values whose distributionsare far from uniform. *Electronic Journal of Statistics*, 7, 2595–2602.

Gelman, A., Meng, X.L., Stern, H. (1996). Posterior predicive assessment of model fitness via realised discrepancies. *Statistica Sinica*, 6, 733–807.

Hastings, W.K. (1970). Monte Carlo sampling methods using Markov chains and their applications. *Biometrika*, 57, 97–109.

Huelsenbeck, J.P., Ronquist, F., Nielsen, R., Bollback, J.P. (2001). Bayesian inference of phylogeny and its impact on evolutionary biology. *Science*, 294(5550), 2310–2314.

Huelsenbeck, J.P., Jain, S., Frost, S.D., Pond, S.L.K. (2006). A Dirichlet process model for detecting positive selection in protein-coding DNA sequences. *Proceedings of the National Academy of Science of the USA*, 103, 6263–6268.

Jeffreys, H. (1935). Some tests of significance, treated by the theory of probability. *Proceedings of the Cambridge Philosophical Society*, 31, 203–222.

Kass, R. and Raftery, A. (1995). Bayes factors and model uncertainty. *Journal of the American Statistical Association*, 90, 773–795.

Lakner, C., Van Der Mark, P., Huelsenbeck, J.P., Larget, B., Ronquist, F. (2008). Efficiency of Markov chain Monte Carlo tree proposals in Bayesian phylogenetics. *Systematic Biology*, 57, 86–103.

Lanave, C., Preparata, G., Saccone, C., Serio, G. (1984). A new method for calculating evolutionary substitution rates. *Journal of Molecular Evolution*, 20, 86–93.

Larget, B. and Simon, D. (1999). Markov chain Monte Carlo algorithms for the Bayesian analysis of phylogenetic trees. *Molecular Biology and Evolution*, 16, 750–759.

Lartillot, N. (2006). Conjugate Gibbs sampling for Bayesian phylogenetic models. *Journal of Computational Biology*, 13, 1701–1722.

Lartillot, N. and Philippe, H. (2004). A Bayesian mixture model for across-site heterogeneities in the amino-acid replacement process. *Molecular Biology and Evolution*, 21, 1095–1109.

Lartillot, N. and Philippe, H. (2006). Computing Bayes factors using thermodynamic integration. *Systematic Biology*, 55, 195–207.

Lartillot, N. and Poujol, R. (2010). A phylogenetic model for investigating correlated evolution of substitution rates and continuous phenotypic characters. *Molecular Biology and Evolution*, 28, 729–744.

Lartillot, N., Brinkmann, H., Philippe, H. (2007). Suppression of long branch attraction artefacts in the animal phylogeny using a site-heterogeneous model. *BMC Evolutionary Biology*, 7(Suppl 1), S4.

Lartillot, N., Rodrigue, N., Stubbs, D., Richer, J. (2013). Phylobayes MPI: Phylogenetic reconstruction with infinite mixtures of profiles in a parallel environment. *Systematic Biology*, 62, 611–615.

Meng, X.-L. (1994). Posterior predictive p-values. *Annals of Statistics*, 22, 1142–1160.

Metropolis, S., Rosenbluth, A.W., Rosenbluth, M.N., Teller, A.H., Teller, E. (1953). Equation of state calculation by fast computing machines. *Journal of Chemical Physics*, 21, 1087–1092.

Nielsen, R. (2002). Mapping mutations on phylogenies. *Systematic Biology*, 51(5), 729–739.

Rambaut, A., Drummond, A.J., Xie, D., Baele, G., Suchard, M.A. (2018). Posterior summarization in Bayesian phylogenetics using tracer 1.7. *Systematic Biology*, 67, 901–904.

Robert, C.P. and Casella, G. (2004). *Monte Carlo Statistical Methods*. Springer, New York.

Robinson, D.M., Jones, D.T., Kishino, H., Goldman, N., Thorne, J.L. (2003). Protein evolution with dependence among codons due to tertiary structure. *Molecular Biology and Evolution*, 18, 1692–1704.

Rodrigue, N. and Aris-Brosou, S. (2011). Fast Bayesian choice of phylogenetic models: Prospecting data augmentation-based thermodynamic integration. *Systematic Biology*, 60, 881–887.

Rodrigue, N. and Lartillot, N. (2016). Detecting adaptation in protein-coding genes using a Bayesian site-heterogeneous mutation-selection codon substitution model. *Molecular Biology and Evolution*, 34, 204–214.

Rodrigue, N., Lartillot, N., Bryant, D., Philippe, H. (2005). Site interdependence attributed to tertiary structure in amino acid sequence evolution. *Gene*, 347, 207–217.

Rodrigue, N., Philippe, H., Lartillot, N. (2006). Assessing site-interdependent phylogenetic models of sequence evolution. *Molecular Biology and Evolution*, 23, 1762–1775.

Rodrigue, N., Philippe, H., Lartillot, N. (2007). Exploring fast computational strategies for probabilistic phylogenetic analysis. *Systematic Biology*, 56, 711–726.

Rodrigue, N., Lartillot, N., Philippe, H. (2008a). Bayesian comparisons of codon substitution models. *Genetics*, 180, 1579–1591.

Rodrigue, N., Philippe, H., Lartillot, N. (2008b). Uniformization for sampling realizations of Markov processes: Applications to Bayesian implementations of codon substitution models. *Bioinformatics*, 24, 56–62.

Rodrigue, N., Philippe, H., Lartillot, N. (2009). Computational methods for evaluating phylogenetic models of coding sequence evolution with dependence between codons. *Molecular Biology and Evolution*, 26, 1663–1676.

Rubin, D.B. (1984). Bayesianly justifiable and relevant frequency calculations for the applied statistician. *Annals of Statistics*, 4, 1151–1172.

Simion, P., Philippe, H., Baurain, D., Jager, M., Richter, D.J., Di Franco, A., Roure, B., Satoh, N., Queinnec, E., Ereskovsky, A. et al. (2017). A large and consistent phylogenomic dataset supports sponges as the sister group to all other animals. *Current Biology*, 27, 958–967.

Xie, W., Lewis, P.O., Fan, Y., Kuo, L., Chen, M.-H. (2011). Improving marginal likelihood estimation for Bayesian phylogenetic model selection. *Systematic Biology*, 60, 150–160.

Yang, Z. (1993). Maximum-likelihood estimation of phylogeny from DNA sequences when substitution rates differ over sites. *Molecular Biology and Evolution*, 10, 1396–1401.

Yang, Z. (1994). Maximum likelihood phylogenetic estimation from DNA sequences with variable rates over sites: Approximate methods. *Journal of Molecular Evolution*, 39, 306–14.

9

Measures of Branch Support in Phylogenetics

Olivier GASCUEL[1] and Frédéric LEMOINE[2]

[1] *ISYEB, CNRS, MNHN, Sorbonne Université, EPHE, UA, Paris, France*

[2] *Unité de bioinformatique évolutive, Hub de bioinformatique et biostatistiques, Institut Pasteur, Université Paris Cité, France*

9.1. Introduction

The first step in phylogeny is to construct a tree from data. In most cases, this is based on a multiple alignment of homologous sequences, DNA or proteins and this is the context that is set for this chapter. Each sequence is associated with a taxon (*taxonomic unit*) represented by a label (for instance, species name). The inferred tree comprises internal nodes or "ancestral nodes", leaves bijectively associated with taxa, and branch lengths measured in expected number of substitutions per site. Most methods produce unrooted and binary trees (each internal node is of degree 3); for simplicity, only this framework will be considered. Every tree branch defines a bipartition of the set of taxa: by removing a branch, the taxon set is split into two. The external branches separate one taxon from all the others, leading to the notion of trivial bipartitions, existing in all phylogenies related to the same taxon set. Conversely, internal bipartitions and branches are specific to the tree under consideration. There is a form of equivalence between a phylogenetic tree and the bipartitions it defines: if its bipartitions are known, the topology of the corresponding tree can be reconstructed without ambiguity. This is the reason why when the support of a tree is the main focus, the support of each of its branches or bipartitions is

Models and Methods for Biological Evolution,
coordinated by Gilles DIDIER and Stéphane GUINDON.

generally considered (see Chapter 1 for further details about trees and their properties).

The reconstruction of a tree from sequences is a form of statistical inference. Similarly to any inference of this type, a major concern is the reliability of the estimates obtained from the data. For example, in a political survey, based on the characteristics of the sample, one calculates a range in between which the estimated percentage of votes for a particular candidate will fall. This range represents the error resulting from the necessarily limited sampling of a very large population. In phylogeny, the estimate is rather more complex, it is a tree and not a simple numerical value. Since we are studying (non-numerical) bipartitions, one will refer to confidence values (or confidence estimates) rather than confidence intervals. Again, a cause for uncertainty is related to the sampling of both the sequences and the sites of the alignment. We shall see that the number s of sequences and the number n of sites have essentially opposite impacts: high values of n and low values of s tend to produce strong supports and vice versa. The choice of sequences can also have a strong impact, and the alignment is a source of uncertainty, since it is the result of a program that is itself uncertain and possibly biased. The whole inference process is also fundamentally based on a more or less explicit model of sequence evolution (see Chapter 2). The tree itself is a model. The global model including the tree and the substitutional mechanisms is necessarily simplifying in light of the complexity of biological evolution. For this reason, one generally cannot merely consider that the confidence values associated with phylogenetic inferences allow us to claim with high probability that a particular evolutionary event did indeed occur in the past. In general, it will be possible to simply say that depending on the method and model under consideration, the result of the analysis is more or less robust and reproducible. If the model and/or the method are inappropriate, this might lead to obtaining high supports for erroneous phylogenetic inferences.

As previously explained (see Chapters 7 and 8), there are essentially four approaches to inferring trees from sequences: parsimony, evolutionary distances, maximum likelihood and Markov Monte Carlo chains (MCMC) in a Bayesian framework. Only the first one (parsimony) does not rely on an explicit probabilistic evolutionary model of the sequences. For parsimony, the only standard approach to calculating branch supports is non-parametric bootstrap introduced in phylogeny by Felsenstein (1985). This approach is applicable to all tree inference methods from multiple sequence alignments. It is very often used, except within the Bayesian framework where the MCMC approach allows for other measures based on a posteriori probabilities to observe or not the bipartitions considered (see Chapter 8). Felsenstein's article is one of the most cited in the history of science with over 45,000 citations (Van Noorden et al. 2014). However, this method is very heavy in computation time because it requires restarting the tree inference with a large number (typically 100–1,000) of pseudo-datasets or “bootstrap samples”. Within the likelihood framework (costly in computational time), algorithms have been proposed to accelerate bootstrapping by combining the resampling specific to the bootstrap with searching for optimal trees in the sense of the likelihood (Stamatakis et al. 2008; Minh et al. 2013), but they still remain computationally expensive and do not ensure the same guarantees as the standard method. This cumbersomeness explains the success of local likelihood-based methods. These methods (Anisimova and Gascuel 2006; Anisimova et al. 2011) do not call into question the whole tree as with the bootstrap, but only the local topological configurations around the branch of interest. Their computational time is low and does not significantly change the time needed to infer a tree in the maximum likelihood sense.

In the following, we introduce fast local methods (*approximate likelihood ratio test* [*aLRT*]) (section 9.2); the Felsenstein bootstrap and a new version based on the transfer distance (*Transfer Bootstrap Expectation*; Lemoine et al. 2018), particularly suited to large datasets with little phylogenetic signal (section 9.3); the posterior probabilities used in a Bayesian framework, including a fast local approach (*aBayes*) (section 9.4); and a discussion and comparison of these supports that rely on very different bases and should be interpreted with caution (section 9.5).

9.2. Local supports: parametric and non-parametric aLRT

9.2.1. *Null branch test and its limitations*

Local methods originate from a test implemented by Felsenstein in several software programs of the PHYLIP package: the “null branch test”. The idea is simple and easy to implement in the maximum likelihood framework, but also for least squares distance-based methods that are similar to the maximum likelihood under the assumption of normality.

Let us consider a tree T, analogous to those returned by programs such as PhyML (Guindon and Gascuel 2003; Guindon et al. 2010) or RAxML (Stamatakis 2014) based on maximum likelihood, and a branch b of T whose support is to be computed. Since these programs are heuristics and that the problem of tree reconstruction is NP-hard (Chor and Tuller 2006), there is little chance that T be the overall optimum among all possible trees, but it is a local optimum in the likelihood sense: its branches have optimal length for the considered topology, and no topology is found in the neighborhood of T that leads to a better likelihood than T. The notion of topological neighborhood in phylogeny is defined by “topological moves”. The simplest of these moves is “nearest neighbor interchange” (NNI). In a binary tree T, an internal branch b defines four subtrees, two at each end of b, denoted by A, B, C

and D (see Figure 9.1). From the current configuration T, two NNIs can be performed around b: swapping B and C and swapping B and D. There are thus three topological configurations resolved around b (A with C, A with D and the current configuration originating from T where A is with B). There is also a unresolved configuration, called star-shaped configuration, where the branch b is of zero length, which will be denoted to simplify $b = 0$. All reasonable software programs ensure that the best likelihood configuration is the current configuration given by T (local optimum) whose likelihood is denoted V_1. By V_2 and V_3, one denotes the likelihoods of the second and third best NNI configurations around b, and V_0 the likelihood of the star-shaped tree $b = 0$. The log-likelihoods are denoted LV_1, LV_2, LV_3 and LV_0. These four likelihoods are defined after maximum likelihood optimization of the branch lengths of T, once the corresponding NNI or zeroing of b is performed.

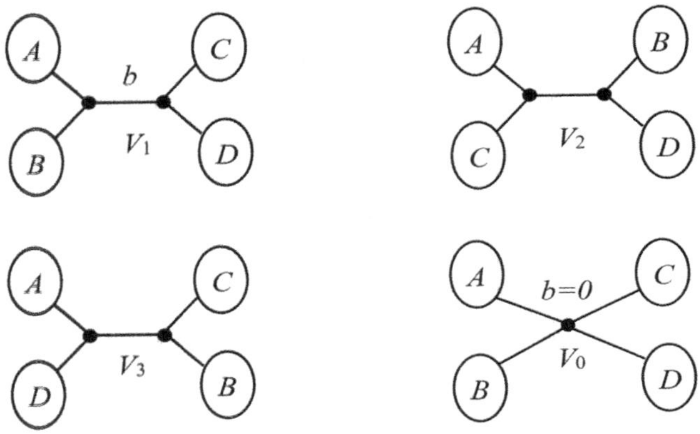

Figure 9.1. *The three NNI configurations, around the branch of interest b, of likelihoods V_1, V_2, V_3. The first configuration (A with B, V_1) is derived from the maximum likelihood tree T, whose branch supports are to be computed. The following NNI configurations correspond to the second and third best likelihoods V_2, V_3. The star configuration, denoted $b = 0$, has a likelihood V_0*

In statistical terms, we have here nested models: each of the three NNI configurations around b contains the star-shaped configuration obtained by setting $b = 0$. And each of the four likelihoods $V_.$ is that of the maximum likelihood of the configuration being considered, once the branch lengths have been optimized. One thus has $V_1 \geq V_2 \geq V_3 \geq V_0$, the last inequality resulting from the nested star configuration in the three NNI alternatives.

The null branch test compares LV_1 and LV_0 and performs a likelihood ratio test based on the $-2(LV_0 - LV_1)$ statistic. Knowing that the star configuration is nested in the current configuration T around b, can the hypothesis H0 be excluded according to which the branch b has zero length? If H0 cannot be rejected, it is considered that the branch b is not supported. In the alternative hypothesis H1, the branch is supported with a confidence level equal to 1 minus the p-value calculated under H0. This is a very simple test between two nested models, one of which has 1 more degree of freedom than the other. A complication now emerges due to the fact that under H0 we are at the border of the possible values for the length of b, which is necessarily positive or zero. Because of this constraint/border, the limit distribution under H0, used to compute the p-value, is a 50|50 mixture between the $chi - square$ distributions with 0 and 1 degree of freedom (Anisimova and Gascuel 2006). Moreover, the multiplicity of the test has to be taken into account since the star configuration is compared to the three alternative configurations. A Bonferroni correction is applied to the p-value with a factor of 3.

This test was quite successful due to its speed and simplicity. It was then realized that with actual data the comparison between the likelihood configurations LV_2 and LV_0, and even LV_3 and LV_0, sometimes yielded significant p-values. In other words, the three NNI configurations around b could be all supported by the data, compared to the star configuration. The test therefore did not answer the question raised, which is to distinguish the current configuration from the two alternative NNI configurations. The null branch test has now been completely forgotten and is no longer available in recent software. It is nonetheless a source of inspiration as we will see.

9.2.2. *Local aLRT test, parametric version*

The idea underlying this test is to make use of the $-2(LV_2 - LV_1)$ statistic which compares the best current NNI configuration derived from T (LV_1) with the second best configuration (LV_2). If the difference is significant, the branch b is supported with a confidence level equal to 1 minus the p-value of the test.

The difficulty is to calculate this p-value. In the first version (Anisimova and Gascuel 2006), we were relying on the inequality $-2(LV_2 - LV_1) \leq -2(LV_0 - LV_1)$, derived from $LV_2 \geq LV_0$, and used the same mixture of χ^2 with a Bonferroni correction as the null-branch test, for computing the p-value associated with this new statistic $-2(LV_2 - LV_1)$. While the latter is significant, null-branch testing is even more significant, but we are now in a position where we are certain of answering the correct question, which is namely to rule out the two alternative NNI configurations. Given the inequality between the two statistics, a conservative test is expected, at least more conservative than the null-branch test. This test directly inspired by the likelihood ratio was called aLRT.

A number of comparisons with the Felsenstein phylogenetic bootstrap have shown that this test is not as conservative as expected and sometimes supports branches that are discarded by the bootstrap, as a matter of fact, itself conservative (see Figure 9.2 for an example). This deviation from the expectation is created by the use of the limit χ^2 distribution, the log-likelihood in phylogeny having a very particular behavior, with considerable differences between the log-likelihoods of each site, with no average site and no normality of the distribution emerging. Therefore, we resorted to a non-parametric approach to calculate the p-value of the test.

9.2.3. *Local aLRT test, SH-like nonparametric version*

Phylogenetic models assume site independence of the multiple alignment given in input to the reconstruction methods. Thus, the data log-likelihood is equal to the sum of the log-likelihoods of each site (see Chapter 7). Therefore, for each statistic LV_1 and LV_2, one has the log-likelihood distribution per site that will be used to calculate the significance of the deviation $(LV_1 - LV_2)$ in the same way as it is done in a mean comparison test with paired observations. Since no normality assumption can be made, one relies on an application of non-parametric bootstrap rather than on a standard deviation calculation or any other parametric approach. Let LV_{1i} and LV_{2i} be the log-likelihoods at site i of the two configurations under study. One has: $(LV_1 - LV_2) = \sum_{i=1}^{n}(LV_{1i} - LV_{2i})$, where n is the number of sites in the alignment.

The method consists of drawing with replacement n site differences $(LV_{1i} - LV_{2i})^{\star}$, to sum these differences, and then to recenter according to the formula:

$$(LV_1 - LV_2)^{\star} = \sum_{i=1}^{n}(LV_{1i} - LV_{2i})^{\star} - (LV_1 - LV_2)$$

The recentering obtained by subtracting the term $(LV_1 - LV_2)$ is used to conform to the null hypothesis that the two configurations are equivalent and the difference observed between their two log-likelihoods would only be due to fluctuations in site log-likelihoods. Therefore, the expectation of the deviation $(LV_1 - LV_2)^{\star}$ in the null hypothesis is indeed equal to zero. By repeating these operations a large number of times, the bootstrap difference distribution $(LV_1 - LV_2)^{\star}$ is obtained to which the observed $(LV_1 - LV_2)$ is compared. If this deviation falls over the tail of the distribution, with a value typically larger than the 95% or 99% quantiles, then the deviation is significant and the branch is supported. The branch support value is given by the empirical distribution function of the bootstrap difference $(LV_1 - LV_2)^{\star}$.

This method is similar to the Shimodaira and Hasegawa test, known as the SH test, for comparing phylogenies (Shimodaira and Hasegawa 1999), hence the name "aLRT SH-like" of this support. As in the SH test, a resampling of the log-likelihoods per site is used here. This is particularly fast, as these values have already been calculated. It is not necessary to go through a phylogenetic inference step again, as in the Felsenstein bootstrap (see section 9.3.2). This is known as RELL bootstrap (Kishino et al. 1990). The method differs from the SH test in its decision statistics. In the SH test, the objective is to list phylogenies that are statistically equivalent to the most likely phylogeny. Here, the objective is to compare the best phylogeny to the second best, but the third NNI configuration is not considered.

There is however a computational cost that we have not mentioned so far: in principle, in these tests, all log-likelihoods (LV_1, LV_2, LV_3 and LV_0) must correspond to the maximum likelihood in terms of branch lengths. This is the case for LV_1 since this is the log-likelihood of the tree T. However, the other log-likelihoods are derived from NNI topological moves applied to T or from zeroing the branch b to 0 for LV_0, and thus, in principle, one would have to re-optimize every branch length of the tree T thus transformed. This involves considerable computation. In practice, we have shown that merely five branches have to be optimized, the central branch and the four adjacent ones (see Figure 9.1), to have an excellent approximation of the log-likelihoods other than LV_1. This is the approach implemented in PhyML. It makes it possible to both compute the branch supports and ensure that the NNI configurations retained for each branch are locally optimal and better than the alternative NNI configurations.

9.2.4. *Comparison with an example of aLRT support and bootstrap*

Figure 9.2 compares the Felsenstein bootstrap and the parametric and non-parametric aLRT supports on a alignment of 54 DNA sequences comprising 1,011 sites with a relatively high phylogenetic signal (extracted from Treebase, reference M2862, see Guindon et al. (2010)). Overall, good correlation between the three approaches is observed. A bootstrap support higher than 75% (a commonly used value) and a SH-like support greater than 90% (as is typical for a test) are essentially equivalent, up to a branch whose bootstrap support is about 60%. Between the two local approaches based on the same decision statistic ($LV_1 - LV_2$), the SH-like (non-parametric) approach is more conservative than the parametric approach based on χ^2. The deviation is even larger between the bootstrap, known to be conservative, and the parametric version of the aLRT support with a dozen branches having a χ^2 support greater than 90%, but a bootstrap support smaller than 75% and sometimes by a lot. Overall, these results are in favor for using the aLRT SH-like support rather than its parametric version. Since these local tests incur a low cost in terms of computing time, they will be used all the more when the dataset is large and contains a large number of taxa.

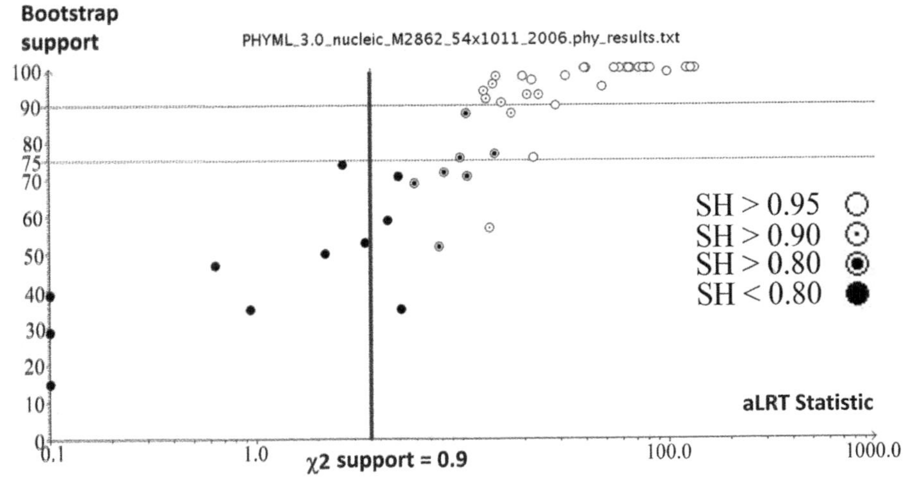

Figure 9.2. *Comparison of bootstrap supports (vertical axis) and aLRT statistics (horizontal axis) with a dataset of 54 sequences and 1,011 nucleic sites, with a tree inferred by PhyML. For a color version of this figure, www.iste.co.uk/didier/models.zip*

COMMENTS ON FIGURE 9.2.– *Each point corresponds to a branch and its various supports. The 90% aLRT support using the mixture of* χ^2 *and Bonferroni correction is indicated by the vertical blue line. The SH-like support is shown with the different types of points. The signal in this dataset is globally high and the supports are well correlated (Guindon et al. 2010).*

An important feature of some implementations of the bootstrap is also clearly visible, corrected as much as possible in PhyML, but still apparent in Figure 9.2: while some branches have a aLRT statistic equal to zero due to an inner branch length of zero, they have non-negligible bootstrap support, which is a form of non-sense. As a matter of fact, a length of zero means that no mutation supports the branch, and thus such a branch should not be supported by any measure and in particular by the bootstrap. The reason for this phenomenon is a hidden determinism in algorithms for tree building, which tend to build the same tree, even when there is no signal in the data. This determinism is, for example, reduced when mixing the order of taxa in bootstrap alignments (as implemented in PhyML). Other deeper factors but of the same order undoubtedly come into play. One will see (see Figure 9.3) that these are very sensitive in the fast implementations of the bootstrap.

9.3. Phylogenetic bootstrap

9.3.1. *Statistic bootstrap*

Bootstrapping is nowadays a widely used computational statistical method. In the context that concerns us here, it is a non-parametric approach (there is also a parametric bootstrap, which is very different). The objective is to study the distribution of a statistic S (e.g. a variance or a quantile) estimated from a n-sample of observations or statistical units denoted by $X^n = \{X^n_{i=1\ldots n}\}$. One denotes by $S(X^n)$ the statistic calculated on this n-sample assuming that it is randomly and independently drawn in the unknown theoretical distribution of observables.

The objective is to quantify the variability of $S(X^n)$. When the model is somewhat complex, analytical formulae for the distribution of $S(X^n)$ are not readily available, which typically leads to resorting to the bootstrap method studied and popularized by Bradley Efron in the late 1970s (Efron 1979; Efron and Tibshirani 1994).

The principle is simple: one draws with replacement n observations of X^n to obtain an n-bootstrap sample or a pseudo-sample $X^{n^\star} = \{X^{n^\star}_{i=1\ldots n}\}$ and $S(X^{n^\star})$ is calculated. By repeating this operation a number of times (typically 100 to 1,000), a distribution is obtained whose variance or quantiles can be calculated, for example, and these values are used for quantifying the variability of $S(X^n)$. It is assumed here that the theoretical distribution of observables is well approximated by the empirical distribution defined by the observed X^n_i. Consequently, not only the variability of $S(X^n)$ can be estimated, but also the bias induced by the use of an n-sample, utilizing the difference $E(S(X^{n^\star})) - S(X^n)$ to estimate the bias $E(S(X^n)) - S(X)$, where $S(X)$ represents the value of the statistic on the set of all observables. This is called the *plug-in principle*. This is a general principle and not a theorem, which cannot be demonstrated without making very specific assumptions about X, S, etc. However, it is clear that the larger n is, the more reliable the approach, because the empirical distribution of observables becomes closer to the theoretical distribution.

9.3.2. *The Felsenstein bootstrap*

The Felsenstein bootstrap follows these main principles. Data are here a multiple alignment of size n (number of sites) × s (number of sequences). Each alignment site (or column) is considered a statistical unit. It is assumed that sites are independant, as in RELL bootstrap (see section 9.2.3) and in many phylogenetic models and programs. The approach consists of drawing with replacement n sites from the initial alignment, in reconstructing a tree from the bootstrap alignment thus obtained and in comparing the bootstrap tree $T^\star$ and the original tree T obtained on the initial alignment. This operation is repeated a large number of times, 100 at least, 1,000 preferably, if computation times are reasonable. In practice, inferring each tree $T^{\star}$ may require a relatively long computation time, but it is possible to parallelize the inference of the different bootstrap trees and utilize fast approaches (Stamatakis et al. 2008; Minh et al. 2013) that employ some trees found during the topological search as bootstrap trees.

Within this framework, the support of a branch (or bipartition) of the tree T is simply the proportion of times where this bipartition is present in the bootstrap trees $T^{\star}$. More formally, the tree T defines a set of bipartitions $B = \{b\}$ where a bipartition b splits the set of taxa into two disjoint non-empty subsets. Similarly, $T^{\star}$ defines $B^{\star} = \{b^{\star}\}$. For a branch b of B and a tree $T^{\star}$, we use the indicator variable $I_{b \in B^{\star}}$, which is equal to 1 if b is again found in $B^{\star}$ and 0 otherwise. The support of b is then:

$$FBP(b) = E_{T^{\star}}(I_{b \in B^{\star}})$$

where the expectation (or average) $E_{T^{\star}}$ is computed over the set of bootstrap trees available and where FBP means *Felsenstein bootstrap proportion.*

This procedure is easily understood intuitively. By bootstrapping pseudo-alignments, one obtains alignments similar to the initial alignment, from which trees are inferred that supposedly look like the initial tree. If they differ significantly, the inference process appears unstable and the support of the initial tree is necessarily low. One thus seeks to measure the robustness of the initial tree and its branches. In other words, it is assumed that the initial alignment is derived from drawing n sites in a larger alignment, typically at the genomic scale, and that by drawing with replacement an n-bootstrap sample, the drawing of n sites is simulated in the full alignment. The reproducibility of the inference of each of the branches of the initial tree is thus measured. It quickly became apparent that this robustness or reproducibility does not guarantee a high probability that the branch under consideration corresponds to a true branch of the evolutionary tree of the taxa considered. As a matter of fact, if the reconstruction method presents bias (for instance, long branch attraction; Felsenstein 1978), the bootstrap trees will suffer from the same bias and a wrong branch resulting from this bias could be strongly supported.

This simple interpretation of FBP supports in terms of robustness was the first. During the 1990s, we witnessed a heated debate about the mathematical meaning of FBP supports, the objective being to make the link between FBP supports and the confidence value of a statistical test in the usual sense (Felsenstein and Kishino 1993; Hillis and Bull 1993). However, nothing is simple here since, to begin with, one must define null and alternative testing hypotheses, which is not obvious when addressing branches and trees. Sophisticated statistical methods have been proposed to correct FBP and better reflect what is commonly meant by testing, p-value, etc.

(Efron et al. 1996; Susko 2010). During the 2000s, these approaches showed their limitations (Susko 2009). They have now been abandoned for most, and the standard Felsenstein approach and the robustness interpretation were again adopted, with the idea that the FBP supports are very conservative, and that a value greater than or equal to 70 % is a solid branch support (Soltis and Soltis 2003).

In recent phylogenomic studies, many genes are combined, sometimes hundreds or even thousands, to constitute a very large alignment in terms of the number of sites n. This very large alignment is intended to infer the tree of the species under study. In this configuration, the FBP branch supports are virtually all equal to 1 because of the very high number of statistical units. Resampling such data hardly changes the result, the same or nearly the same branches are systematically found. However, the branches of the species tree are not necessarily correct and may reflect bias in the inference method. This point and the methods proposed in this framework will be the focus of section 9.5.

9.3.3. *Transfer bootstrap*

Conversely, when addressing very large alignments in terms of the number s of sequences, the opposite problem arises: branch supports become low or very low, even if the data contain a strong phylogenetic signal. Figure 9.3 shows an example of this phenomenon with HIV data (Lemoine et al. 2018). Here, there are over 9,000 sequences of the *pol* gene ($n \approx 1,050$). The sequences were classified into subtypes using a direct method without tree construction (Schultz et al. 2009). In addition to the standard HIV subtypes, this latter detects about 50 recombining sequences in the dataset. Phylogeny separates the subtypes remarkably well, but the clades that correspond to them are however very poorly supported by FBP. For example, the clade corresponding to subtype B contains all sequences assigned to subtype B ($\sim$ 3,500) plus two recombinant sequences, and this almost perfect clade only gets a 3% FBP support.

The explanation is simple. It is related to what are known as *rogue taxa* or unstable taxa. Some sequences (typically recombinant in the HIV example) have unstable positions in the tree and change position in bootstrap trees. Now, for a bootstrap bipartition $b^\star$ induced by $T^\star$ to be registered as present in FBP, it must be strictly identical to the bipartition b being studied. Only one misplaced taxon is enough for $b^\star$ to be unaccounted for. With large datasets (in the sense of s), the presence of such taxa is almost inevitable. It can have multiple causes: recombination, horizontal transfer, erroneous or incomplete sequences, reconstruction error due to numerical and algorithmic approximations, etc.

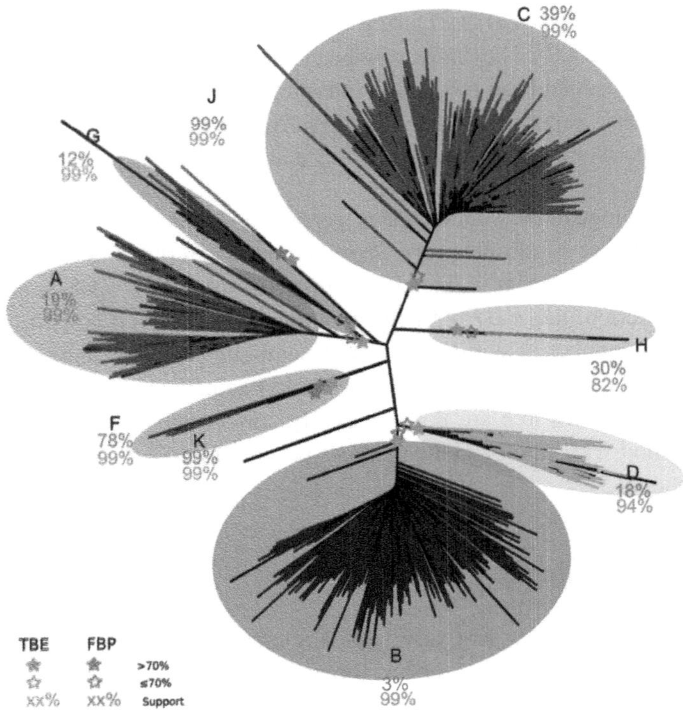

Figure 9.3. *Comparison of the Felsenstein (FBP) and transfer (TBE) supports on a phylogeny of 9,147 strains of HIV-AIDS (FastTree (Price et al. 2009), HIV-1 group M,* pol *gene). For a color version of this figure, www.iste.co.uk/didier/models.zip*

COMMENTS ON FIGURE 9.3.– *The nine HIV1-M subtypes (A, B, C, D, F, G, H, J, K, whose strains are represented in different colors, for example, blue for B) are clearly separated and their branching is consistent with that of previous studies. These nine subtypes have high TBE support, often close to 100%. Conversely, the FPB support is lower, and particularly for the most prevailing subtypes (A, B, C, D, G). For example, FBP does not "see" the clade of subtype B (in majority in Europe and North America, $FBP = 3$ %), whereas this one is very clearly identified (it contains all the blue strains as well as two recombinant strains in black) and obtains a 99% support with TBE.*

The idea behind the transfer bootstrap is natural: instead of using a 0/1 indicator as in the Felsenstein bootstrap, one uses a quasi-continuous measure on $[0, 1]$ of the presence of the studied branch b in the set $B^\star$ of bipartitions induced by $T^\star$. This measure is based on the transfer distance introduced in classification and phylogenetics by Day (1981) and Lin et al. (2011). The transfer distance $\delta(b, b^\star)$

between two bipartitions b and $b^\star$ of the same taxon set is equal to the minimum number of taxa that must be transferred from one side of the bipartition b to the other so that it becomes equal to $b^\star$. The roles are symmetrical, the number of taxa to be transferred into $b^\star$ can be counted so that it becomes equal to b, or the minimum number of taxa that must be removed from the set of all taxa so that the two bipartitions become equal. An alternative definition is based on the Hamming distance denoted H. By ordering the set of taxa, a bipartition can be encoded by a 0/1 vector of length s. Let v be the vector coding for b, and $\underline{v}$ the inverse vector in which 1's become 0's and vice versa; $\underline{v}$ also codes for b; $v^\star$ denotes the vector associated with $b^\star$. It then follows that:

$$\delta(b, b^\star) = \min\{H(v, v^\star), H(\underline{v}, v^\star)\}$$

The transfer index $\phi(b, T^\star)$ measures the presence of a bipartition b of T in the set of bipartitions $B^\star$ induced by $T^\star$ (including trivial bipartitions separating a taxon from $s-1$ other taxa). The transfer index is the minimum of the transfer distances for all bipartitions of $B^\star$:

$$\phi(b, T^\star) = \min_{b^\star \in B^\star} \delta(b, b^\star)$$

This index is normalized in $[0, 1]$ and averaged (as FBP) over the set of bootstrap trees. The normalization is based on the following result. A bipartition b divides the set of taxa into two subsets. Let p represent the cardinal of the smallest of these two subsets, which is denoted P ($p \leq s - p$). We then have $\phi(b, T^\star) \leq p - 1$. Indeed, by taking a trivial bipartition $b^\star$ separating a taxon of P from all other taxa, one has $\delta(b, b^\star) = p - 1$. The normalized form of the transfer index is equal to $1 - \frac{\phi(b,T^\star)}{p-1}$, and the transfer support TBE (*transfer bootstrap expectation*) is defined by:

$$TBE(b) = E_{T^\star}\left(1 - \frac{\phi(b, T^\star)}{p-1}\right)$$

The transfer support TBE thus defined belongs to $[0, 1]$ by construction and proves to be a relevant branch support due to a number of good properties:

– $TBE = 1$ if and only if b belongs to all bootstrap trees $T^\star$ (just as FBP).

– $TBE \approx 0$ when there is no signal in the data (again as FBP). This asymptotic result (Davila Felipe et al. 2019) establishes that if trees $T^\star$ are random, then the TBE support tends to 0 as s increases, and that this convergence is fast.

– $TBE(b) \geq FBP(b)$ when the same set of bootstrap trees is used. The difference between the two supports depends on p and thus on the depth of the branch b. When p is a cherry ($p = 2$), the two supports are equal. Conversely, when the branch is deep and p comes closer to $s/2$, there is often a large gap between the two supports. Figure 9.3 on HIV subtypes is a good illustration thereof. While FBP is 3% for the subtype B, TBE is 99% as intuitively expected given the near perfection of this clade (see above). The same observation can be made about the other subtypes. It should be noted that each support has its own logic here and that each one is "right" in its own way: FBP indicating that clade B is rarely found identical, TBE indicating that clade B is generally found in an almost perfect manner and that there is a strong signal in the data (which FBP is blind to).

– The above property is desirable for large datasets (in number of taxa s), but it has to be verified that TBE does not support largely erroneous branches. This is what the results by Lemoine et al. (2018) show with large simulated datasets and mammalian data for which there is good taxonomic knowledge available on the NCBI website.

– TBE is naturally interpreted as a fraction of taxa to be moved (or removed from the study). It differs radically from FBP which represents the fraction of bootstrap trees that agree with the reference tree for the considered branch. With the examples that we have studied (mammals, HIV, simulated trees), unstable taxa that impact TBE supports are differentiated from others and can be called into question. This is the case, for example, with recombinant HIV sequences that are correctly detected by the approach (Lemoine et al. 2018).

– TBE is computed quickly, in $O(n^2)$ in the worst case, to compare all branches of T to all branches of a given bootstrap tree T^*. The computation is almost as fast as for FBP which uses hashing to speed up comparisons. In practice, for these two branch supports the computation time is negligible compared to that necessary to infer the set of bootstrap trees, even using fast approaches (Stamatakis et al. 2008; Minh et al. 2013).

9.3.4. *Comparison with an example of bootstrap supports*

We reutilize here the data of Figure 9.2 related to 54 DNA sequences of about 1,000 sites with a relatively high phylogenetic signal.

The fast methods available in RAxML and IQ-TREE (Stamatakis et al. 2008; Minh et al. 2013) speed up the computation of bootstrap trees by resampling the sites in the alignment a limited number of times and by proposing as bootstrap trees some of the trees found during the topological research on bootstrap alignments. Figure 9.4 shows clearly that these fast approaches do not lead to the same supports as the slower classical approach, where each bootstrap tree is derived from an independent bootstrap alignment. In particular, IQ-TREE, which is very fast, provides much higher branch supports, where one would expect almost identical supports for the different approaches which are all based on the same statistical principle.

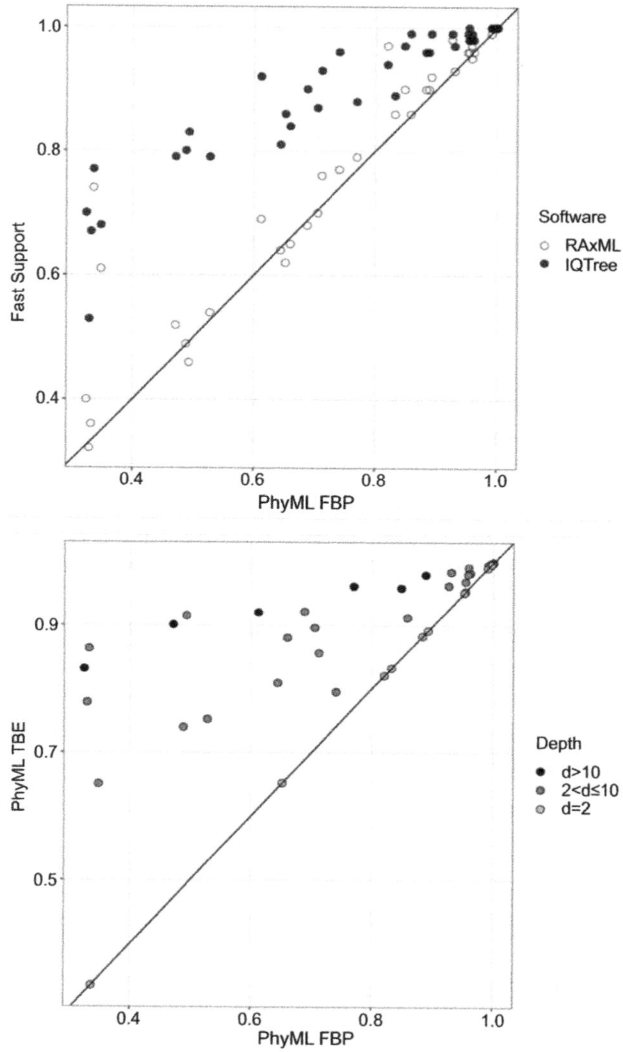

Figure 9.4. *Comparison of different supports*

COMMENTS ON FIGURE 9.4.– *(a) Comparison of the Felsenstein bootstrap supports obtained with PhyML following the classical* FBP *approach, with fast bootstraps proposed by IQ-TREE and RAxML which combine sampling and topological search. Each point represents a branch. (b) Comparison of the* FBP *supports with the* TBE *transfer bootstrap supports in a standard implementation (PhyML).*

Another flaw already pointed out is the hidden determinism of some algorithms, where one would expect a random behavior due to the lack of signal in the data. Thus, a branch of zero or almost zero length, supported by no mutation, induces a trifurcation whose random resolution should receive a bootstrap support of about $\frac{1}{3}$. Three branches have near zero lengths for this set of data and for the three softwares. They properly obtain supports of $\sim 1/3$ with PhyML, which randomizes the sequence order for each bootstrap sample, but much higher supports with fast bootstraps of RAxML and IQ-TREE (respectively, ~ 0.55 and ~ 0.65 on average). This hidden determinism likely explains the very high bootstrap supports of IQ-TREE, and to a lesser extent those of RAxML.

Conversely, when comparing the support of the Felsenstein bootstrap and the transfer bootstrap, the same results would not be expected, since the two supports have very different bases and interpretations. As expected, it can be seen in Figure 9.4 that deep branches tend to be better supported by TBE, while the two supports are identical on the cherries that correspond to a sub-tree comprising only two leaves ($d = p = 2$ where p is the size of the smallest subset induced by the bipartition under consideration, see section 9.3.3). Looking again into the interpretations of these supports, it is equivalent to say that on a cherry the branch is incorrect (FBP) and to say that a taxon has to be moved from the cherry to find a (trivial) bipartition of $T^{\star}$; in both cases, the support is 0, while it is 1 if the cherry in question is found in $T^{\star}$. Conversely, on deep branches ($2 \ll p \leq s/2$) moving a single taxon to make two bipartitions identical may be sufficient, in which case the TBE support will be high while FBP will be zero.

9.4. Bayesian supports

The principles and algorithms of Bayesian methods are described in Chapter 8. We summarize them here to explain the basis of the posterior probabilities that form the branch supports in the Bayesian framework.

9.4.1. *Principle, use of Markov Monte Carlo chains*

The inference methods discussed so far estimate the topology of the tree and the values of the numerical parameters (branch lengths, substitutional model, etc.) by likelihood maximization. Bayesian methods are based on the posterior probability which is the product of the likelihood by the a priori probability on the model parameters. In phylogeny, this a priori may refer to the numerical parameters (e.g. equilibrium frequencies of nucleotides or amino acids), dates of some ancestral nodes when performing molecular dating, on the tree itself sometimes when it is believed to be the result of a well-identified and modeled process (as in epidemiology or ecology, for example). In the absence of information in the data, the a priori and posterior probabilities are identical. Conversely, significant information contained in the data minimizes the role of the a priori. In practice, in most phylogeny problems there is little or no a priori, and maximizing the likelihood or posterior seldom changes the results very little or not at all. In practical terms, using PhyML, RAxML or MrBayes essentially leads to the same tree in most cases, as long as there is a clear signal with respect to the number of taxa under study, and the inferred tree is not the result of random choices carried out by the software programs in question.

The difference between the two approaches lies elsewhere, at least in the context of phylogeny. The usual likelihood methods proceed by direct optimization known as "limited horizon" or "greedy" optimization. This means that they start from an initial tree and change it locally on the topological level or, more globally, for the branch lengths; if the change achieved increases the likelihood, this tree becomes the current tree and the process is repeated; if by means of this mechanism it is impossible to improve the current tree, this latter constitutes the result which is returned to the user (see Chapter 7 for more details).

Within the Bayesian framework as used in phylogeny, the procedure differs by exploring more systematically the space of the trees (there are areas where the methods also follow direct and greedy optimization of the posterior probability). This Bayesian exploration is based on the use of Monte Carlo Markov chains (MCMCs). The current tree undergoes a local modification, for example a change in branch length or an NNI-type topological move (see Figure 9.1). This change is accepted or refused by considering the (Metropolis) ratio of the posterior probabilities of the two configurations, as well as the (Hastings, related to the implementation of the changes and not to the data) ratio of the probability of the change being made to the probability of the reverse change. If the (Metropolis–Hastings) product of the two ratios has a value greater than 1, the new configuration improves the current configuration and is accepted, otherwise it is accepted with a probability equal to the Metropolis–Hastings ratio. This is a logic close to simulated annealing. By doing so, the tree space is more systematically explored and local optima are avoided if the chain is long enough. The great strength of this approach relies on the fact that the frequency of an event observed along the chain approximates the posterior of that event, and this approximation is all the more accurate when the MCMC is long.

In particular, this event can be the presence of a branch in the current tree. With a sufficiently long chain, one will be able to quantify the posterior of all bipartitions of the set of taxa, especially those corresponding to the branches of the tree provided at the method output. The posterior probabilities thus estimated constitute the usual branch supports in a Bayesian framework. They are obtained by sampling the trees within the chain, typically by considering one tree in 1,000 in order to avoid correlations between consecutive trees or trees that are too close in the chain.

An advantage of this approach is that these supports are directly interpretable as the posterior probability of observing any of the branches, including branches that do not belong to the tree provided in the method output. Nonetheless, there are two limitations that must be well understood. The first is that this approach is deeply parametric: by changing the substitution model or the a priori probabilities, branch supports can be significantly changed. It is therefore important to carry out sensitivity studies on the model and on the a priori. Another limitation is that MCMCs are in practice too short to have solid guarantees about the convergence to the posterior. It is therefore necessary to repeat the algorithm several times with different starting points, to use "hot" chains to accelerate convergence, etc., and thus verify the stability of the results.

An abundant literature (see, for example, Douady et al. 2003) addresses the differences between Bayesian supports and the Felsenstein bootstrap. The consensus is that due to the parametric nature and MCMCs necessarily restricted in length, Bayesian supports are generally liberal and tend to over-support certain branches. This is in contrast to the Felsenstein bootstrap, which is considered to be conservative. Many authors produce both supports with the idea that "truth lies somewhere between the two" which is a bit simplistic, because these two supports do not rely on the same bases and do not measure the same thing. It is however a reasonable approach, and so is the one which consists of combining bootstrap and aLRT, each correcting the flaws of the other one (see section 9.2.4).

9.4.2. *Local Bayesian support*

The idea of local Bayesian support, denoted aBayes for *approximate Bayes* (Anisimova et al. 2011), is directly inspired by aLRT support. The three NNI configurations of Figure 9.1 have likelihoods V_1, V_2, V_3, where V_1 is the highest likelihood corresponding to the inferred branch whose support is being evaluated. These are indeed likelihoods (V_i) and not log-likelihoods (LV_i) as in aLRT, and they are again based (see section 9.2.3) on a local optimization of five branch lengths: the branch of interest and its four neighbors (see Figure 9.1). The three likelihoods V_1, V_2, V_3 once normalized can be interpreted as posterior probabilities, because they incorporate the a priori probabilities on the data (typically, the probabilities of each base or each amino acid, see Chapter 2). Thus, an approximation of the Bayesian support is given by the ratio:

$$aBayes(b) = \frac{V_1}{V_1 + V_2 + V_3}$$

As with aLRT, this support is computed remarkably fast without much impact on the total computation time required to the estimation of the tree topology and branch lengths. In PhyML, these local calculations are also used to improve the topology of the inferred tree when an NNI alternative is better than the current configuration.

9.4.3. *Comparison of Bayesian supports with an example*

Again, we use the dataset from Figures 9.2 and 9.4, involving an alignment of 54 DNA sequences of length 1,011, with a relatively high phylogenetic signal.

Figure 9.5(a) clearly shows that the aBayes approximated supports (PhyML) are close to the Bayesian posteriors (MrBayes), without any particular bias and with a high correlation. These approximate supports are, however, significantly faster to compute and can be utilized on very large trees with thousands or even tens of thousands of leaves, which is just impossible for Bayesian methods. It can also be seen in Figure 9.5(b) that Bayesian supports are clearly more liberal than the Felsenstein bootstrap. This point is mentioned in numerous articles in the literature, which also indicate a high sensitivity of Bayesian posteriors to model selection and a priori (Yang and Rannala 2005). By changing the model or the a priori, even by a little, not only a different tree can be obtained, but very different supports on the same branches (or bipartitions).

9.5. Discussion

We have presented three main approaches and their variants for measuring branch supports in phylogeny. The first thing to remember is that these approaches are based on different mathematical principles, and that the supports that they produce cannot be interpreted in the same manner. The local supports (aLRT, aBayes) indicate whether the NNI configuration chosen around the branch of interest is significantly better than both alternatives. The Felsenstein bootstrap tells us whether when resampling the data the branch of interest has a high probability of being found. The transfer bootstrap indicates whether there are a large or small number of taxa that are unstable concerning the branch in question and allow us to identify the most unstable taxa. Bayesian supports give the posterior probability of the branch under study, which depends on the amount of information contained in the data in favor of this branch. In all cases, precautions should be taken, as all these supports depend on the necessarily simplifying substitutional model. Some methods are very sensitive to this, especially Bayesian approaches. In addition, the biases in the models, methods and algorithms lead to the possibility of giving significant supports to erroneous branches, for example, due to the attraction of long branches.

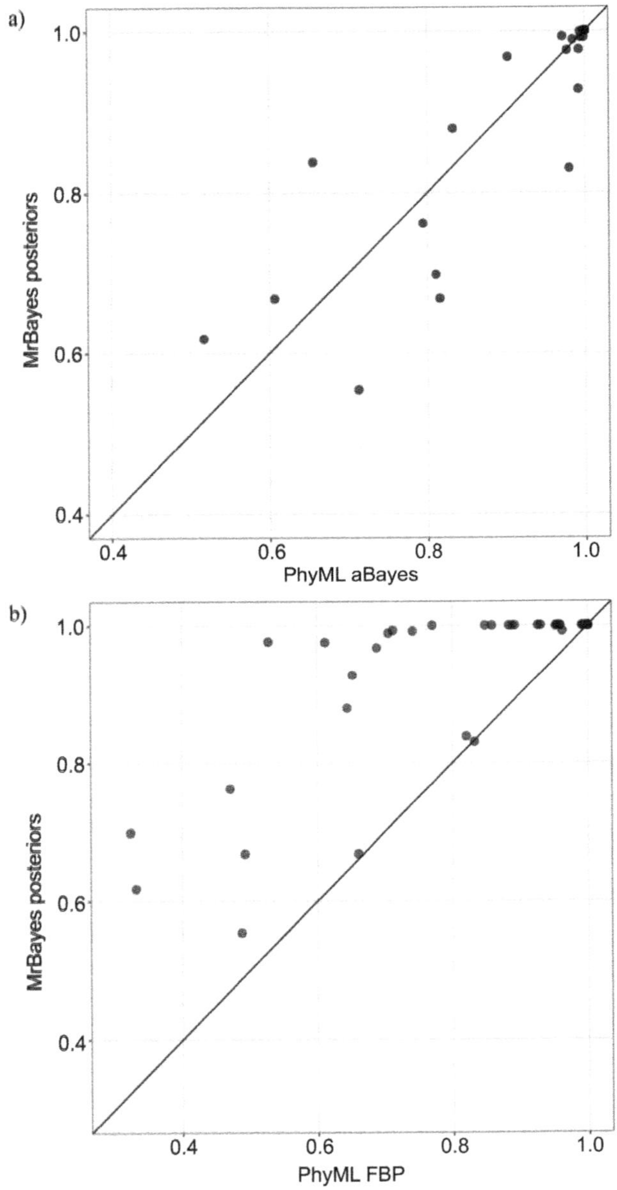

Figure 9.5. *(a) Comparison between the approximated local Bayesian support (PhyML aBayes) and Bayesian posteriors (MrBayes posteriors). Each point represents a branch. (b) Comparison between the Felsenstein bootstrap (PhyML FBP) and Bayesian posteriors (MrBayes posteriors)*

It is becoming increasingly common in publications to provide for each branch the supports from different methods, typically aLRT, Felsenstein bootstrap and a posteriori probability. This is a good practice that makes it possible to overcome the shortcomings of certain methods, for example, the excessive support of Bayesian posteriors, or the hidden determinism of some inference algorithms that induce excessive bootstrap supports, or even the too local nature of aLRT and aBayes. When these supports are weak, the transfer bootstrap allows us to verify whether this is due to a few unstable taxa or if there is simply no signal in the data.

These different supports have very different computational times. A good approach consists of conducting preliminary studies (choice of taxa and sequences, alignment, site cleaning, model selection, etc.) with a rapid local approach (aLRT or aBayes) and then, as the study and data become clearer, in moving to a more global and time-consuming (bootstrap and/or Bayesian) method, which will complete the first local supports.

We have mentioned very large datasets. When the number s of taxa is large but the number of sites is moderate, transfer bootstrap is most likely a good approach. Conversely, in phylogenomic studies involving many genes, n is usually very large, sometimes in the order of a million, while s is moderate. In this case, the focus shifts to the bias induced by the inference method rather than the sampling error, which is almost zero. The issue is one of model sensitivity (Wang et al. 2019) rather than statistical variability. Some authors (Burleigh et al. 2006) have proposed bootstrapping genes rather than sites in the global alignment derived from the concatenation of the alignments specific to each gene. This approach makes it possible to quantify the impact of the choice of the genes used for species tree reconstruction. Some authors have proposed measuring on each branch of the species tree the agreement of the gene trees with respect to the branch being considered (Zhou et al. 2020), thus calling into question the existence of a single tree and taking into account phenomena such as horizontal transfers and gene duplications and losses.

There are many research avenues in this area. The main principles and algorithms of tree inference have changed little in recent years. Implementations are improving, the best ideas are combined, so that the software programs are similar and, in the end, produce similar results with most datasets. The statistical field of branch support is more open. This is a complex area, as shown by the debate in the 1990s on the interpretation of bootstrap supports. This debate is far from being over and the issues raised have not been resolved and remain relevant. Other questions naturally arise, for example, about the use of the transfer distance within a Bayesian framework; the implementation of semi-local but fast supports that would generalize the NNI approach (used in aLRT and aBayes, Figure 9.1) by looking further in the tree; the simple and quick quantification of model and method biases rather than the evaluation of sampling errors, whose impact is increasingly small with the large datasets available today.

9.6. References

Anisimova, M. and Gascuel, O. (2006). Approximate likelihood-ratio test for branches: A fast, accurate, and powerful alternative. *Systematic Biology*, 55(4), 539–552.

Anisimova, M., Gil, M., Dufayard, J.-F., Dessimoz, C., Gascuel, O. (2011). Survey of branch support methods demonstrates accuracy, power, and robustness of fast likelihood-based approximation schemes. *Systematic Biology*, 60(5), 685–699.

Burleigh, J.G., Driskell, A.C., Sanderson, M.J. (2006). Supertree bootstrapping methods for assessing phylogenetic variation among genes in genome-scale data sets. *Systematic Biology*, 55(3), 426–440.

Chor, B. and Tuller, T. (2006). Finding a maximum likelihood tree is hard. *Journal of the ACM (JACM)*, 53(5), 722–744.

Davila Felipe, M., Entfellner, J.-B.D., Lemoine, F., Truszkowski, J., Gascuel, O. (2019). Distribution and asymptotic behavior of the phylogenetic transfer distance. *Journal of Mathematical Biology*, 79(2), 485–508.

Day, W.H. (1981). The complexity of computing metric distances between partitions. *Mathematical Social Sciences*, 1(3), 269–287.

Douady, C.J., Delsuc, F., Boucher, Y., Doolittle, W.F., Douzery, E.J. (2003). Comparison of Bayesian and maximum likelihood bootstrap measures of phylogenetic reliability. *Molecular Biology and Evolution*, 20(2), 248–254.

Efron, B. (1979). Bootstrap methods: Another look at the jackknife. *Annals of Statistics*, 7(1), 1–26.

Efron, B. and Tibshirani, R.J. (1994). *An Introduction to the Bootstrap*. CRC Press, Boca Raton.

Efron, B., Halloran, E., Holmes, S. (1996). Bootstrap confidence levels for phylogenetic trees. *Proceedings of the National Academy of Sciences*, 93(23), 13429–13429.

Felsenstein, J. (1978). Cases in which parsimony or compatibility methods will be positively misleading. *Systematic Zoology*, 27(4), 401–410.

Felsenstein, J. (1985). Confidence limits on phylogenies: An approach using the bootstrap. *Evolution*, 39(4), 783–791.

Felsenstein, J. and Kishino, H. (1993). Is there something wrong with the bootstrap on phylogenies? A reply to Hillis and Bull. *Systematic Biology*, 42(2), 193–200.

Guindon, S. and Gascuel, O. (2003). A simple, fast, and accurate algorithm to estimate large phylogenies by maximum likelihood. *Systematic Biology*, 52(5), 696–704.

Guindon, S., Dufayard, J.-F., Lefort, V., Anisimova, M., Hordijk, W., Gascuel, O. (2010). New algorithms and methods to estimate maximum-likelihood phylogenies: Assessing the performance of PhyML 3.0. *Systematic Biology*, 59(3), 307–321.

Hillis, D.M. and Bull, J.J. (1993). An empirical test of bootstrapping as a method for assessing confidence in phylogenetic analysis. *Systematic Biology*, 42(2), 182–192.

Kishino, H., Miyata, T., Hasegawa, M. (1990). Maximum likelihood inference of protein phylogeny and the origin of chloroplasts. *Journal of Molecular Evolution*, 31(2), 151–160.

Lemoine, F., Entfellner, J.-B.D., Wilkinson, E., Correia, D., Felipe, M.D., De Oliveira, T., Gascuel, O. (2018). Renewing Felsenstein's phylogenetic bootstrap in the era of big data. *Nature*, 556(7702), 452–456.

Lin, Y., Rajan, V., Moret, B.M. (2011). A metric for phylogenetic trees based on matching. *IEEE/ACM Transactions on Computational Biology and Bioinformatics*, 9(4), 1014–1022.

Minh, B.Q., Nguyen, M.A.T., von Haeseler, A. (2013). Ultrafast approximation for phylogenetic bootstrap. *Molecular Biology and Evolution*, 30(5), 1188–1195.

Price, M.N., Dehal, P.S., Arkin, A.P. (2009). FastTree: Computing large minimum evolution trees with profiles instead of a distance matrix. *Molecular Biology and Evolution*, 26(7), 1641–1650.

Schultz, A.-K., Zhang, M., Bulla, I., Leitner, T., Korber, B., Morgenstern, B., Stanke, M. (2009). jpHMM: Improving the reliability of recombination prediction in HIV-1. *Nucleic Acids Research*, 37(Suppl 2), W647–W651.

Shimodaira, H. and Hasegawa, M. (1999). Multiple comparisons of log-likelihoods with applications to phylogenetic inference. *Molecular Biology and Evolution*, 16(8), 1114–1114.

Soltis, P.S. and Soltis, D.E. (2003). Applying the bootstrap in phylogeny reconstruction. *Statistical Science*, 18(2), 256–267.

Stamatakis, A. (2014). RAxML version 8: A tool for phylogenetic analysis and post-analysis of large phylogenies. *Bioinformatics*, 30(9), 1312–1313.

Stamatakis, A., Hoover, P., Rougemont, J. (2008). A rapid bootstrap algorithm for the RAxML web servers. *Systematic Biology*, 57(5), 758–771.

Susko, E. (2009). Bootstrap support is not first-order correct. *Systematic Biology*, 58(2), 211–223.

Susko, E. (2010). First-order correct bootstrap support adjustments for splits that allow hypothesis testing when using maximum likelihood estimation. *Molecular Biology and Evolution*, 27(7), 1621–1629.

Van Noorden, R., Maher, B., Nuzzo, R. (2014). The top 100 papers. *Nature News*, 514(7524), 550.

Wang, H.-C., Susko, E., Roger, A.J. (2019). The relative importance of modeling site pattern heterogeneity versus partition-wise heterotachy in phylogenomic inference. *Systematic Biology*, 68(6), 1003–1019.

Yang, Z. and Rannala, B. (2005). Branch-length prior influences Bayesian posterior probability of phylogeny. *Systematic Biology*, 54(3), 455–470.

Zhou, X., Lutteropp, S., Czech, L., Stamatakis, A., Looz, M.V., Rokas, A. (2020). Quartet-based computations of internode certainty provide robust measures of phylogenetic incongruence. *Systematic Biology*, 69(2), 308–324.

10

Fossils and Phylogeny

Michel LAURIN
CR2P, CNRS, MNHN, Sorbonne Université, Paris, France

10.1. Inferences on topology

Fossils have not always been interpreted as remains of biological organisms that lived in the distant past. This interpretation, which seems obvious today, took a long time to become accepted because, while most people still believed in spontaneous generation, why not also believe that nature was capable of creating, spontaneously in rocks, structures reminiscent of living organisms? Moreover, even among those (quite numerous) who saw in fossils the remains of more or less ancient living beings, various interpretations were proposed, including remains of giants, dragons, unicorns or remains of pilgrims' meals (to explain the presence of mollusc shells or teleost skeletons on hills). Therefore, during the Middle Ages and the Renaissance, many interpretations of fossils existed at the same time (Buffetaut 2017).

10.1.1. *First approaches*

It was with the advent of transformism that the question of taxonomic affinities of fossils began to arise. Lamarck had already integrated fossils into the *Zoological philosophy* (Lamarck 1809), but he regarded evolution as a massively parallel progression toward a higher organization, modulated by adaptations to the habitat and to the lifestyle (hence the great diversity of organizational plans observed in present-day nature). According to Lamarck, these adaptations resulted from the

Models and Methods for Biological Evolution,
coordinated by Gilles DIDIER and Stéphane GUINDON. © ISTE Ltd 2024.

heredity of traits acquired during the life of the organism and explained part of the biodiversity. It should be noted that the heredity of acquired traits was widely accepted long before Lamarck (Burkhardt 2013); it is implicit in Bronze Age myths, and it is well documented by Aristotle (Zirkle 1946). This belief persisted long after Lamarck, whose essential contribution was to incorporate this potential mechanism into an evolutionary theory.

Lamarck thought that very few species had become extinct. According to him, only the large land animals could actually have become extinct as a result of human activity (already!). In most cases where fossils represented unknown forms in present-day faunas, Lamarck believed that the habitats of these taxa had simply not been sufficiently explored to discover them still alive. The principle of plenitude that Plato defined (all possible diversity of living things already exists) suggested that extinction was impossible because it would be against God's will; it was not until the works by Buffon (Clutton-Brock 1995) and Cuvier (1812) that the extinction of ancient species became admitted. This paved the way for a research program into the taxonomic affinities of extinct taxa that became known as systematic paleontology.

Larmack's conception of phylogeny and fossils thus differed significantly from our own. Tassy (2011) states that Lamarck published the first phylogenetic tree, even though the legend of the figure that reproduces it indicates that this is a pre-phylogenetic tree (Tassy 2011, Figure 4). This subtle contradiction probably expresses the fact that Lamarck largely denied the extinction of taxa and, as a result, present-day taxa appeared as ancestors of other present-day taxa. Consequently, in Lamarck's tree, it can be seen that reptiles gave rise to mammals and birds. But can the same thing not be seen much later, with paleontologists such as Romer (1966) and Carroll (1988)? Lamarck thus made some critical advances toward the phylogenetic tree (although his did not integrate evolutionary time, unlike more recent trees), but more modest progress in incorporating fossils into these trees.

It is generally accepted that Darwin played a key role in introducing the evolutionary tree to represent kinship relations between species. Tassy (2006) writes that "the image of the phylogenetic tree is crucial because in terms of fossil evolution the fact of drawing a filiation is a kind of translating words into action, which precisely, neither Lamarck or Geoffroy present". As a matter of fact, Darwin's (1859) hypothetical tree was the first to show the evolutionary lineages in geological time, in stratigraphy. Without it, some of the essential information provided by fossils is missing. Darwin's influence on systematic paleontology, from this point of view, was undoubtedly crucial. However, during the 19th century, no real mention of Darwinian paleontology could be heard of (contrary to Tassy (2006)), because most paleontologists refused to accept natural selection as an evolutionary mechanism (Gaudry 1888, p. 30; Godinot 2011). For Gaudry (1888, p. 30), paleontology could not even be Darwinian, because the evolutionary mechanism could not be known by the fossil record.

Systematic paleontology, after the publication of the *Origin of Species*, increasingly incorporated fossils into phylogenies. Gaudry was a pioneer in this field because he published (Gaudry 1866) the first phylogenies relating extinct taxa (known from the fossil record) and present day taxa (Tassy 2006). These include a phylogeny of Neogene hyenas (p. 36), of Neogene proboscideans (p. 38), Cenozoic rhinoceroses (p. 41), Neogene equids (p. 44) and Cenozoic artiodactyls (p. 46). In short, phylogenetic paleontology was launched!

10.1.2. *Traits usable in paleontology*

Fossilization is selective. Most soft (non-mineralized) tissues seldom fossilize (see Figure 10.1(A–E)). Moreover, DNA is preserved only in the most recent fossils (see section 10.1.7). The available data therefore mainly concern the skeleton, but this is very informative when it is complex, as is the case for vertebrates, arthropods and echinoderms, among others.

The skeleton can be studied morphologically (macroscopically, see Figure 10.1(F)), which provides numerous traits about the presence of various bones, their shape, size and relationships (in contact or not). At a slightly finer scale (under a binocular magnifier, typically), microanatomy can be studied as scale ornamentation (see Figure 10.1(G)), which has been important for the classification of teleosts (Khemiri et al. 2001) and which plays a hydrodynamic role (Burdak et al. 1986). On an even finer scale (microscopic), tissues can studied (see Figure 10.1(H–K)); this is the vast field of paleohistology, which provides traits of systematic and functional interest (Francillon-Vieillot et al. 1990).

The histology of the vertebrate dermoskeleton (dermal skeleton, sometimes erroneously referred to as "exoskeleton", a term that should be kept for arthropods and other organisms with a cuticle superficially located relatively to the epidermis) has provided many important systematic characters, especially among the primitively aquatic Paleozoic taxa (Janvier 1996). Fossils can also be demineralized, often revealing the protein matrix that can fossilize quite often, at least the dense collagen found in many vertebrate mineralized tissues (Zylberberg and Laurin 2011).

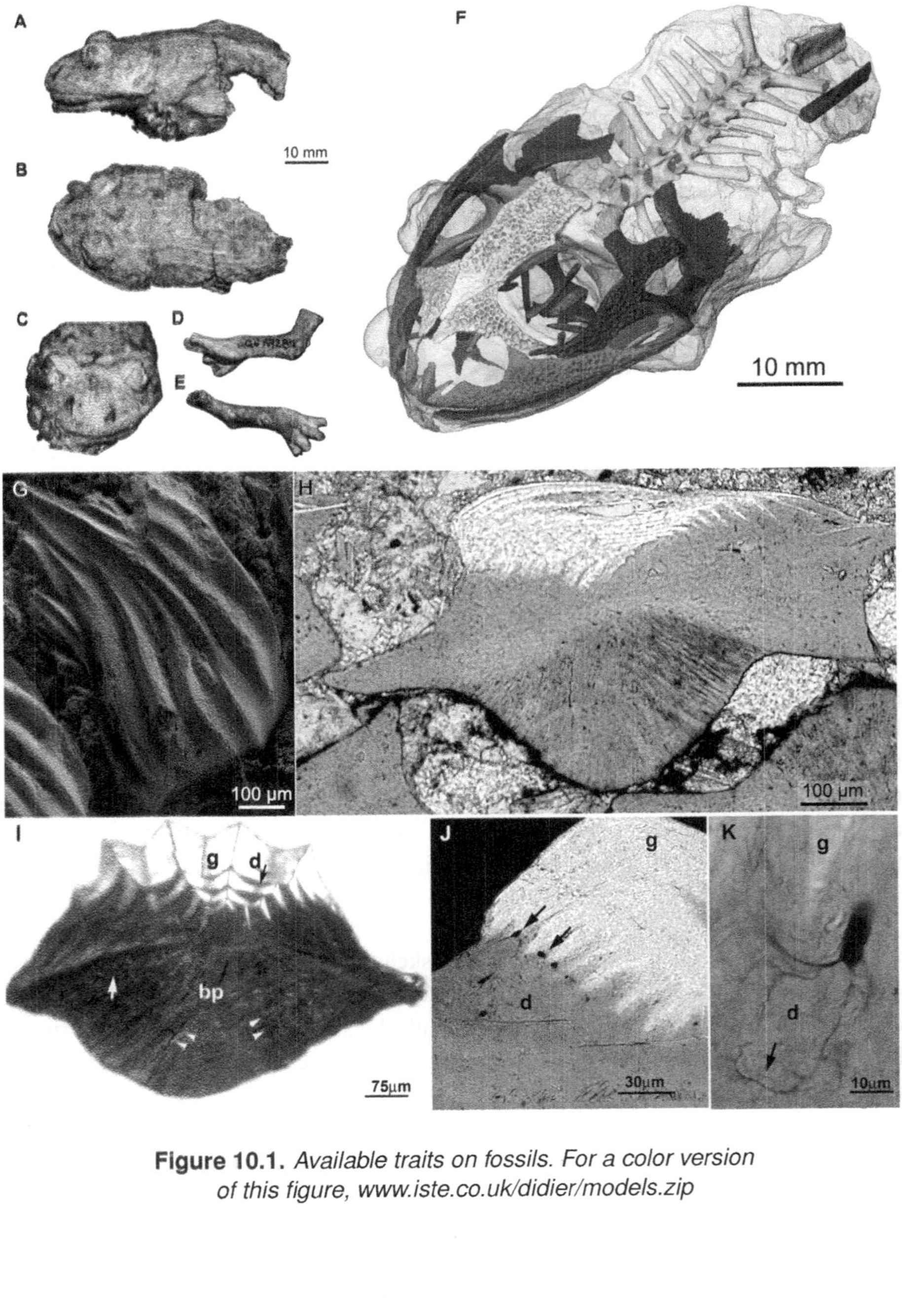

Figure 10.1. *Available traits on fossils. For a color version of this figure, www.iste.co.uk/didier/models.zip*

COMMENTS ON FIGURE 10.1.– *(A–F) Mummy of the anuran* Thaumastosaurus *(Middle or Upper Eocene) showing the skeleton (F) preserved in a natural mummy (A-E). (G–K) Basal Actinopterygian* Cheirolepis canadensis *(Upper Devonian). (G) Superficial view of a scale. Longitudinal (H: front to left) and cross-sections (I) of a scale. Details showing the junction between ganoine and dentin (J) and the vascular channels located at the interface (K). Abbreviations: d, dentin; g, ganoic. Adapted from Laloy et al. (2013) and Zylberberg et al. (2016).*

10.1.3. *First quantitative approach: phenetics*

Until the 1950s, phylogenetics (paleontological or neontological) was based on hypotheses put forward by researchers without relying on algorithms, with no quantification; it was therefore irrefutable. This changed in the 1960s with the advent of two algorithm-based methods: phenetics and cladistics. These two methods appeared more or less simultaneously, but we will address phenetics first because it was less successful and more ephemeral than cladistics in paleontology.

Phenetics is part of numerical taxonomy, which aims at creating objective classifications based on quantification and similarity. It has no historical claim, contrary to cladistics (Sneath 1995), and this is perhaps why it has enjoyed only a limited success in paleontology, the ultimate historical discipline. Phenetics was nevertheless used by a minority of paleontologists, for example, for classifying ichthyosaurs (McGowan 1979). Many other numerical methods were used in systematic paleontology, notably for studying the morphological variability of permo-carboniferous synapsids (Gould 1967; Gould and Littlejohn 1973), and Gould (1967) even made some suggestions on evolution and phylogeny, but these similarity-based methods had little overall impact on paleontological phylogenetics. After the 1970s, these methods were no longer frequently used to infer phylogeny, although various other numerical methods are still very useful for studying patterns of variation, correlations between characters, and the evolution of characters on trees obtained by phylogenetic methods. Phenetics remains useful for delineating species (Anstey and Pachut 2004).

Nevertheless, it should be noted that the quantitative aspect that was introduced by numerical taxonomy, of which phenetics was only one important component among many others, had a great influence on systematic paleontology as we shall see; practically no more phylogenies are produced without using software involving numerical techniques, only these are no longer based on phenetics but rather on parsimony or model-based approaches.

10.1.4. *Stratophenetics*

Stratophenetics represents an evolution of the informal way paleontologists already inferred phylogeny by attempting to "read" the evolutionary tree in fossiliferous strata (Gaudry 1866, 1888; Romer 1966). It differs, however, from this older approach by beginning with measurements and statistics to quantify the phenotype and group specimens into taxa at a given time, which should be more accurate and objective. Moreover, the next step is similar to the traditional approach by hypothesizing evolutionary lineages. These hypotheses can be tested by the subsequent discovery of fossils, because if the lineages and cladogeneses are correctly reconstructed, the fossils found later will fit into these lineages, whereas they might well also fit elsewhere and thus disprove these hypotheses (Gingerich 1979, 1990). However, since the discovery of additional relevant fossils is never guaranteed, the waiting time can be very long. In addition, it is not certain that even such discoveries can test these hypotheses because two different events may result in the same fossil record; as such, a population migration reflecting climatic change could be confused with in situ evolution. More generally, a given fossil record may give rise to several mutually exclusive hypotheses regarding the number of lineages represented and their evolutionary modality (Roopnarine 2005), and the relative weight of stratigraphic and phenotypic arguments is not clear. This approach remained popular for a few decades, but is being gradually abandoned.

10.1.5. *Cladistics*

Similarly to phenetics, cladistics emerged in paleontology in the 1960s and gained considerable momentum from the 1970s onward. The seminal German book by Hennig (1950) was largely ignored; it was not until its English translation (Hennig 1966) that cladistics developed. Hennig's (1966) main innovation in phylogenetics was the distinction between the evolutionary significance of primitive and derived states; only the latter suggest close relationships. Nevertheless, his manual (non-computerized) method for tree searching is no longer used, because homoplasy is arguably more common than what Hennig (1966) expected, and even for relatively small datasets, current phylogenetic software (Swofford 2002; Goloboff and Catalano 2016) are capable of obtaining in seconds much more reliable results than a manual search.

The use of fossils was discussed from the beginning by cladists and not only by Hennig (Kiriakoff 1966; Kluge and Farris 1969). It is therefore not surprising that paleontologists adopted cladistics rather quickly (Kesling and Sigler 1969; Lundberg and Case 1970; MacFadden 1976).

The controversy over the ordering (or not) of multiple character states is not a specifically paleontological problem, but paleontologists were more strongly affected by this controversy than other systematists. As a matter of fact, in the absence of molecular data (available only occasionally in a few relatively recent fossils), they generally use parsimony and thus have to make decisions about the ordering of traits (neontologists, on the contrary, now base their phylogenies mainly on molecular data, usually analyzed by model methods and support for various ordering assumptions can be evaluated). Initially, most systematists ordered multi-state traits into clines (Kesling and Sigler 1969). At that time, the prevailing view stipulated that not ordering states forming clines was not optimal, because this approach did not make use of our knowledge of the plausible evolutionary pattern of such traits (Michevich 1982). In fact, this observation is still valid. Nevertheless, Slowinski (1993) noted that after being popular in the 1970s and early 1980s, ordered traits became rare. Grand et al. (2013) noted that of the articles published in the journal *Cladistics* in 2011 and 2012, half included only unordered traits, a quarter included ordered characters, and for the last quarter of articles, the process was not specified. Yet, simulations demonstrate the importance of ordering traits forming clines (Grand et al. 2013; Rineau et al. 2015, 2018).

Simulations by Rineau et al. (2015) also demonstrate (but not surprisingly) that an extra-group with a short branch (thus phenotypically close to the common intra-group ancestor) performs better than an extra-group located at the end of a long branch. This means that among the available extra-groups, with equal preservation quality, geologically ancient taxa (known from the fossil record) are to be preferred to geologically recent taxa. This effect is stronger in three-element analysis (3ia, an alternative method for cladistic analysis) than in parsimony, but it is also significant in parsimony with ordered or unordered multiple states.

10.1.6. *Model-based approaches: likelihood, Bayesian approaches*

Model-based phylogenetic methods (maximum likelihood and Bayesian analyses) differ from parsimony in many ways and are discussed in detail in Chapters 2, 3, 7 and 8 of this book. The main point that has to be remembered here that when using these methods, branch lengths influence the topology, because the models provide a probability of change of each trait per branch unit length. This is very important in paleontology, because the age of fossils provides useful information about branch lengths that is independent of phenotypic traits; this will be covered in section 10.2.4. Moreover, rapidly evolving traits are automatically downweighted relative to slowly evolving traits, because the probability of change in the latter on a given branch is lower. Thus, the presence of a change in a conservative character has stronger implications than the presence of change in a very labile trait. Unlike parsimony, autapomorphies are informative because they are more likely to occur on long branches than on short branches.

Paleontologists did not begin to adopt model-based phylogenetic methods until well after other systematists, during the 2000s (Müller and Reisz 2006). This is partly because of the unavailability of molecular data in most fossils, as most evolutionary models for discrete traits were developed for nucleotides. For phenotypic traits, only the Mk model (Lewis 2001) is available in a few phylogenetic inference computer programs (Ronquist et al. 2012a, 2012b).

Another reason for the slower and limited adoption of model-based phylogenetic methods in paleontology is that morphological trait evolution is more difficult to model than that of molecular characters. This is partly related to their complexity (a hand being much more complex than a nucleotide or even a sequence of a few hundred nucleotides) and their heterogeneity. Actually, although molecular evolutionary models must essentially determine transition probabilities between four nucleotides (A, T, C and G), phenotypic evolutionary models must be applied to traits that are very different from each other and for which there is no reason to think that they necessarily evolve according to the same pattern. The complexity of phenotypic traits is not surprising, since it is now known that structures are often encoded by several (structural and regulatory) genes, whose interactions can be complex and expression can depend on the environment. They are thus much more complex than nucleotides.

Unfortunately, the development of evolutionary models for phenotypic traits has received much less attention than for molecular traits, even though it is, *a priori*, a more complex problem. Goloboff et al. (2019) analyzed the evolutionary model of 86 empirical data sets. Their results suggest that the Mk model is not a reasonable approximation of the evolution of phenotypic data typically analyzed by systematists. In fact, this model assumes that the probability of change of traits on the tree can be described by a single set of branch lengths that can be transformed (e.g. by multiplication by a constant or exponential transformation), whereas some phenotypic characters often change little or not at all on one part of the tree and frequently change on another, and these areas of invariance or lability vary from trait to trait. Thus, the number of teeth varies little in mammals but significantly in actinopterygians (Farris 1983, p. 15); the number of digits is very constant in turtles but very variable in squamates (frequent reductions accompanying reduction in limb size, even including limb loss); the number of vertebrae is very variable in gymnophiones but fairly constant in mammals and turtles (Carroll 1988). Until further research is achieved in this area and better models for phenotypic traits obtained, parsimony remains an appropriate approach for phylogenetic inference from phenotypic data.

10.1.7. *Fossils and molecular data*

Before the advent of molecular systematics, fossils did not raise any particular problem in phylogenetics, because the latter used phenotypic traits, some of which fossilize. However, since the proliferation of phylogenetic methods specifically designed for molecular data (most of which are model-based methods discussed in other chapters in this book; the reader should refer in particular to Chapter 2), the absence of molecular sequences in the vast majority of fossils has become a problem.

Some of the most recent fossils (less than 2 Myr old, that is, part of the Pleistocene) preserve DNA sequences (Willerslev and Cooper 2005; Prüfer et al. 2010). Recent exploitation of these data has led to significant developments in ancient DNA. Although these DNA sequences are not very old, as they only cover less than 0.4% of the Phanerozoic, they have nevertheless contributed to important progress in the understanding of recent parts of the tree of life, such as the systematic position of the woolly mammoth (Miller et al. 2008) and glyptodonts (Delsuc et al. 2016).

For older fossils (>2 Myr), the absence of molecular sequences restricts phylogenetic inference to using phenotypic data only. This thus limits the number of characters that can be analyzed to a few hundred, or even a few thousand, in some cases. For the moment, there is actually no procedure equivalent to massive DNA sequencing for quickly generating large matrices of phenotypic data in a relatively automated way. It should be noted that the recent trend to increase the size of data matrices does not necessarily represent improvement, because the time that has to be spent for coding these matrices is limited. Thus, there is reason to believe that the larger matrices have been coded with less care than the smaller ones (Laurin and Piñeiro 2017; Simões et al. 2017).

10.2. Dating the tree of life

Initially, the tree of life was dated by means of paleontological trees in which the age of divergences between taxa was either very roughly and subjectively estimated by paleontologists (without algorithms), or assimilated to the age of certain fossils that were thought to be the ancestors of more recent taxa. This approach remained popular until the 1970s through stratophenetics (Gingerich 1990) and more traditional approaches (Romer 1966).

10.2.1. *First qualitative approaches*

Since the emergence of cladistics, the search for direct ancestors in the fossil record has become less of a priority because, as Hennig (1966) suggested, it is impossible to be absolutely sure that fossils indeed represent ancestors. At best, it can be demonstrated that the known phenotype of a given fossil corresponds to the ancestral phenotype that can be inferred on the basis of its presumed descendants (and possibly of other related taxa) and that it is older than its presumed descendants. Among experts on microfossils and more generally on taxa with a very dense fossil record, a large body of work still perpetuates the search for ancestors, because after all, simulations show that in such cases the probabilities of discovering fossils are not negligible (Foote 1996). On the contrary, among vertebrate paleontologists, the search for ancestors has become less frequent, because the vertebrate fossil record, less dense but consisting of organisms showing large numbers of osteological traits (hundreds, even thousands, can be coded on fairly complete and well preserved skeletons) provides fewer opportunities to find plausible direct ancestors.

10.2.2. *First statistical approaches*

To determine the age of the appearance or disappearance of a taxon using its fossil record, indirect approaches must therefore be used. Two distinct cases can be considered: a single lineage and a clade. We will start with the simplest case, the single lineage.

Strauss and Sadler (1989) developed the first approach to provide confidence intervals (CIs) of the age of appearance (and extinction) of species in the simplest case where the potential for fossil discovery is constant in time (over the lifetime of the species in question). For extinctions, they also showed how to test the hypothesis of simultaneous extinction of many lineages, for example, in the context of a mass extinction. In all cases, the 50% CIs are an unbiased estimate of the true age of appearance (or extinction) of species. Species must, however, be conceptualized as phenotypic groups; this is only one of more than 140 concepts of biological species (Lherminier and Solignac 2005), but the most common in paleontology, for obvious practical reasons.

Marshall (1994, 1997) extended the method to clades and to more realistic fossilization potential cases (heterogeneous in time). Marshall (1994) extends the method to cases where the time distribution of fossils is heterogeneous, but where no time trends can be seen. The approach by Marshall (1997) is based on the more general case of a fossilization potential function constructed from criteria external to the data (therefore, without estimating the parameters of the function from the data), such as the inferred depth (below the water surface) of preserved paleoenvironments at fossil sites (appropriate for aquatic, especially benthic, organisms that often have preferences for certain depths) or the surface of exposed fossil sediments that can be used as proxies for fossil searching efforts. However, both of these approaches rely on the implicit (but not always very realistic) assumption that the taxon did not diversify over time or that this diversification did not impact its fossilization potential.

MarjanoviÄĞ and Laurin (2007) proposed incorporating a model of exponential diversification (and the possible impact of major biological crises) into the taxon discovery potential function. For lissamphibians, this works very well and the function thus designed (without relying on data) explains between 86% and 90% of the variance in the time distribution of lissamphibian fossils over time, whereas the use of fossiliferous sediment surface area from various geological periods alone explains only 6% of this variance. For a clade as large and diverse as lissamphibians (and for terrestrial organisms in general), the inferred depth of paleoenvironments would probably not be very useful for calibrating the function (except to eliminate pelagic sites from the analyses), as this group comprises arboreal, terrestrial, amphibious and aquatic taxa; eventually, the distance from the shore (for marine sites) could be useful. These studies provided some interesting empirical results. For example, analysis by MarjanoviÄĞ and Laurin (2007) suggests an appearance of the taxon Lissamphibia during the Permian, between 255 and 300 Myr, whereas previously published molecular ages localized this appearance in the Lower Carboniferous (337 Myr) or even the Upper Devonian (369 Myr) according to various studies (Zhang et al. 2005; Roelants et al. 2007). Some later molecular studies obtained slightly more recent dates (Pyron 2011), but in any case, uncertainties are large and these results are less inconsistent with each other than point estimates alone suggest.

This presentation of paleontological dating methods based on statistical approaches is necessarily simplified, but syntheses expose in more detail the works that have developed this type of methods (Laurin 2012; Marshall 2019; Didier and Laurin 2021).

10.2.3. *Molecular dating*

The proposal of a molecular clock (Zuckerkandl and Pauling 1962, 1965) opened the way not only to molecular phylogenetics, but also to dating the appearance of taxa using molecular data. These methods have been described in detail in a multitude of articles and reviews (Sauquet 2013); we will therefore focus here on the paleontological aspects of this method. In fact, molecular data alone can only provide branch lengths representing a quantity of change. To convert it into time, the molecular clock must be calibrated using the fossil record or geological events, such as the separation between continental plates or the appearance of archipelagos. However, fossil-based calibration is more accurate and reliable than purely geological methods, because the separation between continents is a gradual event and the resulting reproductive isolation depends on the dispersal ability of taxa, which proves difficult to model.

Hasegawa et al. (1989) observed that molecular evolutionary rates varied not only from one gene to another (in addition to varying within the same gene, between positions in codons, introns versus exons, etc.), but also between taxa. Many more recent studies have confirmed the generality of this phenomenon, which has generated a myriad of methods intended to model changes in the rate of molecular evolution on trees. These methods can be grouped into those that rely on autocorrelation of evolutionary rates and those that do not. In the first case, these rates vary slightly between adjacent branches, whereas distant branches may have very different rates (Sanderson 2002). Without autocorrelation, the rates of each branch are drawn from the same statistical distribution over the whole tree, whereas the autocorrelated approach amounts to drawing the evolutionary rate of each branch from an average that is a function of the rate of its parent branch (Drummond et al. 2006). However, in any case, the lack of a global clock requires many calibrations to properly date a tree. The ideal situation would be to have paleontological constraints on most of the nodes and to have to estimate the age of only a few nodes using molecular data (Britton 2005). Unfortunately, in general, the opposite applies; we have only a few paleontological constraints and many nodes to date with molecular data.

A major constraint of molecular dating is that a literal reading of the fossil record provides only minimal ages for the nodes; maximum ages are much more difficult to estimate (Laurin 2012) and the shape of the probability density of these paleontological ages is just beginning to be known (Didier and Laurin 2020). Consequently, molecular trees are typically calibrated with a fairly large number of minimal age constraints, but very few maximal age constraints; for example, Roelants et al. (2007) used minimum constraints for 22 divergences but constrained only one node (Amniota) with one maximum and one minimum constraint. This practice, not differing much from the established practice in molecular dating, probably leads to overestimation of molecular ages (MarjanoviÄĞ and Laurin 2007). Fortunately, recent developments (Heath et al. 2014; Zhang et al. 2015; Didier and Laurin 2020) will soon make it possible to establish numerous constraints on the maximum age and even on the shape of the probability density of these ages (see sections 10.2.4 and 10.2.5).

DNA sequences are available only for present-day taxa and a few fossils dating less than 2 Myr, but some of these few older sequences have been utilized to date much older evolutionary divergences involving extinct taxa. For example, sequences from the woolly mammoth suggest that it shared a last common ancestor with Asian elephants about 7 Myr ago. Similarly, a complete mitochondrial genome of the *Doedicurus* glyptodont dating from about 12,000 years ago suggests that glyptodonts are part of the armadillos, specifically the Chlamyphoridae; its last common ancestor with the clade including *Chlamyphorus* and *Cabassous* (among others) dates from about 35 Myr (Delsuc et al. 2016). This is slightly younger than the oldest known glyptodont which dates from about 36–38 Myr (Woodburne et al. 2014), suggesting that this molecular age is slightly too recent.

Two cases limit the application of molecular dating. First, this method cannot be applied to the numerous clades that are extinct for a long time (more than two Myr), such as ornithischian dinosaurs or seymouriamorphs (Laurin 2008). Second, large present-day clades have no fossil record available or the latter is not very informative because the fossils are not diagnostic. Therefore, we have a small number of fossils of annelids and even fewer nematodes, and among the annelids, fossils are mostly tubes that attest to the very probable presence of this taxon in certain strata, without providing very precise data on their anatomy (beyond the approximate diameter of the body). Worse, other organisms can make such tubes, and several of them, previously attributed to annelids, were probably produced by organisms that belong to another taxon (Taylor and Vinn 2006). In such clades, when no reliable calibrations are possible, the molecular clock must be calibrated using data from fairly distant taxa, which necessarily lowers the accuracy or reliability of estimated molecular ages.

10.2.4. *Tip dating*

Tip dating is often referred to as *total evidence dating* because early applications used both molecular and phenotypic data to produce phylogenies that directly incorporated fossils, without relying on previous (more or less well-supported) assumptions about the systematic position of extinct taxa (Pyron 2011; Ronquist et al. 2012a, 2012b).

Phenotypic data are also used to estimate branch lengths for extinct taxa, whereas branch lengths for extant taxa are estimated using both phenotypic and (especially) molecular data. This allows nodes to be dated without using constraints on nodal ages. However, speaking of *total evidence dating* is inappropriate, because this principle consisting of using fossil ages to constrain the ages of branch tips of extinct taxa and estimating the number of inferred changes (using evolutionary models) on these branches to infer node ages can be used on phenotypic data only (without molecular data), as it has been done in various clades (Wright 2017; Lee and Yates 2018; Paterson et al. 2019). This is particularly useful for dating taxa that became extinct a long ago and for which no molecular information is available.

This method has some advantages and disadvantages over molecular node dating. The main advantage is that the systematic position of the fossils is estimated during the analysis, allowing the impact of uncertainty in the systematic position on the age of the nodes to be evaluated. Nevertheless, this can also be done in the context of node dating, using bootstrapping (Sterli et al. 2013), for example.

A second advantage is that the geological age of the fossils, combined with the use of evolutionary models, influences the topology. Thus, the geological age of the fossils constrains the branch lengths, which changes the likelihood of the topologies. This effect was most clearly demonstrated in the analysis by Lee and Yates (2018) on crocodilians.

A third advantage of this method is that the documented phenotype among the fossils can provide an estimate of branch lengths, which is not the case in node dating. However, the drawback is that the evolutionary models used by this method may not be very realistic (see section 10.1.6). The mismatch between the evolutionary models used (Mk, for the moment) and the phenotypic data analyzed by systematists, identified by Goloboff et al. (2019), is undoubtedly the weakest link of tip dating.

10.2.5. *Birth–death model-based dating*

Another approach to dating the tree of life consists of using birth–death models. Assuming a relatively constant cladogenesis and extinction rate over time (and even by relaxing this assumption with some methods), current and extinct biodiversity data (when available) can provide information on node ages. An early application of this approach in paleontology was that by Wilkinson and Tavaré (2009) who estimated the age of appearance of crown primates and anthropoids using the age of the oldest fossil belonging to these two taxa, the number of known species of these taxa at various times, and simulations. The results suggest that primates may have appeared as early as 98.9 Myr, around middle Cretaceous, which may be surprising given that they consider that no primates belonging to the crown group are known in the fossil record before the Eocene (56 Myr).

Wilkinson and Tavaré's (2009) approach exploited only a small part of the information in the fossil record, because only the fact that a fossil belongs (or not) to the two clades examined (Primates and Anthropoidea) was used. Later approaches utilized more detailed phylogenetic information. Thus, Heath et al. (2014) assign each fossil to a calibrated node ancestral to that fossil (and present-day taxa). More precise information about the fossil's position is ignored (even if it is available). Each fossil can be either on a terminal branch or on an internal branch (it is then the ancestor of other taxa included in the analysis). The approach developed by Didier et al. (2017) and Didier and Laurin (2020) relies on fully resolved trees; phylogenetic uncertainty can then be incorporated by means of populations of trees (in this case, equiparsimonious).

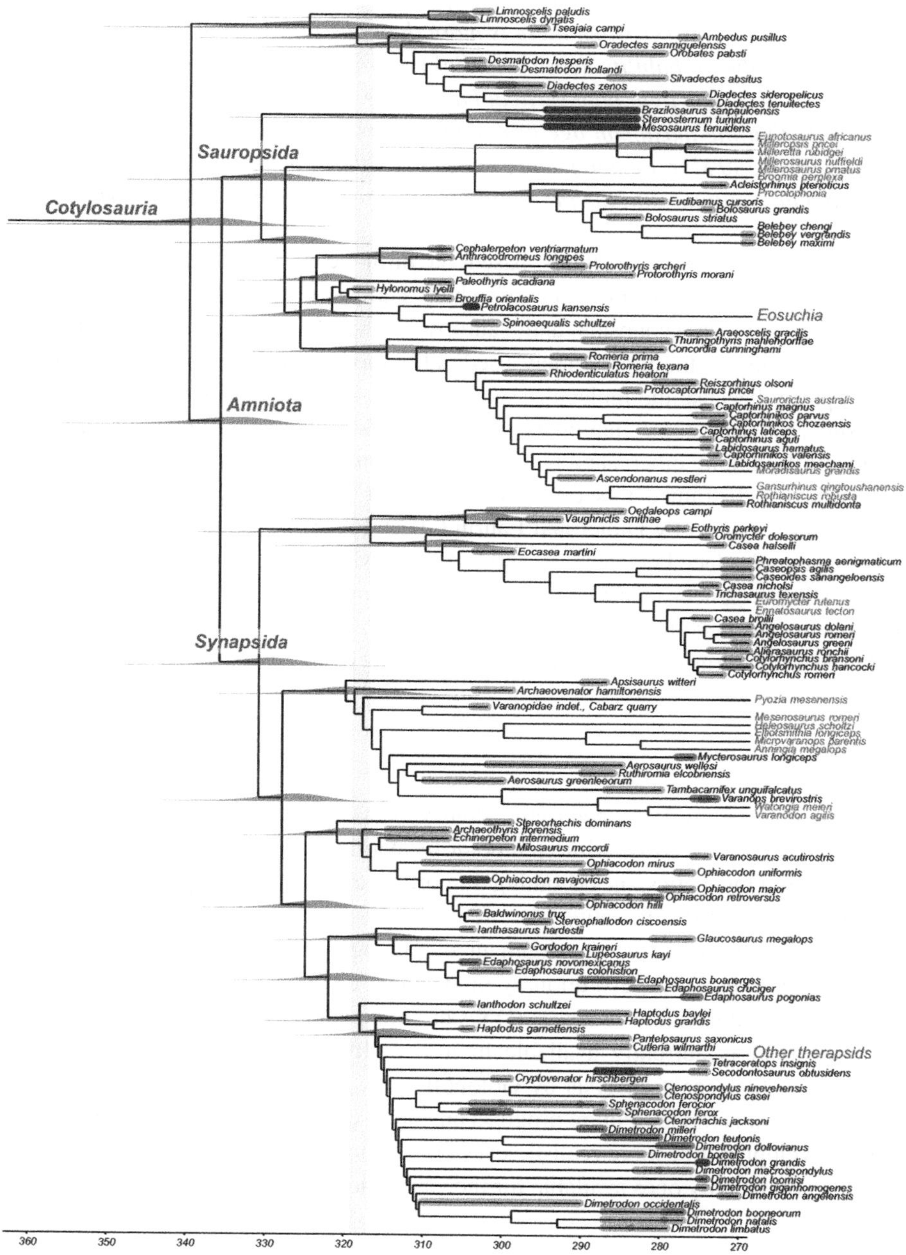

Figure 10.2. *Cotylosaur tree calibrated in time by a birth–death model. For a color version of this figure, www.iste.co.uk/didier/models.zip*

COMMENTS ON FIGURE 10.2.– *The stratigraphic extent of each taxon is denoted by a brown highlight (darker when there is more than one fossiliferous level for a given time interval); the probability density (in pink) is provided for readability reasons for only a few nodes. The light blue vertical band is the age of the oldest known amniote fossil; note that the maximum age probability density of the taxon Amniota is much older, at about 330 Myr. Taxa in red are more recent than the time interval under study. Those in blue continue until the present time (modified from Didier and Laurin 2020).*

These various dating methods based on birth–death models have provided a variety of interesting results, one of which can be presented briefly here. The divergence between synapsids and sauropsids, which marks the appearance of amniotes, is one of the most widely used calibration constraints in molecular node dating, partly because the molecular sequences are so abundant. Actually, synapsids are now represented by mammals (including *Homo sapiens*), while sauropsids include birds, among others, and namely the chicken (*Gallus gallus*). Most molecular studies have constrained this divergence to about 310 or 315 Myr (Hugall et al. 2007), but the results by Didier and Laurin (2020, Figure 8) suggest instead that this divergence took place between 320 and 345 Myr. Moreover, these analyses allow us to date all divergences (see Figure 10.2), which can be very useful for comparative analyses to detect evolutionary trends (Laurin 2004; Bokma et al. 2016) or to identify evolutionary patterns (Ascarrunz et al. 2019; Godoy et al. 2019).

10.3. Conclusion

Phylogenetics has developed tremendously over the last few decades, as this book attests. The role of fossils, while temporarily marginalized following the advent of molecular phylogenetics, is becoming important again, especially because of the importance of fossils in dating the tree of life. Fossils have nevertheless been somewhat neglected in the development of phylogenetic analysis methods. Faced with the sheer number of evolutionary models that can be used to infer a phylogeny on the basis of molecular sequences (DNA or amino acids), for phenotypic data, only the Mk model is currently available in most software packages (Ronquist et al. 2012a). Yet, the recent analysis by Goloboff et al. (2019) clearly demonstrates the limitations of this model. Future progress on dating the tree of life will depend in part on the development of more realistic evolutionary models to manage the phenotypic data that are the only ones available for the vast majority of fossils.

10.4. References

Anstey, R.L. and Pachut, J.F. (2004). Cladistic and phenetic recognition of species in the Ordovician bryozoan genus *Peronopora*. *Journal of Paleontology*, 78(4), 651–674.

Ascarrunz, E., Sánchez-Villagra, M.R., Betancur-R, R., Laurin, M. (2019). On trends and patterns in macroevolution: Williston's law and the branchiostegal series of extant and extinct osteichthyans. *BMC Evolutionary Biology*, 19(117), 1–11.

Bokma, F., Godinot, M., Maridet, O., Ladevèze, S., Costeur, L., Solé, F., Gheerbrant, E., Peigné, S., Jacques, F., Laurin, M. (2016). Testing for Depéret's rule (body size increase) in mammals using combined extinct and extant data. *Systematic Biology*, 65(1), 98–108.

Britton, T. (2005). Estimating divergence times in phylogenetic trees without a molecular clock. *Systematic Biology*, 54(3), 500–507.

Buffetaut, E. (2017). *Fossiles et croyances populaires – Une paléontologie de l'imaginaire*. Le Cavalier Bleu, Paris.

Burdak, V., de Faget, G., Meunier, F., Sire, J.-Y. (1986). Morphologie fonctionnelle du tégument écailleux des poissons. *Cybium*, 10(3), 1–147.

Burkhardt, R.W. (2013). Lamarck, evolution, and the inheritance of acquired characters. *Genetics*, 194(4), 793–805.

Carroll, R. (1988). *Vertebrate Paleontology and Evolution*. W.H. Freeman, New York.

Clutton-Brock, J. (1995). Aristotle, the scale of nature, and modern attitudes to animals. *Social Research*, 62(3), 421–440.

Cuvier, G. (1812). *Recherches sur les ossements fossiles de quadrupèdes, ou l'on rétablit les caractères de plusieurs espèces d'animaux que les révolutions du globe paroissent avoir détruites*. Flammarion, Paris.

Darwin, C. (1859). *On the Origin of Species*. John Murray, London.

Delsuc, F., Gibb, G.C., Kuch, M., Billet, G., Hautier, L., Southon, J., Rouillard, J.-M., Fernicola, J.C., Vizcaíno, S.F., MacPhee, R.D., Poinar, H.N. (2016). The phylogenetic affinities of the extinct glyptodonts. *Current Biology*, 26(4), R155–R156.

Didier, G. and Laurin, M. (2020). Exact distribution of divergence times from fossil ages and tree topologies. *Systematic Biology*, 69(6), 1068–1087.

Didier, G. and Laurin, M. (2021). Distributions of extinction times from fossil ages and tree topologies: The example of some mid-Permian synapsid extinctions. *PeerJ*, 9, e12577.

Didier, G., Fau, M., Laurin, M. (2017). Likelihood of tree topologies with fossils and diversification rate estimation. *Systematic Biology*, 66(6), 964–987.

Drummond, A.J., Ho, S.Y.W., Phillips, M.J., Rambaut, A. (2006). Relaxed phylogenetics and dating with confidence. *PLOS Biology*, 4(5), 699–710 [Online]. Available at: https://doi.org/10.1371/journal.pbio.0040088.

Farris, J. (1983). The logical basis of phylogenetic analysis. In *Advances in Cladistics*, Platnick, N.I. and Funk, V.A. (eds). Columbia University Press, New York.

Foote, M. (1996). On the probability of ancestors in the fossil record. *Paleobiology*, 22(2), 141–151.

Francillon-Vieillot, H., de Buffrénil, V., Castanet, J., Géraudie, J., Meunier, F.J., Sire, J.Y., Zylberberg, L., de Ricqlès, A. (1990). Microstructure and mineralization of vertebrate skeletal tissues. In *Skeletal Biomineralization: Patterns, Processes and Evolutionary Trends*, vol. 5, Carter, J.G. (ed.). Van Nostrand Reinhold, New York.

Gaudry, A. (1866). *Considérations générales sur les animaux fossiles de Pikermi*. F. Savy, Paris.

Gaudry, A. (1888). *Les ancêtres de nos animaux dans les temps géologiques*. J.-B. Baillière et fils, Paris.

Gingerich, P.D. (1979). The stratophenetic approach to phylogeny reconstruction in vertebrate paleontology. In *Phylogenetic Analysis and Paleontology*, Cracraft J. and Elredge, N. (eds). Columbia University Press, New York.

Gingerich, P.D. (1990). The stratophenetic approach to phylogeny reconstruction in vertebrate paleontology. In *Palaeobiology: A Synthesis*, Briggs, D.E.G. and Crowther, P.R. (eds). Blackwell Scientific Publications, Oxford.

Godinot, M. (2011). Avant Teilhard de Chardin: paléontologie, évolution et religion à la fin du XIXe et au début du XXe siècle. In *Théorie de l'évolution et religions*, Portier, M.V.P. and Willaime, J.-P. (eds). Riveneuve éditions, Paris.

Godoy, P.L., Benson, R.B.J., Bronzati, M., Butler, R.J. (2019). The multi-peak adaptive landscape of crocodylomorph body size evolution. *BMC Evolutionary Biology*, 19(167), 1–29.

Goloboff, P.A. and Catalano, S.A. (2016). TNT version 1.5, including a full implementation of phylogenetic morphometrics. *Cladistics*, 32(3), 221–238.

Goloboff, P.A., Pittman, M., Pol, D., Xu, X. (2019). Morphological data sets fit a common mechanism much more poorly than DNA sequences and call into question the Mkv model. *Systematic Biology*, 68(3), 494–504.

Gould, S.J. (1967). Evolutionary patterns in pelycosaurian reptiles: A factor-analytic study. *Evolution*, 21(2), 385–401.

Gould, S.J. and Littlejohn, J. (1973). Factor analysis of caseid pelycosaurs. *Journal of Paleontology*, 47(5), 886–891.

Grand, A., Corvez, A., Duque Velez, L.M., Laurin, M. (2013). Phylogenetic inference using discrete characters: Performance of ordered and unordered parsimony and of three-item statements. *Biological Journal of the Linnean Society*, 110(4), 914–930.

Hasegawa, M., Kishino, H., Yano, T. (1989). Estimation of branching dates among primates by molecular clocks of nuclear DNA which slowed down in Hominoidea. *Journal of Human Evolution*, 18(5), 461–476.

Heath, T.A., Huelsenbeck, J.P., Stadler, T. (2014). The fossilized birthdeath process for coherent calibration of divergence-time estimates. *Proceedings of the National Academy of Sciences*, 111(29), E2957–E2966.

Hennig, W. (1950). *Grundzuge einer Theorie der phylogenetischen Systematik*. Deutscher Zentralverlag, Berlin.

Hennig, W. (1966). *Phylogenetic Systematics*. University of Illinois Press, Urbana/Chicago/London.

Hugall, A.F., Foster, R., Lee, M. S.Y. (2007). Calibration choice, rate smoothing, and the pattern of tetrapod diversification according to the long nuclear gene RAG-1. *Systematic Biology*, 56(4), 543–563.

Janvier, P. (1996). *Early Vertebrates*. Oxford University Press.

Kesling, R.V. and Sigler, J.P. (1969). *Cunctocrinus*, a new Middle Devonian calceocrinid crinoid from the Silica Shale of Ohio. *Museum of Paleontology, University of Michigan*, 22(4).

Khemiri, S., Meunier, F., Laurin, M., Zylberberg, L. (2001). Morphology and structure of the scales in the Gadiformes (Actinopterygii: Teleostei: Paracanthopterygii) and a comparison to the elasmoid scales of other Teleostei. *Cahiers de biologie marine*, 42, 345–362.

Kiriakoff, S.G. (1966). Cladism and phylogeny. *Systematic Biology*, 15(1), 91–93.

Kluge, A.G. and Farris, J.S. (1969). Quantitative phyletics and the evolution of anurans. *Systematic Biology*, 18(1), 1–32.

Laloy, F., Rage, J.-C., Evans, S.E., Boistel, R., Lenoir, N., Laurin, M. (2013). A re-interpretation of the Eocene anuran *Thaumastosaurus* based on microCT examination of a "mummified" specimen. *PLOS ONE*, 8(9), 1–11.

Lamarck, J.-B. (1809). *Philosophie zoologique*. Flammarion, Paris.

Laurin, M. (2004). The evolution of body size, Cope's rule and the origin of amniotes. *Systematic Biology*, 53(4), 594–622.

Laurin, M. (2008). *Systématique, paléontologie et biologie évolutive moderne : l'exemple de la sortie des eaux des vertébrés*. Ellipses, Paris.

Laurin, M. (2012). Recent progress in paleontological methods for dating the tree of life. *Frontiers in Genetics*, 3(130), 1–16.

Laurin, M. and Piñeiro, G. (2017). A reassessment of the taxonomic position of mesosaurs, and a surprising phylogeny of early amniotes. *Frontiers in Earth Science*, 5, 1–13.

Lee, M.S.Y. and Yates, A.M. (2018). Tip-dating and homoplasy: Reconciling the shallow molecular divergences of modern gharials with their long fossil record. *Proceedings of the Royal Society B: Biological Sciences*, 285(1881), 20181071.

Lewis, P.O. (2001). A likelihood approach to estimating phylogeny from discrete morphological character data. *Systematic Biology*, 50(6), 913–925.

Lherminier, P. and Solignac, M. (2005). *De l'espèce*. Syllepse, Paris.

Lundberg, J.G. and Case, G.R. (1970). A new catfish from the Eocene Green River formation, Wyoming. *Journal of Paleontology*, 44(3), 451–457.

MacFadden, B.J. (1976). Cladistic analysis of primitive equids, with notes on other perissodactyls. *Systematic Zoology*, 25(1), 1.

MarjanoviÄĞ, D. and Laurin, M. (2007). Fossils, molecules, divergence times, and the origin of lissamphibians. *Systematic Biology*, 56(3), 369–388.

Marshall, C.R. (1994). Confidence intervals on stratigraphic ranges: Partial relaxation of the assumption of randomly distributed fossil horizons. *Paleobiology*, 20(4), 459–469.

Marshall, C.R. (1997). Confidence intervals on stratigraphic ranges with nonrandom distributions of fossil horizons. *Paleobiology*, 23(2), 165–173.

Marshall, C.R. (2019). Using the fossil record to evaluate timetree timescales. *Front. Gen.*, 10, 1049.

McGowan, C. (1979). A revision of the lower Jurassic ichthyosaurs of Germany with descriptions of two new species. *Palaeontographica Abteilung A*, A166(4–6), 93–135.

Michevich, M.F. (1982). Transformation series analysis. *Systematic Biology*, 31(4), 461–478.

Miller, W., Drautz, D.I., Ratan, A., Pusey, B., Qi, J., Lesk, A.M., Tomsho, L.P., Packard, M.D., Zhao, F., Sher, A. et al. (2008). Sequencing the nuclear genome of the extinct woolly mammoth. *Nature*, 456(7220), 387–390.

Müller, J. and Reisz, R.R. (2006). The phylogeny of early eureptiles: Comparing parsimony and Bayesian approaches in the investigation of a basal fossil clade. *Systematic Biology*, 55(3), 503–511.

Paterson, J.R., Edgecombe, G.D., Lee, M.S.Y. (2019). Trilobite evolutionary rates constrain the duration of the Cambrian explosion. *Proceedings of the National Academy of Sciences*, 116(10), 4394–4399.

Prüfer, K., Stenzel, U., Hofreiter, M., Pääbo, S., Kelso, J., Green, R.E. (2010). Computational challenges in the analysis of ancient DNA. *Genome Biology*, 11(5), R47.

Pyron, R.A. (2011). Divergence time estimation using fossils as terminal taxa and the origins of Lissamphibia. *Systematic Biology*, 60(4), 466–481.

Rineau, V., Grand, A., Zaragüeta, R., Laurin, M., Brazeau, M. (2015). Experimental systematics: Sensitivity of cladistic methods to polarization and character ordering schemes. *Contributions to Zoology*, 84(2), 129–148.

Rineau, V., Bagils, R.Z., Laurin, M., Sluys, R. (2018). Impact of errors on cladistic inference: Simulation-based comparison between parsimony and three-taxon analysis. *Contributions to Zoology*, 87(1), 25–40.

Roelants, K., Gower, D.J., Wilkinson, M., Loader, S.P., Biju, S.D., Guillaume, K., Moriau, L., Bossuyt, F. (2007). Global patterns of diversification in the history of modern amphibians. *Proceedings of the National Academy of Sciences*, 104(3), 887–892.

Romer, A. (1966). *Vertebrate Paleontology*. University of Chicago Press.

Ronquist, F., Klopfstein, S., Vilhelmsen, L., Schulmeister, S., Murray, D.L., Rasnitsyn, A.P. (2012b). A total-evidence approach to dating with fossils, applied to the early radiation of the Hymenoptera. *Systematic Biology*, 61(6), 973–999.

Ronquist, F., Teslenko, M., Van der Mark, P., Ayres, D.L., Darling, A., Höhna, S., Larget, B., Liu, L., Suchard, M.A., Huelsenbeck, J.P. (2012a). MrBayes 3.2: Efficient Bayesian phylogenetic inference and model choice across a large model space. *Systematic Biology*, 61(3), 539–542.

Roopnarine, P.D. (2005). The likelihood of stratophenetic-based hypotheses of genealogical succession. *Special Papers in Palaeontology*, 73, 143–157.

Sanderson, M.J. (2002). Estimating absolute rates of molecular evolution and divergence times: A penalized likelihood approach. *Molecular Biology and Evolution*, 19(1), 101–109.

Sauquet, H. (2013). A practical guide to molecular dating. *Comptes rendus Palevol*, 12(6), 355–367.

Simões, T.R., Caldwell, M.W., Palci, A., Nydam, R.L. (2017). Giant taxon-character matrices: Quality of character constructions remains critical regardless of size. *Cladistics*, 33(2), 198–219.

Slowinski, J.B. (1993). "Unordered" versus "ordered" characters. *Systematic Biology*, 42(2), 155–165.

Sneath, P.H.A. (1995). Thirty years of numerical taxonomy. *Systematic Biology*, 44(3), 281–298.

Sokal, R. and Sneath, P. (1963). *Principles of Numerical Taxonomy*. W. H. Freeman and Company, San Francisco, CA.

Sterli, J., Pol, D., Laurin, M. (2013). Incorporating phylogenetic uncertainty on phylogeny-based palaeontological dating and the timing of turtle diversification. *Cladistics*, 29(3), 233–246.

Strauss, D. and Sadler, P.M. (1989). Classical confidence intervals and Bayesian probability estimates for ends of local taxon ranges. *Mathematical Geology*, 21(4), 411–427.

Swofford, D. (2002). *Paup* Phylogenetic Analysis Using Parsimony (* and Other Methods). Version 4.0b10*. Sinauer Associates, Sunderland, MA.

Tassy, P. (2006). Albert Gaudry et l'émergence de la paléontologie darwinienne au XIXe siècle. *Annales de paléontologie*, 92(1), 41–70.

Tassy, P. (2011). Trees before and after Darwin. *Journal of Zoological Systematics and Evolutionary Research*, 49(2), 89–101.

Taylor, P.D. and Vinn, O. (2006). Convergent morphology in small spiral worm tubes ('*Spirorbis*') and its palaeoenvironmental implications. *Journal of the Geological Society*, 163(2), 225–228.

Wilkinson, R.D. and Tavaré, S. (2009). Estimating primate divergence times by using conditioned birth-and-death processes. *Theoretical Population Biology*, 75(4), 278–285.

Willerslev, E. and Cooper, A. (2005). Ancient DNA. *Proceedings of the Royal Society B: Biological Sciences*, 272(1558), 3–16.

Woodburne, M.O., Goin, F.J., Bond, M., Carlini, A.A., Gelfo, J.N., López, G.M., Iglesias, A., Zimicz, A.N. (2014). Paleogene land mammal faunas of South America; A response to global climatic changes and indigenous floral diversity. *Journal of Mammalian Evolution*, 21(1), 1–73.

Wright, D.F. (2017). Bayesian estimation of fossil phylogenies and the evolution of early to middle Paleozoic crinoids (Echinodermata). *Journal of Paleontology*, 91(4), 799–814.

Zhang, P., Zhou, H., Chen, Y.-Q., Liu, Y.-F., Qu, L.-H. (2005). Mitogenomic perspectives on the origin and phylogeny of living amphibians. *Systematic Biology*, 54(3), 391–400.

Zhang, C., Stadler, T., Klopfstein, S., Heath, T.A., Ronquist, F. (2015). Total-evidence dating under the fossilized birth–death process. *Systematic Biology*, 65(2), 228–249.

Zirkle, C. (1946). The early history of the idea of the inheritance of acquired characters and of pangenesis. *Transactions of the American Philosophical Society*, 35(2), 91.

Zuckerkandl, E. and Pauling, L. (1962). Molecular disease, evolution and genetic heterogeneity. In *Horizons in Biochemistry*, Kasha, M. and Pullman, B. (eds). Academic Press, New York.

Zuckerkandl, E. and Pauling, L. (1965). *Evolutionary Divergence and Convergence in Proteins*. Elsevier, Amsterdam.

Zylberberg, L. and Laurin, M. (2011). Analysis of fossil bone organic matrix by transmission electron microscopy. *Comptes Rendus Palevol*, 10(5), 357–366.

Zylberberg, L., Meunier, F., Laurin, M. (2016). A microanatomical and histological study of the postcranial dermal skeleton of the Devonian actinopterygian *Cheirolepis canadensis* (Whiteaves). *Acta Palaeontologica Polonica*, 61(2), 363–376.

11

Phylodynamics

Samuel ALIZON[1,2]

[1] *MIVEGEC, IRD, CNRS, Université de Montpellier, France*

[2] *CIRB, CNRS, INSERM, Collège de France, Paris, France*

Over the last two decades, some sort of mutual emulation has emerged between the production of microbial genetic sequence data and the development of phylogenetic analysis methods. In effect, infection phylogenies are used to describe the structure of an epidemic and its geographical spread. In turn, the originality of these types of data, in particular their strong anchoring in time, has opened new dimensions for phylogenetic analyses centered around a field known as phylodynamics.

11.1. Reconciling ecology, evolution and mathematics

Population dynamics forms the basis for evolution. Darwin not only understood this but also illustrated it in his book *The Origin of Species*. Nevertheless, mathematical approaches in evolutionary biology have long overlooked these demographic aspects (also known as "ecological" aspects). Therefore, the founding models of population genetics often consider long time scales and infinitely sized populations. The reconciliation between population dynamics and genetics has taken place since the 1970s, first with evolutionary game theory and then, during the 1990s, with adaptive dynamics. Today, modeling in ecology and evolution are inseparable, as detailed in the recent review by Lion (2018).

The study of infectious diseases has played a central role in building bridges between demography, evolution and mathematics. On the one hand, as for evolutionary biology, the tradition of mathematical modeling is old in epidemiology, and models such as those by Kermack and McKendrick (1927) are still utilized as a basis for contemporary studies. On the other hand, at the biological level, microbes generally have short generation times, large population sizes and high mutation rates, which leads to their evolutionary and ecological dynamics overlapping. This is one illustration of the feedback loop between ecology and evolution. For example, an antibiotic resistance mutation (evolution) modifies the spread of the bacteria that carry it (ecology). However, the selection of this mutation (evolution) depends on the environment, because when antibiotic treatments (ecology) are inexisting, antibiotic resistance is rather a cost than a benefit. This has been particularly well captured at the mathematical level by adaptive dynamics approaches (Dieckmann et al. 2002).

Phylodynamics, defined in 2004 by Grenfell et al., has strengthened the links between evolution and epidemiology by providing an increasing volume of data. Until then, the reconciliation between ecology and evolution had been particularly built on inevitably limited phenotypic data. Phylodynamics is primarily based on neutral evolution, which allows it to easily exploit a lot of genetic sequence data. The success of this field is largely based on the hope that a large amount of information relevant to public health will be extracted from the thousands of sequence data routinely generated in laboratories.

In this chapter, we will briefly review the technical developments that contributed to the development of phylodynamics. We will then present infection phylogenies and their initial, almost purely illustrative use, which is still the major one. We will then define phylodynamics itself as well as its basic models. We will next present the spatial analyses or, more generally, in heterogeneous environment that can be performed. Finally, we will discuss more recent extensions coupling infection phenotypes (e.g. viral load) to sequence data. We will conclude by listing some of the existing challenges for the field.

11.2. Data and processors

Phylogenetic analyses have been possible for a long time (Felsenstein 1981), but were initially limited by the availability of biological data and by the computing power that could be dedicated. These limitations are now almost obsolete. Improvements brought to processors coupled with their networked operation enable the simultaneous analysis of tens of thousands of sequence data items. In parallel, advances in sequencing techniques have led to increased quality, length and speed of genetic sequence data production.

11.2.1. *New generation sequencing*

Medicine has often made considerable progress during wars and other health tragedies. The same is true for phylodynamics, if we consider that the deadly human immunodeficiency virus (HIV) pandemic was one of the triggers for using sequencing in clinical routine. As a matter of fact, after the discovery of the first effective antiretroviral molecules against the virus, the problem of the evolution of resistant infections quickly arose. By linking viral sequence data to treatment results, it was possible to develop algorithms predicting the risk of therapeutic failure for a given molecule according to the viral sequence (Beerenwinkel et al. 2003).

These uses for public health have led to the generation of gene sequence data of unprecedented quantity and spatial coverage. This routine sequencing was initially limited to HIV and involved only a few portions of the genome. Over the past decade, it has been extended to other virulent viruses and even to bacterial infections or entire microbiota. Moreover, the percentage of the genome sequenced has exploded and is now moving toward whole genome sequencing. One of the most striking examples was of Quick et al. (2016), who managed to generate 142 whole genome sequences of the Ebola virus using the Oxford Nanopore technology with basic biological and computational material (20 kg) in a country in the midst of a health crisis.

Of these so-called next-generation sequencing techniques, three stand out. The most popular is developed by Illumina Inc. It allows a large quantity of short sequences (200–300 base pairs) to be generated for a low cost with a low error rate. Its main drawback is that the size of the sequences does not make it possible to know whether mutations are present in the same genome and not only in the population of genomes. The second technique by Pacific Biosciences (PacBio) solves this problem to some extent by performing *single-molecule real-time sequencing*. The analyzed fragments are long (potentially more than 30,000 base pairs) but the error rates are higher than for the Illumina. Finally, the Nanopore technique mentioned above is a virtually portable solution, unlike the two others that require expensive equipment to generate very long fragments also with a relatively high error rate. Easier reconstruction of haplotypes are possible with these last two techniques. More generally, new generation sequencing techniques enable low-cost access to microbial genetic diversity.

11.2.2. *PCR and capture*

The power of processors or sequencers is often the main focus, but sample processing techniques have also made progress, especially for managing the overwhelming presence of host DNA. The task of enriching a sample with microbial DNA or RNA is the daily work of many laboratories. If the microbial target is known, DNA amplifications can be carried out by *polymerase chain reaction* (PCR), but these truncate the genetic diversity and require qualitative work. Moreover, the sequencing of an entire genome requires the development of several primers and the performance of several PCR amplifications.

New-generation techniques allow sequencing of all the DNA present without a specific target, using the shotgun technique for example, which uses random DNA targets. Except in special cases (e.g. an extremely high viral load in a medium with little human DNA), it is necessary to enrich the sample with target DNA. This can be done mechanically by ultracentrifugation (to recover smaller bacteria or viruses), chemically by DNAase treatments (to destroy less well-protected human DNA) or even biotechnologically by means of the CRISPR-Cas9 technique to specifically destroy human DNA.

For a few years now, another alternative has been available that allows the generation of extremely good quality sequence data: capture by hybridization. As a result, for less than 50 euros, it is now possible to obtain from a blood sample the viral diversity of HIV along the entire genome but also of flaviviruses and all this without PCR (Bonsall et al. 2018). Ultimately, coverage can be used to estimate viral load, and the sequenced diversity is close to the true diversity. These techniques are opening up new avenues in public health for analyzing transmission networks based on phylogenetic approaches (Wymant et al. 2018).

11.3. Infection phylogenies

11.3.1. *Link to transmission chains*

Phylogenies have invaded the biomedical literature primarily for classification purposes. More precisely, we determine (most often visually) which existing strains a viral (or bacterial) strain infecting a patient is most similar to. From a practical point of view, this may be of interest if the strains differ in terms of virulence or response to treatment (which is the case for hepatitis C virus). From a historical point of view, these classification approaches in molecular epidemiology are very similar to the first microbial phylogenies based on genetic sequences originating from different species. This was the case for endosymbionts (Lake 1988), but also for HIV, for which one of the first phylogenetic works examines its long history and in addition, includes sequences from its simian equivalent (Smith et al. 1988).

A shift took place with the HIV epidemic. This virus differs significantly from most of the organisms sequenced until then due to its rapid evolutionary speed and its epidemiological life cycle (not to mention the quantity of sequences available). This allows the generation of infection phylogenies. One of the first was derived from the data of nine hemophiliac patients infected by the same source (Balfe et al.

1990). Another famous one concerns a chain of transmission linked to a dentist (Ou et al. 1992). In both cases, the ethical and legal implications of these studies are unprecedented for evolutionary biologists who were not used to appearing in courts of law as experts to decide on the sources of transmission. These issues are still relevant in phylodynamics (Coltart et al. 2018; Johnson and Parker 2019).

As a caricature, the shift that took place in the early 1990s meant replacing species with infections on the leaves of phylogenetic trees. Described in this way, the conceptual shift may seem huge. However, not only can most of the techniques be directly transposed, but in addition working on infections rather than species ultimately makes the underlying mechanisms more intuitive. Actually, a parallel between a phylogeny and a transmission chain can easily be drawn. If each individual in the transmission chain is present in the phylogeny, then each leaf corresponds to the end of the infectious period of an individual and each new infection creates a phylogeny branch. Conversely, although species extinctions are obvious, their appearance is much blurrier than a new infection (not to mention, of course, the difficulty of defining a species).

11.3.2. *Dating and evolutionary rates*

For phylogeneticists, infection phylogenies have been truly inspirational, because there is a major difference between them and species phylogenies. Classically, the latter are represented by trees known as "ultrametric trees", because all species are contemporary. Anchoring such a phylogeny in time is a challenge in itself and requires the use of fossil data to calibrate the internal nodes (see Chapter 10). For their part, infection phylogenies are easier to anchor in time, since differences of a few years between two sampling dates are no longer negligible in relation to the date of the phylogeny root. For epidemic emergences, evolutionary rates are such that it is now not uncommon to have good quality phylogenies with some leaves relatively close to the index case of the epidemic (as in the recent Ebola virus epidemic in West Africa described by Holmes et al. 2016).

This strong anchoring in time has led to the development of new techniques for estimating substitution rates, that is, the number of fixed mutations in the genome per unit of time. The simplest one consists of deducing these rates from a regression between the sampling date of the leaves, on the one hand, and their distance (in number of substitutions) from the root of the tree, on the other hand (Rambaut 2000; Drummond et al. 2003). For a given substitution rate, the internal nodes of the tree as well as its root can then be easily dated. Bayesian models have contributed to improve this dating. They have also made it possible to remove the sometimes strong hypothesis of constant evolutionary rate. Consequently, the relaxed molecular clock model allows the evolutionary rates in the different phylogeny branches to be varied by drawing them from a distribution (Drummond et al. 2006), which leads to avoiding having to estimate a value for each phylogeny region as in the local molecular clock models. One of the limitations of these methods stems from their high computational power requirements. More recently, approaches based on maximum likelihood techniques have made it possible to achieve comparable dating qualities in a faster manner (To et al. 2015; Sagulenko et al. 2018).

An emblematic application of time anchoring has been the dating of the origin of HIV in human populations. Shortly after the first sequencing, phylogenetic studies located the common ancestor of the epidemic before 1960 (Li et al. 1988). This estimate was refined with the isolation of the virus in samples from patients infected before 1970 (Jonassen et al. 1997; Korber et al. 2000; Worobey et al. 2008). Over the past decade, inference techniques have advanced and, as it will be seen, both date and geographic origin can now be estimated. For HIV group M, which is the one that has claimed the most victims, they are found in the 1920s in what is now the Democratic Republic of Congo (Faria et al. 2014). This dating is not unique to viruses. For example, the discovery of DNA in human teeth dating back 4,900 years allowed Rascovan et al. (2019) to revise the history of the spread of plague epidemics in Europe since the Neolithic.

Although computational power and inference techniques are important, the discovery of ancient samples remains a driving force in dating studies. However, the biology of viruses constrains phylogenetic inference. For double-stranded DNA viruses evolving at the speed of their vertebrate host, tracing may go back hundreds of thousands of years in the past, most often using internal tree calibrations from host evolution, as in the case of HPV16 (Pimenoff et al. 2016). For more rapidly evolving viruses, the number of substitutions drowns out the phylogenetic signal over long time but it allows inferences over short time scales.

11.3.3. *Biological applications of time calibration*

Dating events that took place during an outbreak obviously involves public health implications. For example, in West Africa in 2014–2016, being able to quickly date the spreading of the Ebola outbreak in Sierra Leone from Guinea helped to understand the magnitude of the outbreak (Gire et al. 2014). At the clinical level, by combining HIV sequence data from an infected mother and child, it is possible to determine whether transmission occurred in utero or at birth (Chaillon et al. 2014).

Evolutionary rates are also informative in their own right. In the case of HIV, this speed is faster at the within-host level than at the inter-host level. By analyzing the whole genome, we have shown that this difference of one order of magnitude in substitution rate is found along the full genome (Alizon and Fraser 2013). The most parsimonious explanation is that the most frequent viruses in the blood are actually in an evolutionary dead end, while those stored in cells in a more latent manner retain their ability to generate new infection. Consequently, "less evolved" viruses are more likely to be passed to the next generation.

These issues also apply to less rapidly evolving viruses such as varicella-zoster, which is a double-stranded DNA (dsDNA) virus and for which the sequence data do not a priori contain enough signal to estimate a reliable substitution rate. However, by studying attenuated vaccine strains, Weinert et al. (2015) were able to estimate this substitution rate. The difference between this strain and the natural strain is that the latter is known to have latency periods during which the virus remains in the host cell without multiplying. This highlighted the importance of taking latency into account when interpreting and dating this virus outbreak.

11.4. Phylodynamics

11.4.1. *A field in search of definition*

The term "phylodynamics" has emerged in a "Perspective" by Grenfell et al. (2004). Their definition of the field is inclusive, to say the least, since, according to them, phylodynamics aims to explain "the wide diversity of pathogen phylogenies observed at scales ranging from an individual host to the population" and is defined as a "mixture of immuno-dynamics, epidemiology and evolutionary biology". What is striking about this definition is the marginal place of phylogenies. Moreover, in the article itself, the use that is made thereof is purely descriptive. Even the question of inference has been evaded in this seminal review.

Since then, many articles have reviewed this expanding field most tend not to face the blur surrounding its definition. For Volz et al. (2013), phylodynamics is "the study of how epidemiological, immunological, and evolutionary processes act and potentially interact to shape [viral] phylogenies". One should however observe that in the last 10 years phylogeny has acquired a central role. The issue of inference is also much more prevalent, and Volz et al. note that different processes can lead to the same phylogeny, generating problems of parameter identifiability.

Perhaps the clearest way to define phylodynamics is to invoke the function used in Bayesian inference models Kühnert et al. (2011) to capture the likelihood of having a phylogenetic tree T, a population dynamics model with parameters η and a sequence evolution model with parameters θ best explain the data (i.e. a sequence alignment):

$$f\left[T,\eta,\theta|\text{data}\right] = \frac{f\left[\text{data}|T,\theta\right]\ f\left[T|\eta\right]\ f\left[\eta,\theta\right]}{f\left[\text{data}\right]}$$

where $f\,[\text{data}|T,\theta]$ is Felsenstein's likelihood function, $f\,[T|\eta]$ is the transmission model likelihood, $f\,[\eta,\theta]$ is the a priori parameter distributions and $f\,[\text{data}]$ is the normalization term. The originality of phylodynamics is the $f\,[T|\eta]$ term.

For the sake of clarity, we will here define phylodynamics as the link between genetic sequence data (or phylogenies) and population dynamics.

11.4.2. *As closely as possible to epidemiology*

Phylodynamics is one of the many extensions of the founding work by Felsenstein (1981) for calculating the probability of observing a sequence alignment from a phylogenetic tree, also called phylogenetic likelihood (see Chapter 7). The pioneering work in phylodynamics dates back to the 1990s and consisted of extracting information about the demographic history of microbial populations contained in the sequences of (Holmes et al. 1995; Rodrigo and Felsenstein 1999). They have shown that variations in population size affect the distribution of phylogenies of individuals originating from the same population (also called genealogy in population genetics, see Chapter 1) allowing periods of growth or decrease in population size to be detected (Nee et al. 1995).

Microbial sequences have driven the rise of phylodynamics, partly because of the intuitive parallel that can be drawn between transmission chains and infection phylogenies. With that respect, it is necessary to formalize a little more what is meant by population dynamics or, in our case, epidemiological dynamics. The SIR model is one of the most commonly used epidemiological models (Anderson and May 1991; Keeling and Rohani 2008). Its name is derived from the three possible host states: susceptible, infected and recovered (i.e. immune, dead or cured). It is formalized by means of the following system of differential equations:

$$\frac{\mathrm{d}S(t)}{\mathrm{d}t} = -\beta\, S(t)\, I(t) + \sigma\, R(t) \quad [11.1]$$

$$\frac{\mathrm{d}I(t)}{\mathrm{d}t} = \beta\, S(t)\, I(t) - \gamma\, I(t) \quad [11.2]$$

$$\frac{\mathrm{d}R(t)}{\mathrm{d}t} = \gamma\, I(t) - \sigma\, R(t) \quad [11.3]$$

where $S(t)$, $I(t)$ and $R(t)$ are the densities of susceptible, infected and immune individuals at time t, whereas β, γ and σ are the transmission, recovery and loss of immunity rates. Susceptible individuals become infected following their encounter with infected individuals according to the mass action law hypothesis. They are then infected for an average of $1/\gamma$ units of time, after which they become immune. Immunity is assumed here to last for an average of $1/\sigma$ units of time.

For field epidemiologists, one of the key parameters is the basic reproduction number, denoted by R_0. It corresponds to the average number of secondary infections generated by an infected individual in a fully susceptible population (Anderson and May 1991). In a deterministic model, for an epidemic to occur, it is necessary that $R_0 > 1$. This R_0 can be easily calculated from any system of differential equations (Diekmann and Heesterbeek 2000; Hurford et al. 2010). For example, according to equation [11.2], one has $R_0 = \beta S(0)/\gamma$, where $S(0)$ is the number of individuals initially present in the population. Public health interventions aim to decrease this R_0 by virtue of therapeutic or quarantine methods.

Although simple in appearance, the SIR model involves nonlinear phenomena. For example, at the beginning of an epidemic, the number of new infections per unit of time $\beta \cdot S(t) \cdot I(t)$, also called "incidence", increases as $I(t)$ increases, and gradually slows down as the number of susceptible hosts ($S(t)$) decreases. Infection times are also difficult to estimate using incidence data alone, because they have a limited time component. This illustrates both the challenges and the potential of phylodynamics: to be able to make use of the temporal signal present in genetic sequences while taking into account the constraints inherent to epidemiology, such as nonlinear processes.

11.4.3. *Coalescent*

Phylodynamics was first based on the coalescent theory by Kingman (1982), especially by analyzing the number of lineages through time (or LTT) (Ong et al. 1996; Pybus et al. 1999). In practice, this amounts to summing the number of active branches of a dated phylogeny at a given time (Nee et al. 1995; Rambaut et al. 1997). It is important to underline this point because it illustrates the fact that the structure of the phylogeny itself (in particular possible asymmetries) have little incidence. For more details, one should refer to the review by Kühnert et al. (2011) on phylodynamic inference that contains detailed aspects on the coalescent theory as well as to Chapter 1 in this book.

The development of phylodynamic methods has followed the increasing availability of molecular sequence data for well-studied human viruses such as HIV, hepatitis C virus (HCV) or influenza. As noted above, unlike most population genetic data, viral sequences are often sampled at different points in time. Inference methods using such dated phylogenies allow, among other things, the estimation of variations in pathogen population size over time from sequences and sampling dates (Drummond et al. 2002, 2005; Lemey et al. 2003; Rambaut et al. 2008). More general methods, such as Bayesian Skyline (Drummond et al. 2005) and Skygrid (Gill et al. 2013), have made possible the reconstruction of increasingly rich population dynamics.

Coalescent models have long been based only on simple demographic cases that best correspond to the epidemiological susceptible infected (SI) model with two host classes. There are exceptions with phylodynamics coalescent models analyzing structured populations (Hudson 1990; Notohara 1990), for example, to describe the dynamics of different influenza strains (Koelle et al. 2006; Vaughan et al. 2014).

The relevance of standard coalescent models to infectious pathogens has therefore been progressively questioned. As a matter of fact, in most of these models, the coalescence rate is inversely proportional to the effective population size ($N_e(t)$), which is different from the absolute population size. This problem arises even within an individual infected by a diverse virus population (as discussed by Kouyos et al. (2006) in the case of HIV). It is all the more acute at the inter-host level when it is assumed that $N_e(t)$, which is derived from microbial population sequence data, is proportional to the number of infected hosts (Pybus et al. 2000; Drummond et al. 2002). Studies have first pointed out that the phylodynamic reconstruction of $N_e(t)$ should not be interpreted as a dynamic of the number of infections but as a measure of relative genetic diversity, itself influenced by absolute population size, the population structure and reproductive variability (Pybus et al. 2001; Rambaut et al. 2008). More recently, research around the structured coalescent (Hudson 1990) has looked deeper into the link between genealogy and infection dynamics (Volz et al. 2009; Bedford et al. 2010; Volz 2012; Volz and Siveroni 2018). The underlying assumption is that each lineage in the genealogy corresponds to a single infected host (thus neglecting within-host evolution), which implies that coalescence events correspond to transmission events. The coalescence rate is therefore proportional to the incidence of the microbe under study (i.e. $\beta \cdot S(t) \cdot I(t)$ in the SIR model). It thus varies in a nonlinear way over time, except under conditions where the number of susceptible individuals can be considered to be constant, for example, at the very beginning of an epidemic (Frost and Volz 2010; Koelle and Rasmussen 2012).

This research has led to the development of new Bayesian inference approaches based on the calculation of the likelihood of a coalescent model associated with the epidemiological model. For example, Rasmussen et al. (2011) used the expression of the likelihood of the coalescent model associated with an epidemic of the SIR type established by Volz et al. (2009) to develop an approach enabling the use of both dated phylogeny and time series data. This type of approach was then extended to a wide range of epidemiological models (Volz 2012; Rasmussen et al. 2014). These approaches are based on the simulation of epidemiological trajectories and then proceed by calculating the likelihood of the dated phylogeny under the assumption of a coalescent model by integrating over the possible population size variations over time described by the trajectories. Nevertheless, they often require large computational resources (Ratmann et al. 2017).

11.4.4. *Birth–death models*

Among the approaches making use of phylogenetic likelihood, an alternative to coalescent models is based on the birth and death process (Kendall 1948). These models, called BD (for *birth and death*), have been applied to the inference of epidemiological parameters from genotyping data (Tanaka et al. 2006), as well as to the study of the antigenic dynamics of the influenza virus (Koelle et al. 2009). In its simplest formulation, the birth–death model is the exact equivalent of an SI model where the population size of susceptible hosts is constant, which may be the case at the beginning of an epidemic (because the whole population is susceptible). Furthermore, if the birth rate is higher than the death rate, this model can be compared to a classical coalescent model with exponential population size growth (Stadler 2009).

One of the major differences relative to classical coalescent models is that in BD models the population size can stochastically vary over time, whereas in classical coalescent models the population size varies deterministically. Moreover, Stadler et al. (2015) have compared coalescent models (Volz 2012; Rasmussen et al. 2014) to the BD model for phylodynamic inference and showed that a classical coalescent model fails to infer the correct transmission tree coalescence times corresponding to an SI-type epidemic (with dynamics identical to the stochastic BD model). However, this bias is corrected in the new versions of the coalescent model that take into account epidemiological trajectories, but only for small values of R_0 and large population sizes.

As with coalescent ones, phylodynamic approaches based on birth–death models have been extended to capture more detailed epidemiological patterns. One of the advances was thus to obtain a *skyline*-based approach not to estimate variations in effective population size over time, but variations in epidemiological parameters such as R_0 or the duration of the infection (Stadler et al. 2013; Kühnert et al. 2014). More recently, Kühnert et al. (2018) have generalized the birth–death model to heterogeneous host populations, making it possible to measure the effect of resistance mutations to antiretrovirals on the spread of HIV.

11.4.5. *Limitations of likelihood approaches*

Most existing phylodynamic methods require that a phylogenetic likelihood function be differentiated. Although some models, such as the structured coalescent, are capable of capturing a wide range of patterns, in most of them the addition of even the smallest biological detail requires an enormous amount of analytical work. And even when it is possible to express analytically the likelihood function, its numerical resolution is not always possible. Finally, the computational time required to find the parameter values that maximize the likelihood remains an issue. While Bayesian approaches have made it possible to save time, the convergence times of Markov chain Monte Carlo methods increase rapidly with the details of the model and, above all, with the number of sequences (Ratmann et al. 2017).

More generally, the field seems to be undergoing significant disruption. Since its origin, phylodynamics has aimed to extract the maximum signal from a limited amount of data. For this purpose, the estimation of phylogeny while inferring the demographic model from a dated sequence alignment resulted in maximizing the use of information. This is also what led to the success of the BEAST (Bayesian evolutionary analysis by sampling trees) (Drummond and Rambaut 2007) and BEAST2 (Bouckaert et al. 2014) software programs: by allowing the generation of consistent results in a relatively simple manner, the user has the possibility of performing parameter estimations from a given model (typically involving coalescence or birth and death processes). Nowadays, the problem has been inverted and the objective now consists of sorting the information from large volumes of data. Due to this last point, algorithms estimating phylogeny and demographic models simultaneously become slow. This was particularly striking during the Pangea "contest", whose purpose was to compare the ability of different approaches to infer epidemiological parameters from simulated sequence alignments. The first step for all participants was to infer a phylogeny using fast maximum likelihood methods (with software such as PhyML or fastML) and then apply phylodynamic methods (Ratmann et al. 2017).

11.4.6. *ABC phylodynamics*

One response to the limitations outlined in section 11.4.5 was to abandon the analytical differentiation of a likelihood function and simply approximate it by numerical simulations. Therefore, the so-called *approximate Bayesian computation* (ABC) methods consist of simulating a large number of data sets using an underlying model with known parameter values to determine the parameter values that lead to generating the objects closest to the observed objects. In the case of phylogenies, this approach is particularly suitable, because while phylogenies are difficult objects to compare two by two, their simulation proves quite easy because of this parallel with a transmission chain. In practice, an algorithm such as the Gillespie algorithm, commonly used in epidemiology (Keeling and Rohani 2008), is enough. The difficulty lies more in the comparison between phylogenetic trees. To this end, a conventional method involves decomposing the trees into summary statistics such as the average branch length, the variance of the branch length or the asymmetry of the tree. Then, comparing two trees through their summary statistics values is easy.

ABC in phylodynamics is not such a new idea and Ratmann et al. (2012) had already attempted to use it to combine incidence data with phylogenetic data. The mixed success of their approach may be explained by the fact that the method used imposed a strict limit on the number of summary statistics that could be used. More recently, we have developed an approach using ABC regression (Csilléry et al. 2012; Blum 2018) to instead maximize the number of summary statistics used. Tests on SIR models showed performance for inference comparable to that obtained by birth–death models in BEAST (Saulnier et al. 2017). Moreover, one of the strengths of these approaches is that their performance increases with the size of the phylogeny, without slowing down the computations in a prohibitive way. This formalism was also extended to capture an epidemic of hepatitis C spreading in a population with two types of hosts (Danesh et al. 2021). However, analyses on more detailed models are needed to evaluate the true potential of the method. Another advantage is that the regression models used in these ABC approaches, such as random forests for instance, offer enhanced possibilities for model comparison (Pudlo et al. 2016).

11.5. Infection phylogeography

Phylogeography as such is an old discipline (Avise et al. 1987) whose statistical methods were well developed before the advent of phylodynamics (Knowles and Maddison 2002). Nevertheless, infection phylogenies have unquestionably renewed the field. As with time calibration, the main reason is that rapid microbial evolution allows the generation of non-ultrametric trees, namely for which some leaves are contemporary to internal nodes. Keeping in line with our definition of phylodynamics, we will present here only those approaches that incorporate population dynamics. Most of these are detailed in the summary by Lemey et al. (2009) with more recent work summarized by Baele et al. (2018).

The first type of approach considers spatial location as a discrete trait. From a practical point of view, the underlying models developed are the same continuous-time Markov chains employed for sequence evolution. This parallel with nucleotide base mutations has led to the model being often referred to as a "mugration" model, where migrations are seen as random events that can occur at different frequencies. Bayesian approaches allow us to infer these frequencies that correspond to the connectivity between locations, while inferring the evolutionary history (Lemey et al. 2009). These approaches have been applied in many contexts, in part due to their implementation integrated in the BEAST and BEAST2 software programs. However, the recent Ebola virus outbreak has probably led to the most extensive developments given the large number of sequences and associated geographic information (Dudas et al. 2017).

A limitation of these discrete models is that they are sensitive to sampling disparities (a recurring theme in phylodynamics). De Maio et al. (2015) have developed a BEAST package called BASTA that relies on a structured coalescent approximation to make phylogeographic inference more robust to sampling, thus allowing more localities and migration flows to be generated. Recently, Müller et al. (2018) proposed a faster program based on an exact solution of the structured coalescent.

The second type of approach consists of modeling the geographic location using two continuous traits, typically latitude and longitude (Lemey et al. 2009). An obvious advantage of this method is that it makes it possible to infer the presence of individuals in unsampled locations. It also provides a significant visual dimension in terms of attractiveness. Finally, it allows the calculation of a rate parameter for the spread of the epidemic. This spatial modeling can easily be implemented by means of classical Brownian motion models that can vary according to the branches. This has greatly contributed to its popularization (Pybus et al. 2012). However, this diffusion model can of course prove to be simplistic when considering an epidemic. Moreover, external factors may affect the spread rate, which can be tested a posteriori using robust statistical models (Dellicour et al. 2016).

Finally, while not strictly addressing phylogeography, several studies have attempted to infer the structure of an underlying contact network between individuals from genetic sequence data. For example, Leventhal et al. (2012) conducted a simulation study to determine the extent to which different network families could be characterized by different summary statistics. On the other hand, Rasmussen et al. (2017) implemented a maximum likelihood model using the technique of pairwise approximations to obtain an analytical solution. For their part, Giardina et al. (2017) used an approach based on the ABC computation described in section 11.4.6 to perform this inference. Finally, for some networks such as sexual contact networks, it should be noted that these inferences can be further complicated by the dynamics of the network itself (Metzig et al. 2019).

11.6. Infection and viral life history traits

Comparative phylogeny is a means to determine to what extent the phylogenetic structure between populations explains the trait distribution measured in these populations (Felsenstein 1985). To overlook this (phylogenetic) non-independence in data is often tantamount to overestimating the potential correlations between traits. Surprisingly, this notion was only belatedly applied to life history traits of infections. In 2010, we analyzed the distribution of viral loads measured in pre-treatment HIV-positive individuals and found that, in one subpopulation, nearly half of the variance was explained by phylogeny (Alizon et al. 2010). For a human infectious disease, this result has a particular implication, because by assuming that viruses are the only link between infected individuals, one demonstrates genetic control of the virus on the trait under study.

The link between the notions of phylogenetic signal and heritability had already been emphasized for species phylogenies (Housworth et al. 2004). For infection phylogenies, it takes on even more meaning when we consider that we are working with a population of individuals and not of species. Nonetheless, there are several drawbacks that have to be considered when applying it to traits such as viral load, including within-host evolution or the bottleneck associated with transmission (Leventhal and Bonhoeffer 2016). From a phylodynamic perspective, Vrancken et al. (2014) developed a Bayesian approach for estimating this phylogenetic signal along with phylogeny. In addition, Mitov and Stadler (2018) meanwhile relaxed the neutral evolution assumption by using an Ornstein–Uhlenbeck process.

Another direct application of this trait evolution has focused not on an infection trait but on a viral trait. One of the reasons that influenza virus re-emerges every year is related to the rapid evolution of surface proteins presented to the immune system (Koelle et al. 2006). This evolution can be captured *via* inhibition analyses performed by infecting ferrets, which are an ideal model animal for influenza virus. Some kind of matrix is then obtained that describes the ability of antibodies synthesized by a ferret following infection with an influenza virus X to prevent the agglutination of red blood cells by an influenza virus Y. Therefrom, a distance between viruses can be calculated. Since the research by Smith et al. (2004), it was known that two-dimensional clustering made it possible to visualize the evolution of influenza antigens with a very strong temporal structure (the viruses isolated in 1995 being closer to those isolated in 1992 or 1998 than to those isolated in 1987 or 2002). The device employed by Bedford et al. (2014) consisted of combining this trait related to inhibition analyses with genetic sequences. By combining molecular and antigenic sources of information, they were able to better explain the variations in prevalence observed from one year to the next.

In the end, the combination of phylogeny and life history traits is very similar to phylogeography. However, this analogy should be regarded with caution, as traits cause their own biological issues and their evolution differs greatly from spatial propagation.

11.7. Perspectives and challenges

Phylodynamics was built on promises, the main one being to provide a real application to the thousands of microbial sequences routinely generated every day in laboratories. Although significant progress has been made in the last 20 years, the field remains marginal in public health. In the case of the 2013–2016 Ebola epidemic in West Africa, admittedly, the first two large-scale publications contained phylogenetic trees (Baize et al. 2014; Gire et al. 2014), but the phylodynamic dimension remained very limited (with no estimation of population dynamics, for example). Moreover, for their part the analyses carried out by the WHO working group, resorted merely to traditional methods based on incidence surveys and contact tracing (WHO Ebola Response Team 2014). This reluctance may have two origins. The first is that phylodynamics, similarly to most phylogenetic inferences, relies on strong assumptions such as the absence of recombination or the neutrality of molecular evolution. Some studies certainly include natural selection in the generation of phylogenies (Neher and Hallatschek 2013), but for the moment they are the sole exception.

Another less discussed limitation is that phylodynamics involving likelihood functions has difficulty incorporating heterogeneous data. Admittedly, using only genetic sequence data to make an inference is a considerable feat in itself, but, from an applied point of view, why not consider other sources of data, such as incidence data in particular? This data aggregation from different sources has been somewhat explored, especially by means of ABC approaches that facilitate this integration of heterogeneous data (Ratmann et al. 2012; Saulnier 2017; Smith et al. 2017). They could also help to correct for sampling biases (Volz and Frost 2014). Nonetheless, a lot remains to be done, especially for widespread use.

Despite the fact that phylodynamics has to overcome obstacles in order to make its mark in epidemiology, it undergoes more success in places where it was less expected. For example, the explosion in the number of sequences has contributed to an increase in within-host phylogenies (Hartfield et al. 2014). The term phyloanatomy has even been introduced to characterize these phylodynamic approaches applied to virus propagation between organs (Bons and Regoes 2018).

An even more promising aspect concerns the analysis of transmission clusters that correspond to areas of the phylogeny densely populated with leaves. One of the contributions of NGS has been to allow the identification of transmission pairs within these clusters and even the direction of transmission to be inferred (Romero-Severson et al. 2016; Wymant et al. 2018). The latter is particularly important in public health as it makes it possible to identify the risk factors associated with the donor or recipient (Le Vu et al. 2019).

Finally, although quite young, the field of phylodynamics has seen its philosophy undergo major upheavals. When its initial goal was to extract the maximum amount of information from a limited amount of genetic data, it is now confronted with an avalanche of data (genetic or not). As a result, tools such as BEAST find themselves at odds, since their computation times increase nonlinearly with the number and length of sequences, even if there are existing libraries that allow these slowdowns to be mitigated (Ayres et al. 2019). In this context, techniques based on machine learning and likelihood-free methods open interesting perspectives for the field.

This chapter is a translation of a French version written in 2019. The author slightly edited the translation performed by the publisher but did not make major changes, which is why the SARS-CoV-2 research is absent.

11.8. References

Alizon, S. (2016). *C'est grave docteur Darwin ? L'évolution, les microbes et nous*. Le Seuil, Paris.

Alizon, S. and Fraser, C. (2013). Within-host and between-host evolutionary rates across the HIV-1 genome. *Retrovirology*, 10, 49.

Alizon, S., von Wyl, V., Stadler, T., Kouyos, R.D., Yerly, S., Hirschel, B., Böni, J., Shah, C., Klimkait, T., Furrer, H. et al. (2010). Phylogenetic approach reveals that virus genotype largely determines HIV set-point viral load. *PLOS Pathogens*, 6(9), e1001123.

Anderson, R.M. and May, R.M. (1991). *Infectious Diseases of Humans: Dynamics and Control*. Oxford University Press.

Avise, J.C., Arnold, J., Ball, R.M., Bermingham, E., Lamb, T., Neigel, J.E., Reeb, C.A., Saunders, N.C. (1987). Intraspecific phylogeography: The mitochondrial DNA bridge between population genetics and systematics. *Annual Review of Ecology and Systematics*, 18(1), 489–522.

Ayres, D.L., Cummings, M.P., Baele, G., Darling, A.E., Lewis, P.O., Swofford, D.L., Huelsenbeck, J.P., Lemey, P., Rambaut, A., Suchard, M.A. (2019). BEAGLE 3: Improved performance, scaling, and usability for a high-performance computing library for statistical phylogenetics. *Systematic Biology*, 68(6), 1052–1061.

Baele, G., Dellicour, S., Suchard, M.A., Lemey, P., Vrancken, B. (2018). Recent advances in computational phylodynamics. *Current Opinion in Virology*, 31, 24–32.

Baize, S., Pannetier, D., Oestereich, L., Rieger, T., Koivogui, L., Magassouba, N., Soropogui, B., Sow, M.S., Keïta, S., De Clerck, H. et al. (2014). Emergence of Zaire Ebola virus disease in Guinea. *New England Journal of Medicine*, 371(15), 1418–1425.

Balfe, P., Simmonds, P., Ludlam, C.A., Bishop, J.O., Brown, A.J. (1990). Concurrent evolution of human immunodeficiency virus type 1 in patients infected from the same source: Rate of sequence change and low frequency of inactivating mutations. *Journal of Virology*, 64(12), 6221–6233.

Bedford, T., Cobey, S., Beerli, P., Pascual, M. (2010). Global migration dynamics underlie evolution and persistence of human influenza A (H3n2). *PLOS Pathogens*, 6(5), e1000918.

Bedford, T., Suchard, M.A., Lemey, P., Dudas, G., Gregory, V., Hay, A.J., McCauley, J.W., Russell, C.A., Smith, D.J., Rambaut, A. (2014). Integrating influenza antigenic dynamics with molecular evolution. *eLife*, 3, e01914.

Beerenwinkel, N., Däumer, M., Oette, M., Korn, K., Hoffmann, D., Kaiser, R., Lengauer, T., Selbig, J., Walter, H. (2003). Geno2pheno: Estimating phenotypic drug resistance from HIV-1 genotypes. *Nucleic Acids Research*, 31(13), 3850–3855.

Blum, M.G.B. (2018). Regression approaches for ABC. In *Handbook of Approximate Bayesian Computation*, Sisson, S.A., Fan, Y., Beaumont, M.A. (eds). Chapman and Hall/CRC Press, Boca Raton, FL.

Bons, E. and Regoes, R.R. (2018). Virus dynamics and phyloanatomy: Merging population dynamic and phylogenetic approaches. *Immunological Reviews*, 285(1), 134–146.

Bonsall, D., Golubchik, T., Di Cesare, M., Limbada, M., Kosloff, B., MacIntyre-Cockett, G., Hall, M., Wymant, C., Ansari, M.A., Abeler-Dörne, L. et al. (2018). A comprehensive genomics solution for HIV surveillance and clinical monitoring in a global health setting. *bioRxiv*, 397083.

Bouckaert, R., Heled, J., Kühnert, D., Vaughan, T., Wu, C.-H., Xie, D., Suchard, M.A., Rambaut, A., Drummond, A.J. (2014). BEAST 2: A software platform for Bayesian evolutionary analysis. *PLOS Computational Biology*, 10(4), e1003537.

Chaillon, A., Samleerat, T., Zoveda, F., Ballesteros, S., Moreau, A., Ngo-Giang-Huong, N., Jourdain, G., Gianella, S., Lallemant, M., Depaulis, F., Barin, F. (2014). Estimating the timing of mother-to-child transmission of the human immunodeficiency virus type 1 using a viral molecular evolution model. *PLOS One*, 9(4), e90421.

Coltart, C.E.M., Hoppe, A., Parker, M., Dawson, L., Amon, J.J., Simwinga, M., Geller, G., Henderson, G., Laeyendecker, O., Tucker, J.D. et al. (2018). Ethical considerations in global HIV phylogenetic research. *Lancet HIV*, 5(11), e656–e666.

Csilléry, K., Olivier, F., Blum, M.G.B. (2012). abc: An R package for approximate Bayesian computation (ABC). *Methods in Ecology and Evolution*, 3(3), 475–479.

Danesh, G., Virlogeux, V., Ramière, C., Charre, C., Cotte, L., Alizon, S. (2021). Quantifying transmission dynamics of acute hepatitis C virus infections in a heterogeneous population using sequence data. *PLoS Pathog*., 17(9), e1009916.

De Maio, N., Wu, C.-H., OâĂŹReilly, K.M., Wilson, D. (2015). New routes to phylogeography: A Bayesian structured coalescent approximation. *PLOS Genetics*, 11(8), e1005421.

Dellicour, S., Rose, R., Faria, N.R., Lemey, P., Pybus, O.G. (2016). SERAPHIM: Studying environmental rasters and phylogenetically informed movements. *Bioinformatics*, 32(20), 3204–3206.

Diekmann, O. and Heesterbeek, J. (2000). *Mathematical Epidemiology of Infectious Diseases: Model Building, Analysis, and Interpretation*. Wiley, New York.

Dieckmann, U., Metz, J.A.J., Sabelis, M.W., Sigmund, K. (eds) (2002). *Adaptive Dynamics of Infectious Diseases: In Pursuit of Virulence Management*. Cambridge University Press.

Drummond, A.J. and Rambaut, A. (2007). BEAST: Bayesian evolutionary analysis by sampling trees. *BMC Evolutionary Biology*, 7, 214.

Drummond, A.J., Nicholls, G.K., Rodrigo, A.G., Solomon, W. (2002). Estimating mutation parameters, population history and genealogy simultaneously from temporally spaced sequence data. *Genetics*, 161(3), 1307–1320.

Drummond, A.J., Pybus, O.G., Rambaut, A., Forsberg, R., Rodrigo, A.G. (2003). Measurably evolving populations. *Trends in Ecology & Evolution*, 18(9), 481–488.

Drummond, A.J., Rambaut, A., Shapiro, B., Pybus, O.G. (2005). Bayesian coalescent inference of past population dynamics from molecular sequences. *Molecular Biology and Evolution*, 22(5), 1185–1192.

Drummond, A.J., Ho, S.Y.W., Phillips, M.J., Rambaut, A. (2006). Relaxed phylogenetics and dating with confidence. *PLOS Biology*, 4(5), e88.

Dudas, G., Carvalho, L.M., Bedford, T., Tatem, A.J., Baele, G., Faria, N.R., Park, D.J., Ladner, J.T., Arias, A., Asogun, D. et al. (2017). Virus genomes reveal factors that spread and sustained the Ebola epidemic. *Nature*, 544(7650), 309–315.

Faria, N.R., Rambaut, A, Suchard, M.A., Baele, G., Bedford, T., Ward, M.J., Tatem, A.J., Sousa, J.D., Arinaminpathy, N., Pépin, J. et al. (2014). The early spread and epidemic ignition of HIV-1 in human populations. *Science*, 346(6205), 56–61.

Felsenstein, J. (1981). Evolutionary trees from DNA sequences: A maximum likelihood approach. *Journal of Molecular Evolution*, 17(6), 368–376.

Felsenstein, J. (1985). Phylogenies and the comparative method. *The American Naturalist*, 125(1), 1–15.

Frost, S.D.W. and Volz, E.M. (2010). Viral phylodynamics and the search for an "effective number of infections". *Philosophical Transactions of the Royal Society B*, 365(1548), 1879–1890.

Giardina, F., Romero-Severson, E.O., Albert, J., Britton, T., Leitner, T. (2017). Inference of transmission network structure from HIV phylogenetic trees. *PLOS Computational Biology*, 13(1), e1005316.

Gill, M.S., Lemey, P., Faria, N.R., Rambaut, A., Shapiro, B., Suchard, M.A. (2013). Improving Bayesian population dynamics inference: A coalescent-based model for multiple loci. *Molecular Biology and Evolution*, 30(3), 713–724.

Gire, S.K., Goba, A., Andersen, K.G., Sealfon, R.S.G., Park, D.J., Kanneh, L., Jalloh, S., Momoh, M., Fullah, M., Dudas, G. et al. (2014). Genomic surveillance elucidates Ebola virus origin and transmission during the 2014 outbreak. *Science*, 345(6202), 1369–1372.

Grenfell, B.T., Pybus, O.G., Gog, J.R., Wood, J.L., Daly, J.M., Mumford, J.A., Holmes, E.C. (2004). Unifying the epidemiological and evolutionary dynamics of pathogens. *Science*, 303(5656), 327–32.

Hartfield, M., Murall, C.L., Alizon, S. (2014). Clinical applications of pathogen phylogenies. *Trends in Molecular Medicine*, 20(7), 394–404.

Holmes, E.C., Nee, S., Rambaut, A., Garnett, G.P., Harvey, P.H., Leigh Brown, A.J., Smith, J.M. (1995). Revealing the history of infectious disease epidemics through phylogenetic trees. *Philosophical Transactions of the Royal Society B*, 349(1327), 33–40.

Holmes, E.C., Dudas, G., Rambaut, A., Andersen, K.G. (2016). The evolution of Ebola virus: Insights from the 2013-2016 epidemic. *Nature*, 538(7624), 193–200.

Housworth, E.A., Martins, E.P., Lynch, M. (2004). The phylogenetic mixed model. *The American Naturalist*, 163(1), 84–96.

Hudson, R.R. (1990). Gene genealogies and the coalescent process. *Oxford Surveys in Evolutionary Biology*, 7, 1–44.

Hurford, A., Cownden, D., Day, T. (2010). Next-generation tools for evolutionary invasion analyses. *Journal of the Royal Society Interface*, 7(45), 561–71.

Johnson, S.B., Parker, M. (2019). The ethics of sequencing infectious disease pathogens for clinical and public health. *Nature Reviews Genetics*, 20(6), 313–315.

Jonassen, T.O., Stene-Johansen, K., Berg, E.S., Hungnes, O., Lindboe, C.F., Frøland, S.S., Grinde, B. (1997). Sequence analysis of HIV-1 group O from Norwegian patients infected in the 1960. *Virology*, 231(1), 43–47.

Keeling, M.J. and Rohani, P. (2008). *Modeling Infectious Diseases in Humans and Animals*. Princeton University Press.

Kendall, D.G. (1948). On the generalized "birth-and-death" process. *The Annals of Mathematical Statistics*, 19(1), 1–15.

Kermack, W.O. and McKendrick, A.G. (1927). A contribution to the mathematical theory of epidemics. *Proceedings of the Royal Society A*, 115, 700–721.

Kingman, J.F.C. (1982). The coalescent. *Stochastic Processes and their Applications*, 13(3), 235–248.

Knowles, L.L. and Maddison, W.P. (2002). Statistical phylogeography. *Molecular Ecolology*, 11(12), 2623–2635.

Koelle, K. and Rasmussen, D.A. (2012). Rates of coalescence for common epidemiological models at equilibrium. *Journal of the Royal Society Interface*, 9(70), 997–1007.

Koelle, K., Cobey, S., Grenfell, B., Pascual, M. (2006). Epochal evolution shapes the phylodynamics of interpandemic influenza A (H3n2) in humans. *Science*, 314(5807), 1898–1903.

Koelle, K., Kamradt, M., Pascual, M. (2009). Understanding the dynamics of rapidly evolving pathogens through modeling the tempo of antigenic change: Influenza as a case study. *Epidemics*, 1(2), 129–137.

Korber, B., Muldoon, M., Theiler, J., Gao, F., Gupta, R., Lapedes, A., Hahn, B.H., Wolinsky, S., Bhattacharya, T. (2000). Timing the ancestor of the HIV-1 pandemic strains. *Science*, 288(5472), 1789–1796.

Kouyos, R.D., Althaus, C.L., Bonhoeffer, S. (2006). Stochastic or deterministic: What is the effective population size of HIV-1? *Trends in Microbiology*, 14(12), 507–511.

Kühnert, D., Wu, C.-H., Drummond, A.J. (2011). Phylogenetic and epidemic modeling of rapidly evolving infectious diseases. *Infection, Genetics and Evolution*, 11(8), 1825–1841.

Kühnert, D., Stadler, T., Vaughan, T.G., Drummond, A.J. (2014). Simultaneous reconstruction of evolutionary history and epidemiological dynamics from viral sequences with the birth-death SIR model. *Journal of the Royal Society Interface*, 11(94), 20131106.

Kühnert, D., Kouyos, R., Shirreff, G., Pečerska, J., Scherrer, A.U., Böni, J., Yerly, S., Klimkait, T., Aubert, V., Günthard, H.F. et al. (2018). Quantifying the fitness cost of HIV-1 drug resistance mutations through phylodynamics. *PLOS Pathogens*, 14(2), e1006895.

Lake, J.A. (1988). Origin of the eukaryotic nucleus determined by rate-invariant analysis of rRNA sequences. *Nature*, 331(6152), 184–186.

Le Vu, S., Ratmann, O., Delpech, V., Brown, A.E., Gill, O.N., Tostevin, A., Dunn, D., Fraser, C., Volz, E.M. (2019). HIV-1 transmission patterns in men who have sex with men: Insights from genetic source attribution analysis. *AIDS Research and Human Retroviruses*, 35(9), 805–813.

Lemey, P., Pybus, O.G., Wang, B., Saksena, N.K., Salemi, M., Vandamme, A.-M. (2003). Tracing the origin and history of the HIV-2 epidemic. *Proceedings of the National Academy of Sciences of the USA*, 100(11), 6588–6592.

Lemey, P., Rambaut, A., Drummond, A.J., Suchard, M.A. (2009). Bayesian phylogeography finds its roots. *PLOS Computational Biology*, 5(9), e1000520.

Leventhal, G.E. and Bonhoeffer, S. (2016). Potential Pitfalls in Estimating Viral Load Heritability. *Trends in Microbiology*, 24(9), 687–698.

Leventhal, G.E., Kouyos, R., Stadler, T., von Wyl, V., Yerly, S., Böni, J., Cellerai, C., Klimkait, T., Günthard, H.F., Bonhoeffer, S. (2012). Inferring epidemic contact structure from phylogenetic trees. *PLOS Computational Biology*, 8(3), e1002413.

Li, W.H., Tanimura, M., Sharp, P.M. (1988). Rates and dates of divergence between AIDS virus nucleotide sequences. *Molecular Biology and Evolution*, 5(4), 313–330.

Lion, S. (2018). Theoretical approaches in evolutionary ecology: Environmental feedback as a unifying perspective. *The American Naturalist*, 191(1), 21–44.

Metzig, C., Ratmann, O., Bezemer, D., Colijn, C. (2019). Phylogenies from dynamic networks. *PLOS Computational Biology*, 15(2), e1006761.

Mitov, V. and Stadler, T. (2018). A practical guide to estimating the heritability of pathogen traits. *Molecular Biology and Evolution*, 35(3), 756–772.

Müller, N.F., Rasmussen, D., Stadler, T. (2018). MASCOT: Parameter and state inference under the marginal structured coalescent approximation. *Bioinformatics*, 34(22), 3843–3848 [Online]. Available at: https://academic.oup.com/bioinformatics/article/34/22/3843/5001387.

Nee, S., Holmes, E.C., Rambaut, A., Harvey, P.H., Leigh Brown, A.J., Smith, J.M. (1995). Inferring population history from molecular phylogenies. *Philosophical Transactions of the Royal Society B*, 349(1327), 25–31.

Neher, R.A. and Hallatschek, O. (2013). Genealogies of rapidly adapting populations. *Proceedings of the National Academy Sciences of the USA*, 110(2), 437–442.

Notohara, M. (1990). The coalescent and the genealogical process in geographically structured population. *Journal of the Mathemetical Biology*, 29(1), 59–75.

Ong, C.-K., Nee, S., Rambaut, A., Harvey, P.H. (1996). Inferring the population history of an epidemic from a phylogenetic tree. *Journal of Theoretical Biology*, 182(2), 173–178.

Ou, C.-Y., Ciesielski, C.A., Myers, G., Bandea, C.I., Luo, C.C., Korber, B.T., Mullins, J.I., Schochetman, G., Berkelman, R.L., Economou, A.N. et al. (1992). Molecular epidemiology of HIV transmission in a dental practice. *Science*, 256(5060), 1165–1171.

Pimenoff, V.N., de Oliveira, C.M., Bravo, I.G. (2016). Transmission between archaic and modern human ancestors during the evolution of the oncogenic human papillomavirus 16. *Molecular Biology and Evolution*, 34(1), 4–19.

Pudlo, P., Marin, J.-M., Estoup, A., Cornuet, J.-M., Gautier, M., Robert, C.P. (2016). Reliable ABC model choice via random forests. *Bioinformatics*, 32(6), 859–866.

Pybus, O.G. and Rambaut, A. (2009). Evolutionary analysis of the dynamics of viral infectious disease. *Nature Reviews Genetics*, 10(8), 540–550.

Pybus, O.G., Holmes, E.C., Harvey, P.H. (1999). The mid-depth method and HIV-1: A practical approach for testing hypotheses of viral epidemic history. *Molecular Biology and Evolution*, 16(7), 953–959.

Pybus, O.G., Rambaut, A., Harvey, P.H. (2000). An integrated framework for the inference of viral population history from reconstructed genealogies. *Genetics*, 155(3), 1429–1437.

Pybus, O.G., Charleston, M.A., Gupta, S., Rambaut, A., Holmes, E.C., Harvey, P.H. (2001). The epidemic behavior of the hepatitis C virus. *Science*, 292(5525), 2323–2325.

Pybus, O.G., Suchard, M.A., Lemey, P., Bernardin, F.J., Rambaut, A., Crawford, F.W., Gray, R.R., Arinaminpathy, N., Stramer, S.L., Busch, M.P. et al. (2012). Unifying the spatial epidemiology and molecular evolution of emerging epidemics. *Proceedings of the National Academy of Sciences of the USA*, 109(37), 15066–15071.

Quick, J., Loman, N.J., Duraffour, S., Simpson, J.T., Severi, E., Cowley, L., Bore, J.A., Koundouno, R., Dudas, G., Mikhail, A. et al. (2016). Real-time, portable genome sequencing for Ebola surveillance. *Nature*, 530(7589), 228–232.

Rambaut, A. (2000). Estimating the rate of molecular evolution: Incorporating non-contemporaneous sequences into maximum likelihood phylogenies. *Bioinformatics*, 16(4), 395–399.

Rambaut, A., Harvey, P.H., Nee, S. (1997). End-Epi: An application for inferring phylogenetic and population dynamical processes from molecular sequences. *Bioinformatics*, 13(3), 303–306.

Rambaut, A., Pybus, O.G., Nelson, M.I., Viboud, C., Taubenberger, J.K., Holmes, E.C. (2008). The genomic and epidemiological dynamics of human influenza A virus. *Nature*, 453(7195), 615–619.

Rascovan, N., Sjögren, K.-G., Kristiansen, K., Nielsen, R., Willerslev, E., Desnues, C., Rasmussen, S. (2019). Emergence and spread of basal lineages of Yersinia pestis during the Neolithic decline. *Cell*, 176(1), 295–305.e10.

Rasmussen, D.A., Ratmann, O., Koelle, K. (2011). Inference for nonlinear epidemiological models using genealogies and time series. *PLOS Computational Biology*, 7(8), e1002136.

Rasmussen, D.A., Volz, E.M., Koelle, K. (2014). Phylodynamic inference for structured epidemiological models. *PLOS Computational Biology*, 10(4), e1003570.

Rasmussen, D.A., Kouyos, R., Günthard, H.F., Stadler, T. (2017). Phylodynamics on local sexual contact networks. *PLOS Computational Biology*, 13(3), e1005448.

Ratmann, O., Donker, G., Meijer, A., Fraser, C., Koelle, K. (2012). Phylodynamic inference and model assessment with approximate Bayesian computation: Influenza as a case study. *PLOS Computational Biology*, 8(12), e1002835.

Ratmann, O., Hodcroft, E.B., Pickles, M., Cori, A., Hall, M., Lycett, S., Colijn, C., Dearlove, B., Didelot, X., Frost, S. et al. (2017). Phylogenetic tools for generalized HIV-1 epidemics: Findings from the PANGEA-HIV methods comparison. *Molecular Biology and Evolution*, 34(1), 185–203.

Rodrigo, A.G. and Felsenstein, J. (1999). Coalescent approaches to HIV population genetics. In *The Evolution of HIV*, Crandall, K.A. (ed.). John Hopkins University Press, Baltimore.

Romero-Severson, E.O., Bulla, I., Leitner, T. (2016). Phylogenetically resolving epidemiologic linkage. *Proceedings of the National Academy of Sciences of the USA*, 113(10), 2690–2695.

Sagulenko P., Puller V., Neher R.A. (2018) TreeTime: Maximum-likelihood phylodynamic analysis. *Virus Evolution*, 4(1), vex042.

Saulnier, E. (2017). Phylodynamique des pathogènes viraux par calcul bayésien approché. Thesis, Université de Montpellier.

Saulnier, E., Gascuel, O., Alizon, S. (2017). Inferring epidemiological parameters from phylogenies using regression-ABC: A comparative study. *PLOS Computational Biology*, 13(3), e1005416.

Smith, T.F., Srinivasan, A., Schochetman, G., Marcus, M., Myers, G. (1988). The phylogenetic history of immunodeficiency viruses. *Nature*, 333(6173), 573–575.

Smith, D.J., Lapedes, A.S., deJong, J.C., Bestebroer, T.M., Rimmelzwaan, G.F., Osterhaus, A.D.M., Fouchier, R.A.M. (2004). Mapping the antigenic and genetic evolution of influenza virus. *Science*, 305(5682), 371–376.

Smith, R.A., Ionides, E.L., King, A.A. (2017). Infectious disease dynamics inferred from genetic data via sequential Monte Carlo. *Molecular Biology and Evolution*, 34(8), 2065–2084.

Stadler, T. (2009). On incomplete sampling under birth-death models and connections to the sampling-based coalescent. *Journal of the Theoretical Biology*, 261(1), 58–66.

Stadler, T., Kühnert, D., Bonhoeffer, S., Drummond, A.J. (2013). Birth-death skyline plot reveals temporal changes of epidemic spread in HIV and hepatitis C virus (HCV). *Proceedings of the National Academy of Sciences of the USA*, 110(1), 228–233.

Stadler, T., Vaughan, T.G., Gavryushkin, A., Guindon, S., Kühnert, D., Leventhal, G.E., Drummond, A.J. (2015). How well can the exponential-growth coalescent approximate constant-rate birth–death population dynamics? *Proeedings of the Royal Society B*, 282(1806), 20150420.

Tanaka, M.M., Francis, A.R., Luciani, F., Sisson, S.A. (2006). Using approximate Bayesian computation to estimate tuberculosis transmission parameters from genotype data. *Genetics*, 173(3), 1511–1520.

To, T.-H., Jung, M., Lycett, S., Gascuel, O. (2015). Fast dating using least-squares criteria and algorithms. *Systematic Biology*, 65(1), 82–97.

Vaughan, T.G., Kühnert, D., Popinga, A., Welch, D., Drummond, A.J. (2014). Efficient Bayesian inference under the structured coalescent. *Bioinformatics*, 30(16), 2272–2279.

Volz, E.M. (2012). Complex population dynamics and the coalescent under neutrality. *Genetics*, 190(1), 187–201.

Volz, E.M. and Frost, S.D.W. (2014). Sampling through time and phylodynamic inference with coalescent and birth–death models. *Journal of the Royal Society Interface*, 11(101), 20140945.

Volz, E.M. and Siveroni, I. (2018). Bayesian phylodynamic inference with complex models. *PLOS Computational Biology*, 14(11), e1006546.

Volz, E.M., Kosakovsky Pond, S.L., Ward, M.J., Leigh Brown, A.J., Frost, S.D.W. (2009). Phylodynamics of infectious disease epidemics. *Genetics*, 183(4), 1421–1430.

Volz, E.M., Koelle, K., Bedford, T. (2013). Viral phylodynamics. *PLOS Computational Biology*, 9(3), e1002947.

Vrancken, B., Lemey, P., Rambaut, A., Bedford, T., Longdon, B., Günthard, H.F., Suchard, M.A. (2014). Simultaneously estimating evolutionary history and repeated traits phylogenetic signal: Applications to viral and host phenotypic evolution. *Methods in Ecology and Evolution*, 6(1), 67–82.

Weinert, L.A., Depledge, D.P., Kundu, S., Gershon, A.A., Nichols, R.A., Balloux, F., Welch, J.J., Breuer, J. (2015). Rates of vaccine evolution show strong effects of latency: Implications for varicella zoster virus epidemiology. *Molecular Biology and Evolution*, 32(4), 1020–1028.

WHO Ebola Response Team (2014). Ebola virus disease in West Africa – The first 9 months of the epidemic and forward projections. *New England Journal of Medicine*, 371(16), 1481–1495.

Worobey, M., Gemmel, M., Teuwen, D.E., Haselkorn, T., Kunstman, K., Bunce, M., Muyembe, J.-J., Kabongo, J.-M.M., Kalengayi, R.M., Van Marck, E. et al. (2008). Direct evidence of extensive diversity of HIV-1 in Kinshasa by 1960. *Nature*, 455(7213), 661–664.

Wymant, C., Hall, M., Ratmann, O., Bonsall, D., Golubchik, T., de Cesare, M., Gall, A., Cornelissen, M., Fraser, C. (2018). PHYLOSCANNER: Inferring transmission from within- and between-host pathogen genetic diversity. *Molecular Biology and Evolution*, 35(3), 719–733.

Yoder, A.D. and Yang, Z. (2000). Estimation of primate speciation dates using local molecular clocks. *Molecular Biology and Evolution*, 17(7), 1081–1090.

12

Inference of Demographic Processes in Human Populations

Frédéric AUSTERLITZ
Éco-anthropologie, CNRS, MNHN, Université Paris Cité, France

12.1. Introduction

Demographic phenomena are among the major forces affecting the distribution of genetic diversity within and among populations. For example, a recently expanded population will not have the same genetic diversity distribution among its members as a population of constant size (Slatkin 1991). Another important parameter is migration between populations. The simplest way to take into account this migration is through a global estimate of the differentiation between populations by way of the fixation index F_{ST} (Wright 1949). Under the hypothesis of an island-based model that assumes a set of populations of the same size N exchanging a proportion m of migrants randomly, independently of any geographical distribution, the following equation could be established: $F_{ST} = 1/(1 + 4Nm)$ (Wright 1949), which makes it possible to estimate the product Nm from the value of F_{ST}.

Furthermore, research by Wright (1943, 1946), Malécot (1948) and Kimura and Weiss (1964) explored the relationships between spatial distances and genetic distances between individuals, using so-called isolation-by-distance models, where the dispersal of individuals within their habitat during evolution governs the spatial

Models and Methods for Biological Evolution,
coordinated by Gilles DIDIER and Stéphane GUINDON.

distribution of genetic diversity. This has led to the development of methods for estimating dispersal that take into account the spatial position of populations (Slatkin 1993; Rousset 1997). Studies have also been done on the impact of other processes such as extinction-recolonization mechanisms in the context of metapopulations (Wade and McCauley 1988; Whitlock and McCauley 1990), but without leading to concrete methods for detecting and estimating these processes.

However, demographic processes have long been considered as a "black box" by population geneticists: it involved only the calculation of an effective size, that is the constant size of an ideal panmictic population that would have the same genetic diversity as the population under study, and then to reason as if the population or populations under study were indeed panmictic and of constant size. While this can be justified in situations of equilibrium (Charlesworth 1994), the same cannot be said for transient situations. Therefore, Nei et al. (1975) have demonstrated the impact of a bottleneck (a temporary reduction in the size of a population) on neutral genetic diversity, which as such is strongly reduced, and this loss of diversity persists for a fairly long period after the return to demographic equilibrium.

The coalescent theory (Kingman 1982) is an extremely effective tool for inferring demographic processes from population genetics data (for a summary, see Hudson (1990) and Nordborg (2001)). This theory studies the probability distributions of coalescence times, that is, the times that separate two or more genes from their common ancestor. The coalescent tree is the gene family tree. Coalescent time distributions can be analytically inferred in a number of cases. For example, in the case of a constant size population, these times follow an exponential distribution (Kingman 1982a, 1982b; Tavaré 1984). With the knowledge of the probability distributions of these coalescent trees one is able to deduce those of the expected genetic diversity in a population, since the divergence between two genes at the genetic level comes from mutations that have accumulated in the lineages since their common ancestor.

This coalescent theory is particularly useful for addressing the genetics of populations undergoing demographic changes. This was first achieved in simple cases, such as exponentially growing populations (Slatkin 1991), the complete theory having become available later on (Griffiths and Tavare 1994). In parallel, this was also extended to subdivided populations (see among others (Slatkin 1991)). This allowed predictions to be made about the distribution of the expected neutral genetic diversity in populations subject to demographic processes, which could be verified experimentally (Johnson 1988; Excoffier 1990; Westerbergh and Saura 1994; Lavery et al. 1996).

As a result of these theoretical developments, computationally intensive methods have been developed to infer past demographic processes from genetic diversity. The first category of these methods is based on Markov chain Monte Carlo (MCMC) methods, specifically based on algorithms such as Metropolis–Hastings (Metropolis et al. 1953; Hastings 1970). This consists of simultaneously reconstructing the probability distributions of coalescent trees and the underlying demographic parameters. To this end, the algorithm explores the space of coalescent trees and parameters until it obtains a stationary distribution of the latter. This has allowed the development of parameter estimators such as population growth rates (Kuhner et al. 1998), inter-population migration rates (Bahlo and Griffiths 2000; Beerli and Felsenstein 2001) and population separation times (Wilson et al. 2003).

MCMC methods generally work well for models with few parameters for which there is a strong signal existing in the data. Nonetheless, they are very greedy in computational terms, since the time required to compute them increases significantly as the number of parameters increases, to the point where convergence to a stationary state may become impossible in a reasonable time. Moreover, it becomes extremely complex to derive likelihood in models with many parameters. In order to be able to use these complex models, another category of methods has been developed, namely *approximate Bayesian computation* (ABC; Beaumont et al. 2002; Csilléry et al. 2010) methods. These methods are based on an intensive simulation process. Several possible demographic models are assumed and a large number of simulations will be performed by drawing demographic (population size, growth rate, migration rate, etc.) and genetic (mutation rate, recombination rate, etc.) parameters from a priori distributions that will generally be considered as uninformative. For each simulation, simulated data will be obtained to be compared to real data.

In order to make this comparison, summary statistics on the simulated data are computed. The summary statistics are quantities calculated from the genetic data, they may be, for example, the expected genetic diversity H_e in a population (Nei 1973) which corresponds to the probability that two alleles of the same locus taken from a population are different, or to the parameter F_{ST} (Wright 1949) which measures the genetic differentiation between populations. The values of these statistics will then be compared to the values obtained from the real data. As a result, it will be possible to determine which of the assumed demographic models is the most compatible with the real data, and also to estimate the a posteriori parameter distribution of this model, in both cases by analyzing which simulations generated summary statistics on the simulated data closest to those obtained from the observed data.

In this chapter, we will first see application examples for these methods with classical population genetics data, namely polymorphism data for microsatellites or short DNA sequences (section 12.2). We will then see how these methods have been extended to high-throughput sequence polymorphism as it is currently developed (section 12.3). Finally, we will see how genetic polymorphism can be combined with cultural data, such as linguistic data, in order to reconstruct the recent history of human populations (section 12.4).

12.2. Demographic inferences from population genetics data

12.2.1. *Reconstruction of the history of Central African Pygmies*

Central Africa is a region of the world where two types of human populations coexist, hunter-gatherer Pygmies and farmer non-Pygmies (in particular, Bantus). These populations with substantially different statures and lifestyles generally coexist in close proximity to each other. The question therefore arose as to whether these populations have been separated for a long time and whether they have continued to exchange migrants since the time of their separation. To this end, Verdu et al. (2009) analyzed population genetic data acquired from DNA samples of individuals originating from 21 populations in West Central Africa (Gabon, Cameroon) genotyped for 28 autosomal microsatellite markers. These 21 populations included nine Pygmy populations (from five different groups) and 12 neighboring non-Pygmy populations. In particular, this study used ABC methods performed with the DIYABC program (Cornuet et al. 2008) in order to determine the most adequate demographic model for these populations, assuming models characterized either by a common origin of all Pygmy populations derived from an ancestral population that would have separated from the ancestral non-Pygmy population, or a separate origin where each Pygmy population would have independently derived from the non-Pygmy population. The ABC method was then used to estimate the parameters (separation times, effective population sizes, etc.) of this scenario.

The model that proved to be the most probable is that of a separation between a Pygmy and a non-Pygmy ancestral population at a time between 50,000 and 90,000 years before our current time (see Figure 12.1). Much more recently, this ancestral Pygmy population would have subdivided into several populations which gave birth, about 3,000 years ago to the present groups. This period coincides with the Neolithic expansion of non-Pygmy populations in Central Africa. It seems likely that the fragmentation of Pygmy populations is linked to this massive arrival of non-Pygmies, who greatly disrupted the Pygmies' environment and developed complex socioeconomic relationships with them. This study also revealed that introgression phenomena between these two populations (the arrival of genes from one population into the other) is significantly more frequent from non-Pygmies into Pygmies than in the opposite direction, which is consistent with what is known about the sociocultural relations between these two groups. Some Pygmy groups are more affected by this introgression than others.

This work continued with the study of male–female differences in these introgression dynamics (Verdu et al. 2013). This study was based on the use of

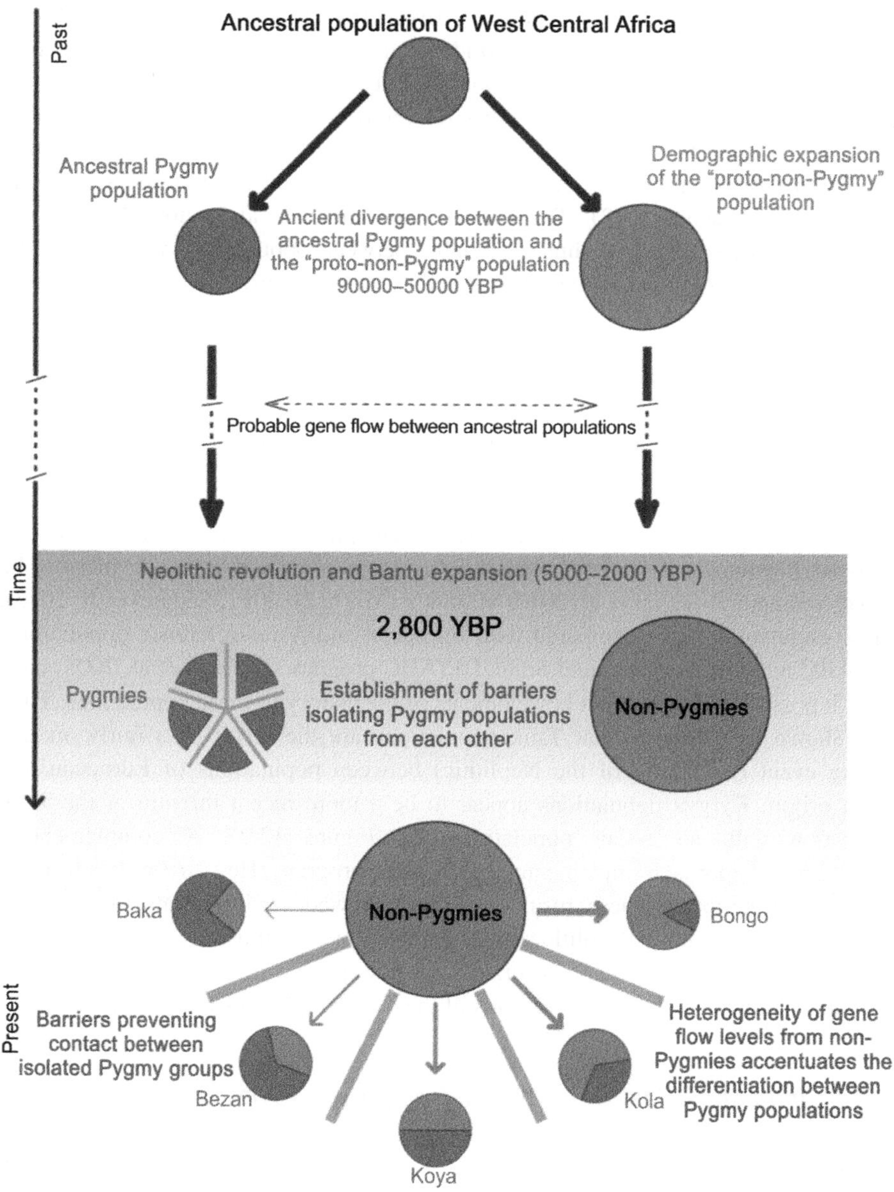

Figure 12.1. *Diagram representing the population history of West Central Africa inferred from genetic data. Adapted from Verdu et al. (2009). For a color version of this figure, see www.iste.co.uk/didier/models.zip*

markers located in the mitochondrial DNA and on the Y chromosome, which are transmitted in a uniparental manner, respectively, from mother to daughter and from father to son. In particular, it has led to identifying a much higher degree of genetic differentiation between Pygmies and non-Pygmies for mitochondrial DNA than for the Y chromosome, suggesting that introgression occurs mainly through males. This result is consistent with the observations of ethnologists that indicate that a non-Pygmy man can marry a Pygmy woman, while the opposite is virtually impossible (Bahuchet 2012). The use of ABC methods also showed that these sex-specific differences were highly variable between populations, which could be related to the ethnological characteristics of these populations.

12.2.2. *Inference of the history of populations in Central Asia*

Palstra et al. (2015) have studied the history of Tajik and Kyrgyz populations in Central Asia. These populations are distinguished by their way of life: Tajiks are sedentary farmers while Kyrgyzes are traditionally nomadic herders. Several individuals from these populations had been previously typed for a set of microsatellite loci and sequences taken away from any gene, allowing them to be considered neutral (Laval et al. 2010; Martínez-Cruz et al. 2011; Ségurel et al. 2013). These populations were compared with European and Asian reference populations. The ABC methods implemented in the DIYABC program (Cornuet et al. 2008) have made it possible to reconstruct key elements of the history of these populations. They have shown in particular that Tajik populations are the result of a fairly ancient mixing event (beginning of the Neolithic) between populations of European and Asian origin. Kyrgyz populations appear to be a more recent mixture of the Tajik population with an Asian population (see Figure 12.2). A complementary MCMC-based approach implemented in the Ima2 program (Hey 2010b) has resulted in the confirmation of these results and also showed earlier events of east-west migration during the Palaeolithic. This work properly illustrates the impact of historical processes on genetic diversity and how powerful statistical methods are capable of inferring substantial elements of the demographic history of populations.

12.2.3. *Impact of lifestyle on population growth dynamics*

Human populations are characterized by very different lifestyles. In particular, three major groups of populations can be distinguished that differ in their diet and exposure to pathogens: farmers, herders and hunter-gatherers. Researchers studied a set of Eurasian and African populations in order to determine whether differences in lifestyle affect the patterns of population expansion (Aimé et al. 2013, 2014, 2015).

For this purpose, they used sequence polymorphism from the HV1 hypervariable region of mitochondrial DNA and autosomal sequences from regions far from any gene, from autosomal microsatellites and from the Y chromosome. They applied a Bayesian method based on an MCMC algorithm to these data that led to estimating the demographic regime of populations (BEAST; Drummond and Rambaut 2007). This method allows for the consideration of various scenarios of population growth (constant size or growing populations according to logistic or exponential models). These researchers have adapted it in order to include a criterion for choosing the best scenario. This choice criterion is based on the DIC (Deviance Information Criterion) (Spiegelhalter et al. 2002), which is suitable for model selection in a Bayesian framework.

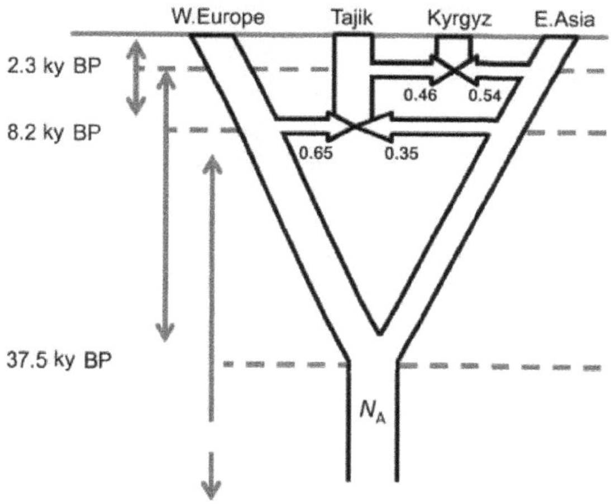

Figure 12.2. *Diagram representing the history of Central Asian populations inferred from genetic data. Adapted from Palstra et al. (2015)*

The results show that hunter-gatherer populations present no expansion signal, whereas herder populations present a limited expansion signal, and farmer populations generally show a high rate of expansion (see Figure 12.3). In other words, sedentary populations with a diet based on cereals have experienced demographic expansions that are much more substantial than nomadic populations with a diet based on animal substances (milk, meat). This can be related to the very high prevalence of agricultural populations on the planet today. Moreover, these studies have shown that certain regions present exceptional characteristics, such as Central Asia, where the expansion signal is generally higher for herders than for farmers, which could be linked to local climatic constraints and/or to particular historical processes.

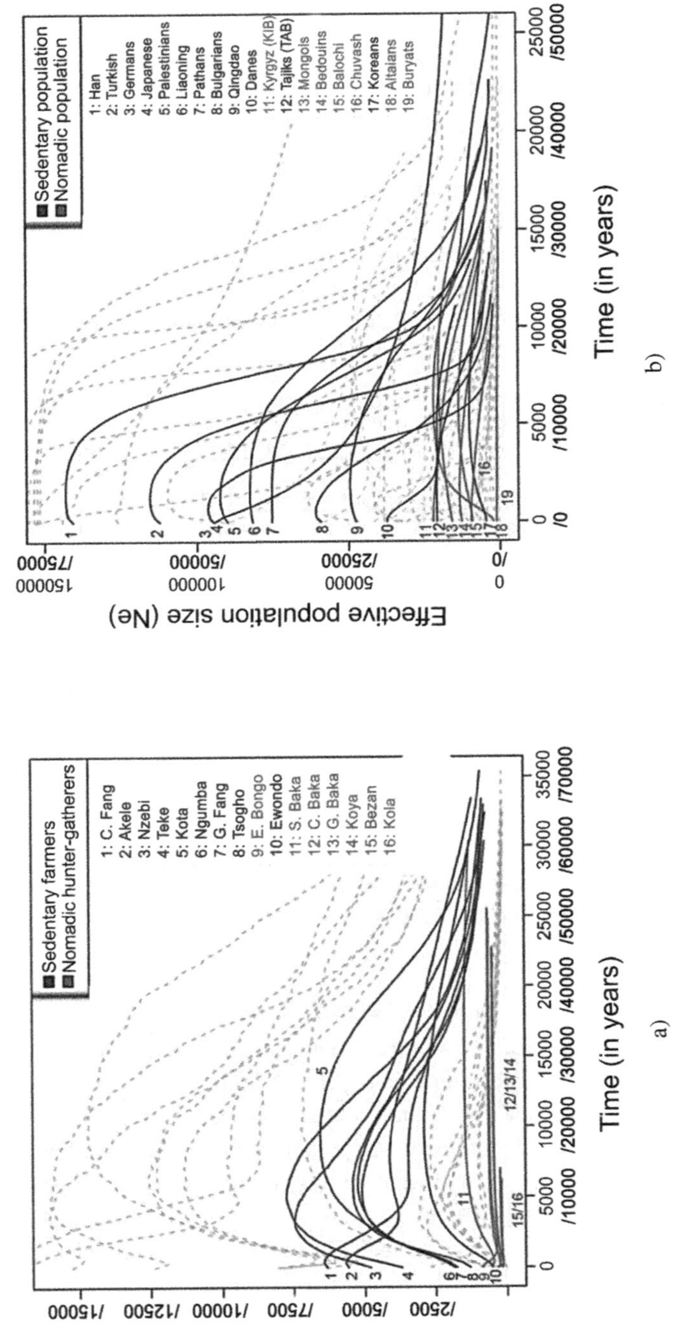

Figure 12.3. *Demographic expansions of farmer (green) and hunter-gatherer or pastoralist (red) populations for African (a) and Eurasian (b) populations inferred using the BEAST software (Drummond and Rambaut, 2007), applied to data from the HV1 hypervariable mitochondrial region. For a color version of this figure, see www.iste.co.uk/didier/models.zip*

COMMENTS ON FIGURE 12.3.– *As is usual with coalescence software, the present is on the left-hand side and the past on the right-hand side. The two scales of values on the axes correspond to two different assumptions for the mutation rate. The figure is inspired by Aimé et al. (2013).*

Finally, it is interesting to observe that the use of markers exhibiting different mutation rates makes it possible to infer expansions at different times. In effect, when using DNA sequence type markers, the ages of expansion of the populations obtained are situated in the Paleolithic. These expansions would therefore not result from the invention of agriculture, but from technological changes that brought, for example, better hunting techniques or environmental changes (Aimé et al. 2013). Conversely, with microsatellite markers, the inferred expansions date back to the Neolithic period (Aimé et al. 2014). One plausible explanation is that there were in fact two successive expansions of human populations, one during the Paleolithic and the other during the Neolithic, and that markers with low mutation rates such as DNA sequences detect the former, while those with high mutation rates detect the latter, as a simulation-based study has shown (Aimé and Austerlitz 2017). Actually, markers with a low mutation rate cannot detect recent events, because the time since their appearance is too short; conversely, markers with a high mutation rate cannot detect older events because their trace has faded or disappeared. In the end, the use of two different types of markers makes it possible to detect two successive waves in the demographic expansion of human agricultural populations, one during the Palaeolithic and the other during the Neolithic.

12.3. Inferring human expansions from next-generation sequence data

In recent time, we have witnessed the development of *next-generation sequencing (NGS)* data, which consists of sequences of full genomes or significant portions thereof. Sequences are now available for samples of individuals from many human populations with, in some cases, several tens of individuals per population as is the case for the 1000 Genomes Project (The 1000 Genomes Project Consortium, 2010, 2015) where about 4,000 complete genomes have been deposited to date. These data are characterized by the contribution of a very large amount of information, the challenge being to extract from these very large data sets the relevant information for a given type of analysis. They therefore require tools that are fast enough to process very large data sets. Furthermore, another layer of complexity is added to sequence analysis by errors due to DNA fragmentation, repeated regions of the genome, alignment, sequencing and contamination. Statistical methods for demographic inference from these data are currently being developed, in particular for the reconstruction of past population sizes and the dating of expansion or contraction events of these populations.

In particular, an ABC package (named DemoSeq) has been produced to reconstruct population histories from large genetic polymorphism datasets (Jay et al. 2019), such as NGS sequencing data. DemoSeq considers demographic models characterized by successive bottleneck and expansion events. These models are characterized by a given number of parameters such as population sizes, the times at which expansion or contraction events occur or even the duration of these events. The first study was conducted with simulated data (fastsimcoal2 software; Excoffier et al. 2013) in order to test the effectiveness of different statistics in inferring these phenomena. These statistics were of two types: (i) "classical", that is commonly calculated for polymorphism data originating from low-density SNPs arrays or microsatellite markers (average heterozygosity, average differentiation between individuals, Tajima's (1983) D, number of polymorphic sites, linkage disequilibrium, etc.) and (ii) "adapted to dense data" such as DNA sequences, in other words, statistics that use distance information between SNPs (for instance, the distribution of haplotypes length shared by multiple individuals or the exact nature of the decline in linkage disequilibrium according to the distance between SNPs).

The authors were thus able to develop an ABC method capable of exploiting high-throughput polymorphism data to be able to reconstruct the demographic population history from an intermediate number of individuals (of the order of 5–20). This is a considerable advantage over methods such as MSMC (Schiffels and Durbin 2014) which are limited to a small number of individuals for computational reasons, while methods like fastNeutrino (Bhaskar et al. 2015) based on the allelic frequency spectrum are very fast but require about a hundred individuals to finely reconstruct the demography. Moreover, this last method does not take into account parameters such as the decline of linkage disequilibrium according to the distance between SNPs, thus ignoring a precious part of the information contained in complete genomes.

It was also possible to study the conditions under which the correct demographic scenario can be identified among several alternative scenarios and, once the choice is made, to infer the parameters corresponding to that scenario. This study also made it possible to determine which method was best for analyzing the summary statistics tables produced by ABC, in this case a method using neural networks to improve parameter estimation (Blum and François 2010), although methods such as ABC with random forests should also be tested (Pudlo et al. 2016; Raynal et al. 2019). It was also shown that it was preferable to co-estimate the sequencing error with the other parameters rather than just merely filtering the data to try to remove errors. These results reflect that it is possible to detect population growth events from a limited number of individuals if correlation patterns at different genome scales are used.

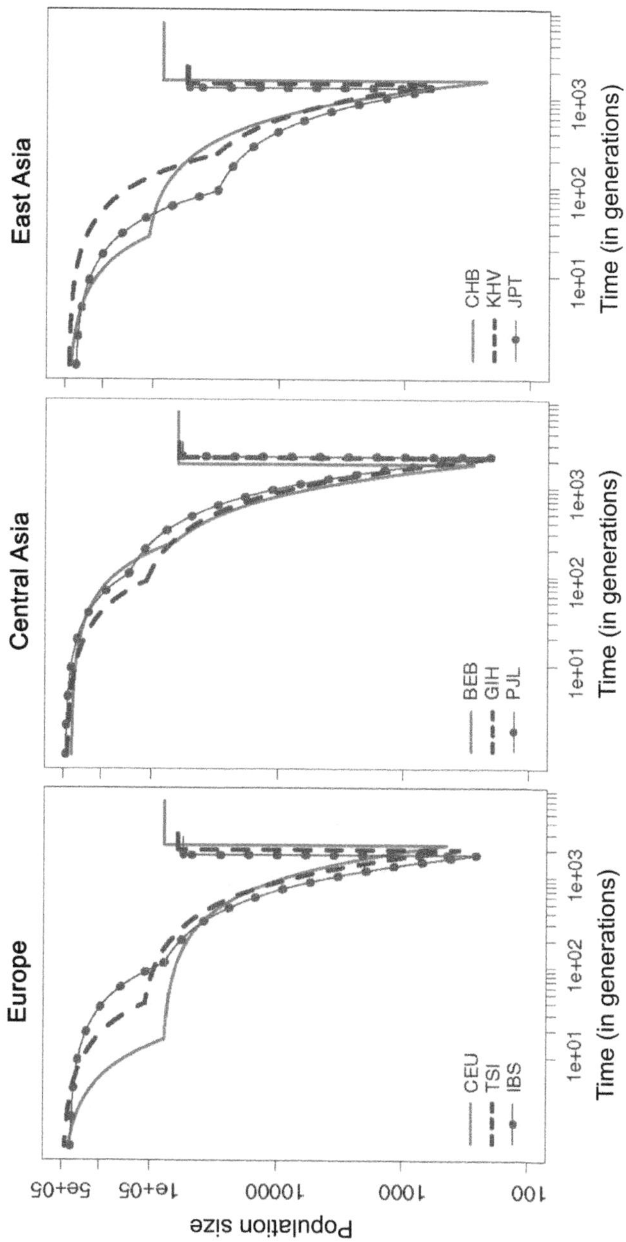

Figure 12.4. *Demographic history inferred by the ABC method for nine human populations. For a color version of this figure, see www.iste.co.uk/didier/models.zip*

COMMENTS ON FIGURE 12.4.– *From left to right: European (CEU, TSI and IBS), Central Asian (BEB, GIH and PJL) and East Asian (CHB, KHV and JPT) populations. For each population, the ABC method was applied to nine individuals drawn from the 1000 Genomes Project database. Bottlenecks and two extensions were independently identified as the best model for each of these nine populations. Adapted from Jay et al. (2019).*

By applying the DemoSeq package to complete genome data from public datasets, Complete Genomics (Drmanac et al. 2010) and 1000 Genomes Project, it was shown that African and non-African populations had contrasting demographic histories (Jay et al. 2019). In particular, it has resulted in the observation that different Eurasian populations (e.g. Italian, Indian, Chinese, and so on) provided very similar signals and allowed to infer two successive growth phases. It was thus possible to jointly detect Paleolithic and Neolithic expansions, as well as the earlier bottleneck related to the exit from Africa (see Figure 12.4). In Africa, a single expansion has been detected in the different populations under study, and it also follows a bottleneck.

12.4. Reconstructing population dynamics from genetic and cultural data

12.4.1. *Simultaneous analysis of genetic and linguistic diversity*

The evolutionary history of humans is marked by cultural and biological exchanges and interbreeding. Researchers have developed methods for analyzing genetic and linguistic data within a unified framework to reconstruct past human migrations (Thouzeau et al. 2017). They developed a computer tool (PopLingSim) that can simulate gene and language evolution for a large number of possible scenarios. These simulations are then compared to real genetic and linguistic data using ABC methods in order to identify the most realistic historical scenario and estimate its parameters. The scenarios were chosen using an ABC approach coupled with random forests implemented in the R abcrf package (Pudlo et al. 2016), whereas the parameters of the chosen model were estimated by an ABC approach with a neural network implemented in the R abc package (Csilléry et al. 2012). A cross-validation approach (simulated datasets analysis as if they were real datasets) has shown that it is indeed possible to use genetic and linguistic data together, and that this allows the choice between several scenarios and estimating the parameters of these scenarios.

This approach has been applied to human populations in Central Asia, being the historical crossroads of Turko-Mongolian and Indo-Iranian language migrations, based on genetic and linguistic data available for this region. For example, in the case of an Uzbek population in the region, the study showed that it was the result of a mixing event between Indo-European and Turko-Mongolian populations, both at the linguistic and genetic level. However, the mixing appears to be much more recent regarding linguistics, and also much more unbalanced: it mainly occurred in favor of the Turko-Mongolian languages. The method has also led to showing that for one particular Indo-Iranian population, the Yagnobes, known for its isolation from other Indo-Iranian populations, this isolation is more significant at the genetic level than at the linguistic level.

The work then continued with the development of an individual-centered method (Thouzeau et al. 2017). This method (PolLingSim2) allows the simulation of populations of individuals transmitting both genes and linguistic elements. It has also been coupled with an ABC method, this time by random forest, including parameter estimation (Raynal et al. 2019). This has allowed demonstrating that this method can be used to determine which linguistic and genetic model best explains a genetic and linguistic data set, and to estimate parameters such as the effective number of speakers of a given language.

12.4.2. *Detecting the intergenerational transmission of reproductive success*

This transmission occurs when the number of offspring an individual of a given generation leaves is positively correlated with the number of offspring his or her parents had themselves. This can be linked to cultural causes, such as the transmission of social status between generations, or to genetic causes. In this context, it was shown that this transmission leads to a decrease in genetic diversity and the possibility for certain genes to rapidly increase in frequency (Austerlitz and Heyer 1998). Moreover, the impact of this transmission on the shape of coalescent trees was demonstrated (Sibert et al. 2002), which enabled developing a method to detect this transmission from genetic data (Blum et al. 2006; Brandenburg et al. 2012). When applied to human mitochondrial DNA data, this method has enabled conclusions that concerning the matrilineal pathway, transmission is much more common in hunter-gatherer populations than in agricultural populations.

This method was then applied to populations of Central Asia (Heyer et al. 2015). These populations are either patrilineal populations where the transmission of social status is done through the father, or cognatic populations where this transmission is done via both parents. This study focused on mitochondrial DNA and the Y chromosome transmitted by females and males, respectively. By reconstructing coalescent trees for the two types of genetic markers in each population, the authors showed that the transmission of reproductive success is much stronger for the male pathway in patrilineal populations than for cognatic populations. Conversely, they found no difference between the two types of populations for the female pathway (see Figure 12.5). This difference may be related to a cultural phenomenon, namely a transmission of social status by males in patrilineal populations, which causes a transmission of reproductive success. However, it cannot be excluded that the phenomenon also has in part genetic causes.

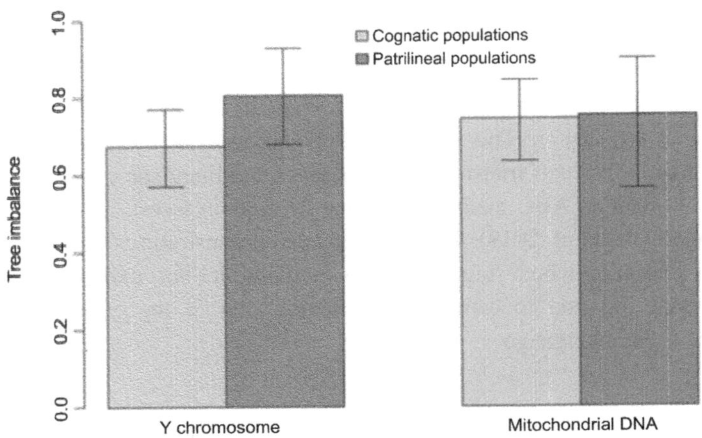

Figure 12.5. *Inferred coalescent tree imbalance for Y chromosome and mitochondrial DNA in cognate and patrilineal populations of Central Asia. The imbalance is all the more significant when the transmission of reproductive success is high. Adapted from Heyer et al. (2015)*

12.5. Conclusion

ABC and MCMC methods based on coalescent theory are very powerful in reconstructing the demographic history of human populations. By using only data from current populations, they are capable of reconstructing past events such as bottlenecks and demographic expansion phenomena by dating these events, which has resulted in highlighting the occurrence of two successive expansions, one in the Palaeolithic and the other in the Neolithic. These methods have also made it possible to identify genetic exchanges (possibly asymmetrical) between populations, such as between Pygmy and non-Pygmy populations in Equatorial Africa, or between Indo-Iranian and Turkic populations in Central Asia. The emergence of high-throughput sequencing data, such as those from the 1000 Genomes Project, leads to obtaining increasingly accurate inferences on the history of populations by relying on parallel computing machines capable of processing these types of data and performing the simulations necessary for the ABC and MCMC approaches.

An interesting avenue for the future is the integration of data extracted from fossil individuals. Genetic sequences are actually currently available from such individuals dating back up to 45,000 years for complete genomes (Fu et al. 2014) and even up to 430,000 years for mitochondrial DNA (Meyer et al. 2016). This will allow for better parameter estimates such as separation times between populations, but also for determining to what extent current populations are the result of mixtures of populations and in what proportions.

Moreover, we have chosen to focus here on the human species, but it should be obvious that the approaches presented have been applied to many other species, including animal species (Hey 2010a; Laurent et al. 2011; Fontaine et al. 2012; Boitard et al. 2016), plants (Cornille et al. 2013), fungi (Barres et al. 2012) or oomycetes (Fontaine et al. 2013).

12.6. References

1000 Genomes Project Consortium (2010). A map of human genome variation from population-scale sequencing. *Nature*, 467(7319), 1061.

1000 Genomes Project Consortium (2015). A global reference for human genetic variation. *Nature*, 526(7571), 68–74.

Aimé, C. and Austerlitz, F. (2017). Different kinds of genetic markers permit inference of paleolithic and neolithic expansions in humans. *European Journal of Human Genetics*, 25(3), 360–365.

Aimé, C., Laval, G., Patin, E., Verdu, P., Ségurel, L., Chaix, R., Hegay, T., Quintana-Murci, L., Heyer, E., Austerlitz, F. (2013). Human genetic data reveal contrasting demographic patterns between sedentary and nomadic populations that predate the emergence of farming. *Molecular Biology and Evolution*, 30(12), 2629–2644.

Aimé, C., Verdu, P., Ségurel, L., Martinez-Cruz, B., Hegay, T., Heyer, E., Austerlitz, F. (2014). Microsatellite data show recent demographic expansions in sedentary but not in nomadic human populations in Africa and Eurasia. *European Journal of Human Genetics*, 22(10), 1201–1207.

Aimé, C., Heyer, E., Austerlitz, F. (2015). Inference of sex-specific expansion patterns in human populations from Y-chromosome polymorphism. *American Journal of Physical Anthropology*, 157(2), 217–225.

Austerlitz, F. and Heyer, E. (1998). Social transmission of reproductive behavior increases frequency of inherited disorders in a young-expanding population. *Proceedings of the National Academy of Sciences*, 95(25), 15140–15144.

Bahlo, M. and Griffiths, R. (2000). Inference from gene trees in a subdivided population. *Theoretical Population Biology*, 57(2), 79–95.

Bahuchet, S. (2012). Changing language, remaining Pygmy. *Human Biology*, 84(1), 11–43.

Barres, B., Carlier, J., Seguin, M., Fenouillet, C., Cilas, C., Ravigné, V. (2012). Understanding the recent colonization history of a plant pathogenic fungus using population genetic tools and approximate Bayesian computation. *Heredity*, 109(5), 269–279.

Beaumont, M.A., Zhang, W., Balding, D.J. (2002). Approximate Bayesian computation in population genetics. *Genetics*, 162(4), 2025–2035.

Beerli, P. and Felsenstein, J. (2001). Maximum likelihood estimation of a migration matrix and effective population sizes in n subpopulations by using a coalescent approach. *Proceedings of the National Academy of Sciences*, 98(8), 4563–4568.

Bhaskar, A., Wang, Y.R., Song, Y.S. (2015). Efficient inference of population size histories and locus-specific mutation rates from large-sample genomic variation data. *Genome Research*, 25(2), 268–279.

Blum, M.G. and François, O. (2010). Non-linear regression models for approximate Bayesian computation. *Statistics and Computing*, 20(1), 63–73.

Blum, M.G., Heyer, E., François, O., Austerlitz, F. (2006). Matrilineal fertility inheritance detected in hunter–gatherer populations using the imbalance of gene genealogies. *PLoS Genetics*, 2(8), e122.

Boitard, S., Rodriguez, W., Jay, F., Mona, S., Austerlitz, F. (2016). Inferring population size history from large samples of genome-wide molecular data – An approximate Bayesian computation approach. *PLoS Genetics*, 12(3), e1005877.

Brandenburg, J.-T., Austerlitz, F., Toupance, B. (2012). Impact of fertility transmission and other sociodemographic factors on reproductive success and coalescent trees. *Genetics Research*, 94(3), 121–131.

Charlesworth, B. (1994). *Evolution in Age-structured Populations*, vol. 2. Cambridge University Press.

Cornille, A., Giraud, T., Bellard, C., Tellier, A., Le Cam, B., Smulders, M., Kleinschmit, J., Roldan-Ruiz, I., Gladieux, P. (2013). Postglacial recolonization history of the European crabapple (*Malus sylvestris* M ill.), a wild contributor to the domesticated apple. *Molecular Ecology*, 22(8), 2249–2263.

Cornuet, J.-M., Santos, F., Beaumont, M.A., Robert, C.P., Marin, J.-M., Balding, D.J., Guillemaud, T., Estoup, A. (2008). Inferring population history with DIY ABC: A user-friendly approach to approximate Bayesian computation. *Bioinformatics*, 24(23), 2713–2719.

Csilléry, K., Blum, M.G., Gaggiotti, O.E., François, O. (2010). Approximate Bayesian computation (ABC) in practice. *Trends in Ecology & Evolution*, 25(7), 410–418.

Csilléry, K., François, O., Blum, M.G. (2012). An R package for approximate Bayesian computation (ABC). *Methods in Ecology and Evolution*, 3(3), 475–479.

Drmanac, R., Sparks, A.B., Callow, M.J., Halpern, A.L., Burns, N.L., Kermani, B.G., Carnevali, P., Nazarenko, I., Nilsen, G.B., Yeung, G. et al. (2010). Human genome sequencing using unchained base reads on self-assembling DNA nanoarrays. *Science*, 327(5961), 78–81.

Drummond, A.J. and Rambaut, A. (2007). BEAST: Bayesian evolutionary analysis by sampling trees. *BMC Evolutionary Biology*, 7(1), 1–8.

Excoffier, L. (1990). Evolution of human mitochondrial DNA: Evidence for departure from a pure neutral model of populations at equilibrium. *Journal of Molecular Evolution*, 30(2), 125–139.

Excoffier, L., Dupanloup, I., Huerta-Sánchez, E., Sousa, V.C., Foll, M. (2013). Robust demographic inference from genomic and SNP data. *PLoS Genetics*, 9(10), e1003905.

Fontaine, M.C., Snirc, A., Frantzis, A., Koutrakis, E., Öztürk, B., Öztürk, A.A., Austerlitz, F. (2012). History of expansion and anthropogenic collapse in a top marine predator of the Black Sea estimated from genetic data. *Proceedings of the National Academy of Sciences*, 109(38), E2569–E2576.

Fontaine, M.C., Austerlitz, F., Giraud, T., Labbé, F., Papura, D., Richard-Cervera, S., Delmotte, F. (2013). Genetic signature of a range expansion and leap-frog event after the recent invasion of Europe by the grapevine downy mildew pathogen *Plasmopara viticola*. *Molecular Ecology*, 22(10), 2771–2786.

Fu, Q., Li, H., Moorjani, P., Jay, F., Slepchenko, S.M., Bondarev, A.A., Johnson, P.L., Aximu-Petri, A., Prüfer, K., de Filippo, C. et al. (2014). Genome sequence of a 45,000-year-old modern human from western Siberia. *Nature*, 514(7523), 445–449.

Griffiths, R.C. and Tavare, S. (1994). Sampling theory for neutral alleles in a varying environment. *Philosophical Transactions of the Royal Society of London. Series B: Biological Sciences*, 344(1310), 403–410.

Hastings, W.K. (1970). Monte Carlo sampling methods using Markov chains and their applications. *Biometrika*, 57(1), 97–109

Hey, J. (2010a). The divergence of chimpanzee species and subspecies as revealed in multipopulation isolation-with-migration analyses. *Molecular Biology and Evolution*, 27(4), 921–933.

Hey, J. (2010b). Isolation with migration models for more than two populations. *Molecular Biology and Evolution*, 27(4), 905–920.

Heyer, E., Brandenburg, J.-T., Leonardi, M., Toupance, B., Balaresque, P., Hegay, T., Aldashev, A., Austerlitz, F. (2015). Patrilineal populations show more male transmission of reproductive success than cognatic populations in Central Asia, which reduces their genetic diversity. *American Journal of Physical Anthropology*, 157(4), 537–543.

Hudson, R.R. (1990). Gene genealogies and the coalescent process. *Oxford Surveys in Evolutionary Biology*, 7, 1–44.

Jay, F., Boitard, S., Austerlitz, F. (2019). An ABC method for whole-genome sequence data: Inferring paleolithic and neolithic human expansions. *Molecular Biology and Evolution*, 36(7), 1565–1579.

Johnson, M.S. (1988). Founder effects and geographic variation in the land snail *Theba pisana*. *Heredity*, 61(1), 133–142.

Kimura, M. and Weiss, G.H. (1964). The stepping stone model of population structure and the decrease of genetic correlation with distance. *Genetics*, 49(4), 561–576.

Kingman, J.F.C. (1982). The coalescent. *Stochastic Processes and their Applications*, 13(3), 235–248.

Kuhner, M.K., Yamato, J., Felsenstein, J. (1998). Maximum likelihood estimation of population growth rates based on the coalescent. *Genetics*, 149(1), 429–434.

Laurent, S.J., Werzner, A., Excoffier, L., Stephan, W. (2011). Approximate Bayesian analysis of *Drosophila melanogaster* polymorphism data reveals a recent colonization of Southeast Asia. *Molecular Biology and Evolution*, 28(7), 2041–2051.

Laval, G., Patin, E., Barreiro, L.B., Quintana-Murci, L. (2010). Formulating a historical and demographic model of recent human evolution based on resequencing data from noncoding regions. *PloS One*, 5(4), e10284.

Lavery, S., Moritz, C., Fielder, D. (1996). Genetic patterns suggest exponential population growth in a declining species. *Molecular Biology and Evolution*, 13(8), 1106–1113.

Malécot, G. (1948). *Les mathématiques de l'hérédité*. Masson, Paris.

Martínez-Cruz, B., Vitalis, R., Ségurel, L., Austerlitz, F., Georges, M., Théry, S., Quintana-Murci, L., Hegay, T., Aldashev, A., Nasyrova, F., Heyer, E. (2011). In the heartland of Eurasia: The multilocus genetic landscape of Central Asian populations. *European Journal of Human Genetics*, 19(2), 216–223.

Metropolis, N., Rosenbluth, A.W., Rosenbluth, M.N., Teller, A.H., Teller, E. (1953). Equation of state calculations by fast computing machines. *The Journal of Chemical Physics*, 21(6), 1087–1092.

Meyer, M., Arsuaga, J.-L., de Filippo, C., Nagel, S., Aximu-Petri, A., Nickel, B., Martínez, I., Gracia, A., de Castro, J.M.B., Carbonell, E. et al. (2016). Nuclear DNA sequences from the Middle Pleistocene Sima de los Huesos hominins. *Nature*, 531(7595), 504–507.

Nei, M. (1973). Analysis of gene diversity in subdivided populations. *Proceedings of the National Academy of Sciences*, 70(12), 3321–3323.

Nei, M., Maruyama, T., Chakraborty, R. (1975). The bottleneck effect and genetic variability in populations. *Evolution*, 29, 1–10.

Nordborg, M. (2001). Coalescent theory. In *Handbook of Statistical Genetics*, Balding, D., Bishop, M., Cannings, C. (eds). John Wiley & Sons, Chichester.

Palstra, F.P., Heyer, E., Austerlitz, F. (2015). Statistical inference on genetic data reveals the complex demographic history of human populations in Central Asia. *Molecular Biology and Evolution*, 32(6), 1411–1424.

Pudlo, P., Marin, J.-M., Estoup, A., Cornuet, J.-M., Gautier, M., Robert, C.P. (2016). Reliable ABC model choice via random forests. *Bioinformatics*, 32(6), 859–866.

Raynal, L., Marin, J.-M., Pudlo, P., Ribatet, M., Robert, C P., Estoup, A. (2019). ABC random forests for Bayesian parameter inference. *Bioinformatics*, 35(10), 1720–1728.

Rousset, F. (1997). Genetic differentiation and estimation of gene flow from F-statistics under isolation by distance. *Genetics*, 145(4), 1219–1228.

Schiffels, S. and Durbin, R. (2014). Inferring human population size and separation history from multiple genome sequences. *Nature Genetics*, 46(8), 919–925.

Ségurel, L., Austerlitz, F., Toupance, B., Gautier, M., Kelley, J.L., Pasquet, P., Lonjou, C., Georges, M., Voisin, S., Cruaud, C. et al. (2013). Positive selection of protective variants for type 2 diabetes from the Neolithic onward: A case study in Central Asia. *European Journal of Human Genetics*, 21(10), 1146–1151.

Sibert, A., Austerlitz, F., Heyer, É. (2002). Wright–Fisher revisited: The case of fertility correlation. *Theoretical Population Biology*, 62(2), 181–197.

Slatkin, M. (1991), Inbreeding coefficients and coalescence times. *Genetics Research*, 58(2), 167–175.

Slatkin, M. (1993). Isolation by distance in equilibrium and non-equilibrium populations. *Evolution*, 47(1), 264–279.

Tajima, F. (1983). Evolutionary relationship of DNA sequences in finite populations. *Genetics*, 105(2), 437–460.

Thouzeau, V., Mennecier, P., Verdu, P., Austerlitz, F. (2017). Genetic and linguistic histories in Central Asia inferred using approximate Bayesian computations. *Proceedings of the Royal Society B: Biological Sciences*, 284(1861), 20170706.

Verdu, P., Austerlitz, F., Estoup, A., Vitalis, R., Georges, M., Théry, S., Froment, A., Le Bomin, S., Gessain, A., Hombert, J.-M. et al. (2009). Origins and genetic diversity of Pygmy hunter-gatherers from Western Central Africa. *Current Biology*, 19(4), 312–318.

Verdu, P., Becker, N.S., Froment, A., Georges, M., Grugni, V., Quintana-Murci, L., Hombert, J.-M., Van der Veen, L., Le Bomin, S., Bahuchet, S. et al. (2013). Sociocultural behavior, sex-biased admixture, and effective population sizes in Central African Pygmies and non-Pygmies. *Molecular Biology and Evolution*, 30(4), 918–937.

Wade, M.J. and McCauley, D.E. (1988). Extinction and recolonization: Their effects on the genetic differentiation of local populations. *Evolution*, 42(5), 995–1005.

Westerbergh, A. and Saura, A. (1994). Genetic differentiation in endemic Silene (Caryophyllaceae) on the Hawaiian Islands. *American Journal of Botany*, 81(11), 1487–1493.

Whitlock, M.C. and McCauley, D.E. (1990). Some population genetic consequences of colony formation and extinction: Genetic correlations within founding groups. *Evolution*, 44(7), 1717–1724.

Wilson, I.J., Weale, M.E., Balding, D.J. (2003). Inferences from DNA data: Population histories, evolutionary processes and forensic match probabilities. *Journal of the Royal Statistical Society: Series A (Statistics in Society)*, 166(2), 155–188.

Wright, S. (1943). Isolation by distance. *Genetics*, 28(2), 114–138.

Wright, S. (1946). Isolation by distance under diverse systems of mating. *Genetics*, 31(1), 39–59.

Wright, S. (1949). The genetical structure of populations. *Annals of Eugenics*, 15(1), 323–354.

List of Authors

Guillaume ACHAZ
SMILE Group
CIRB, CNRS, INSERM
Collège de France
Paris
and
Éco-anthropologie, CNRS
MNHN
Université Paris Cité
France

Samuel ALIZON
MIVEGEC, IRD, CNRS
Université de Montpellier
and
CIRB, CNRS, INSERM
Collège de France
Paris
France

Frédéric AUSTERLITZ
Éco-anthropologie, CNRS
MNHN
Université Paris Cité
France

Paul BASTIDE
IMAG, CNRS
Université de Montpellier
France

Anne BERGERON
LACIM
Université du Québec à Montréal
Canada

Gilles DIDIER
IMAG, CNRS
Université de Montpellier
France

Julien Y. DUTHEIL
RG Molecular Systems Evolution
MPI for Evolutionary Biology
Plön
Germany

Olivier GASCUEL
ISYEB, CNRS
MNHN
Sorbonne Université
EPHE, UA
Paris
France

Laurent GUÉGUEN
LBBE
Université Claude Bernard Lyon 1
Villeurbanne
France
and
Swedish Collegium for Advanced Study
Uppsala
Sweden

Stéphane GUINDON
LIRMM, CNRS
Université de Montpellier
France

Michel LAURIN
CR2P, CNRS
MNHN
Sorbonne Université
Paris
France

Frédéric LEMOINE
Unité de bioinformatique évolutive
Hub de bioinformatique et biostatistiques
Institut Pasteur
Université Paris Cité
France

Mahendra MARIADASSOU
MaIAGE, INRAE
Université Paris-Saclay
Jouy-en-Josas
France

Fabio PARDI
LIRMM, CNRS
Université de Montpellier
France

Étienne PARDOUX
I2M, CNRS
Aix-Marseille Université
France

Stéphane ROBIN
LPSM
Sorbonne Université
Paris
France

Nicolas RODRIGUE
Carleton University
Ottawa
Canada

Krister M. SWENSON
LIRMM, CNRS
Université de Montpellier
France

Index

Q, R

S

T

W, Y

Printed and bound by CPI Group (UK) Ltd, Croydon, CR0 4YY

13/09/2026

14960996-0005